ROUTLEDGE LIBRARY EDITIONS: GEOLOGY

Volume 14

GEOMORPHOLOGY AND ENGINEERING

GEOMORPHOLOGY AND ENGINEERING

Binghamton Geomorphology Symposium 7

Edited by
DONALD R. COATES

Routledge
Taylor & Francis Group

LONDON AND NEW YORK

First published in 1976 by Dowden, Hutchinson & Ross, Inc.

This edition first published in 2020
by Routledge
2 Park Square, Milton Park, Abingdon, Oxon OX14 4RN

and by Routledge
52 Vanderbilt Avenue, New York, NY 10017

Routledge is an imprint of the Taylor & Francis Group, an informa business

British Library Cataloguing in Publication Data
A catalogue record for this book is available from the British Library

ISBN: 978-0-367-18559-6 (Set)
ISBN: 978-0-429-19681-2 (Set) (ebk)
ISBN: 978-0-367-46446-2 (Volume 14) (hbk)
ISBN: 978-0-367-46452-3 (Volume 14) (pbk)
ISBN: 978-1-00-302882-6 (Volume 14) (ebk)

Publisher's Note
The publisher has gone to great lengths to ensure the quality of this reprint but points out that some imperfections in the original copies may be apparent.

Disclaimer
The publisher has made every effort to trace copyright holders and would welcome correspondence from those they have been unable to trace.

GEOMORPHOLOGY AND ENGINEERING

Edited by

Donald R. Coates

Dowden, Hutchinson & Ross, Inc.

STROUDSBURG, PENNSYLVANIA

Distributed by

HALSTED PRESS

A Division of John Wiley & Sons, Inc.

Library of Congress Cataloging in Publication Data
Geomorphology Symposium, 7th, Binghamton, N.Y., 1976.
    Geomorphology and engineering.
    Held in the Dept. of Geological Sciences at the State
University of New York, Sept. 24–25, 1976.
    Includes index.
    1. Engineering geology—Congresses. 2. Geomorpho-
logy—Congresses. I. Coates, Donald Robert, 1922–
II. Title.
TA705.G424 1976   624'.151   76-16504
ISBN 0-87933-244-1

Exclusive distributor: **Halsted Press**
A Division of John Wiley & Sons, Inc.
ISBN: 0-470-98930-0

To Nathaniel and Morgan
and the new generation

PREFACE

This book is the proceedings volume of the Seventh Annual Geomorphology Symposium held in the Department of Geological Sciences at the State University of New York at Binghamton on September 24–25, 1976. The organization of this meeting, "geomorphology and engineering," was set in motion more than three years ago in correspondence with Professor John Orsborn, then chairman, Department of Civil and Environmental Engineering at Washington State University, who suggested the idea for combining the two disciplines. With the exception of the first symposium on environmental geomorphology, the other five have dealt largely with landforms, surface processes, and techniques of analysis of primary interest to geomorphologists. Thus it is appropriate to provide data, examples, and methods that are applicable on the interdisciplinary level and to show the joint involvement of earth scientists with the engineering profession.

The potential subject matter that comprises the total interface of geomorphology and engineering is so vast that an attempt at complete coverage in a single volume would be impractical. Therefore, it is not intended or implied that this book is comprehensive or includes all important subject areas. Instead, the various chapters are suggestive and illustrative of the broad range of subject matter. Indeed, several of the other symposiums contain pertinent data on some of the fields, and the interested reader is referred to them. [For example, elements of coastal engineering are discussed in *Coastal Geomorphology* (1973) and highway engineering is covered in *Environmental Geomorphology* (1971) (both volumes edited by D. R. Coates and published by State University of New York, Binghamton, N.Y.]

Several criteria were considered when selecting a team of authors–speakers for the symposium. Since one purpose of the symposium is to expand communications between the geomorphology and engineering disciplines and to allow for interaction, there is an even split; half the articles are by geomorphologists and half by engineers and engineering geologists. The authors represent a broad spectrum of professional pursuits, including federal and state governmental agencies (U.S. Department of Agriculture's Forest Service and Soil Conservation Service, Corps of Engineers, U.S. Geological Survey, New York State Geological Survey), private business (Hershey) and private consulting firms, and a variety of universities and colleges. The authors come from broad geographic areas in North America, and they describe a range of field localities from Alaska to Canada, and from Washington, Oregon, California, and Colorado to the Mississippi River Basin. In the east, a variety of case histories are developed in such states as North Carolina, Maryland, Pennsylvania, and New York. The themes and topical areas that comprise the subject matter of the volume have been grouped into five parts for reader convenience and organizational style.

Part I: Methods and Mapping. Chapter 1, by Coates, sets the stage by describing the nature of the techniques, subject matter, and philosophies of the geomorphology and engineering disciplines. Chapter 2, by Olson, discusses the importance of mapping and understanding soils and how this relates to wise landuse decisions. Examples in New York and Central and South America are provided. In Chapter 3, Schmidt and Pierce describe the soil mapping techniques being used by the U.S. Geological Survey in the Denver region. Kreig and Reger, in Chapter 4, analyze the terrain classification system that was used to aid in routing the Trans-Alaska Pipeline and its network of support roads.

Part II: River Engineering. Although many other chapters contain elements of river planning and management, the papers in this section are exclusively devoted to an understanding of the dynamics and properties of the river. In Chapter 5, Noble reviews manmade structures of the Mississippi River, the great flood of 1973, and engineering implications. The companion Chapter 6, by Kolb, discusses the importance of the recognition and interpretation of alluvial landforms and soils for levee construction to reduce sand boils during high-water stages of the Mississippi River. Keller, in Chapter 7, analyzes the controversial engineering topic of channelization and suggests techniques that can be used to minimize environmental damage to streams. In Chapter 8, Orsborn provides quantitative methods that can be used by engineers for design purposes on rivers in ungaged watersheds.

Part III: Resource Engineering. Part III deals with the engineering and geomorphic relations involved when man plans and develops resources. Water is a recreational and scenic resource, and its impoundment creates new economies and the utilization of raw materials. Construction of Kinzua Dam and the rerouting of communication facilities are discussed by Philbrick in Chapter 9, wherein he shows the relation of geomorphology and engineering. Swanston and Swanson in Chapter 10 discuss the large array and chain reactions in erosion and sedimentation that result from improper roads and harvest methods for such living resources as timber in the Pacific Northwest. Chapter 11, by Fakundiny, shows the importance of landuse planning and resource assessment in the mining and transport of such nonrenewable resources as sand and gravel for the greater Rochester, New York, region.

Part IV: Urbanization Effects. Different aspects of man's impact on natural systems in urban areas are discussed. The effect that urbanization has on major floods, erosion, and sedimentation in stream channels in Maryland is analyzed by Fox in Chapter 12. Landslides provide geomorphic hazards in many urbanizing areas of California; in Chapter 13, Leighton provides several case histories and discusses how geomorphology proved effective in cost and hazard reduction. Urban areas produce enormous waste products that must be disposed of; Foose and Hess, in Chapter 14, provide a case history of a landfill site in Pennsylvania where imaginative methods were employed to eliminate contamination of the land–water ecosystem.

Part V: Geomorphic Synthesis. Legget, in Chapter 15, summarizes the role of geomorphology in engineering planning by using case studies drawn from Canada and Europe of landslides, floods, and subsurface features. The concluding Chapter 16, by Palmer, shows the broad range of environmental and management considerations that are necessary to assess the entire river corridor. By using such an approach, the maintenance and integrity of the river process can be preserved.

I wish to thank the Geomorphology Group and the Department of Geological Sciences for their support of the symposium. Special thanks are reserved for Dr. Jorge Rabassa of the Barloche Foundation of Argentina for his assistance in editorial review during the year he spent at Binghamton while on a Fulbright Scholarship.

DONALD R. COATES

CONTENTS

LIST OF CONTRIBUTORS

Donald R. Coates
    Department of Geological Sciences, State University of New York at Binghamton

Robert H. Fakundiny
    New York Geological Survey, Museum and Science Service

Richard M. Foose
    Department of Geology, Amherst College

Helen L. Fox
    Soil Conservation Service, U.S. Department of Agriculture, College Park, Maryland

Paul W. Hess
    Hershey Foods Corporation, Hershey, Pennsylvania

Edward A. Keller
    Department of Geological Sciences, University of California, Santa Barbara

Charles R. Kolb
    3314 Highland Drive, Vicksburg, Mississippi

Raymond A. Kreig
    D. J. Belcher & Associates, Inc., Ithaca, New York

Robert F. Legget
    531 Echo Drive, Ottawa, Canada

F. Beach Leighton
    Leighton and Associates, Irvine, California

Charles C. Noble
    Charles T. Main, Inc., Boston, Massachusetts

Gerald W. Olson
    Department of Agronomy, New York State College of Agriculture and Life Sciences

John F. Orsborn
    Department of Civil and Environmental Engineering, Washington State University

Leonard Palmer
    Earth Sciences Department, Portland State University

Shailer S. Philbrick
    117 Texas Lane, Ithaca, New York

Kenneth L. Pierce
U.S. Geological Survey, Denver, Colorado

Richard D. Reger
Alaska Division of Geological and Geophysical Surveys, College, Alaska

Paul W. Schmidt
U.S. Geological Survey, Denver, Colorado

Frederick J. Swanson
Department of Geology, University of Oregon

Douglas N. Swanston
Forest Service, U.S. Department of Agriculture, Corvallis, Oregon

I
METHODS
AND MAPPING

1

GEOMORPHIC ENGINEERING

Donald R. Coates

INTRODUCTION

Many dictionaries describe engineering as the business of planning, designing, constructing, and managing machinery, roads, bridges, highways, dams, tunnels, etc. Others have defined it as the art, or even the science, of using power and materials most effectively in ways that are valuable and necessary to man. Engineering has been divided into many subfields, which represent nearly all the activities of man. Some of the older and more traditional fields include civil, mining, structural, hydraulic, sanitary, electrical, chemical, mechanical, military, agricultural, and related engineering specialities. Some newer fields in which the appellation "engineering" is commonly used include traffic, communications, illuminating, hydroelectric, aeronautical, automotive, heating, ventilation, acoustical, electronics, marine, and nuclear. New terms and combinations are continually instituted. For example, the U.S. Army Corps of Engineers, which was traditionally oriented along civil engineering lines, has begun to use such new terms as "coastal engineering," as in their Coastal Engineering Research Center (CERC) near Washington, D.C., and many of their works in the Mississippi River Basin are described as being "river engineering." Thus, the geomorphic process is increasingly being recognized as a significant component of engineering interest.

Geomorphology is the science of the study of landforms and the processes that create them. That part of the discipline which comes into contact with the engineering profession is the area that Coates has called "environmental geomorphology." He has defined this area as follows:

> Environmental geomorphology is the practical use of geomorphology for the solution of problems where man wishes to transform landforms or to use and change surficial processes (Coates, 1971, p. 6).

It thus involves

> (1) study of geomorphic processes and terrain that affect man, including hazard phenomena such as floods and landslides; (2) analysis of problems where man plans to disturb or has already degraded the land—water ecosystem; (3) man's utilization of geomorphic agents or products as resources, such as water or sand and gravel; and (4) how the science of geomorphology can be used in environmental planning and management (Coates, 1972, p. 3).

It therefore follows that

> The goal for geomorphic environmental studies is to minimize topographic distortions and to understand the interrelated processes necessary in restoration, or maintenance, of the natural balance (Coates, 1971, p. 6).

3

In summary, environmental geomorphology treats man as a physical process in changing the terrain, in the same manner that other surficial forces transform the landscape, such as rivers, oceans, winds, gravity movements, etc. Since man lives, works, and plays on the surface of the earth, many of his activities are designed to modify the land–water ecosystem. This habitat is the province of the geomorphologist, who is trained in understanding what constitutes process and landform equilibrium.

The activities of man in civilized society can be broadly grouped into water resources, living resources, nonrenewable resources, and services.

Relation of Geology, Geomorphology, and Engineering

Because geology, the science of the earth, is so broad and complex, it has been subdivided into many fields. A variety of names has been used for those disciplines that have some overlap with engineering: engineering geology, economic geology, environmental geology, environmental geomorphology, urban geology, urban geomorphology, etc. It is the thesis of this chapter that because of the vast spread in competences—and yet the need exists for coordinated expertise—the area of *geomorphic engineering* is a significant and integral part of man's stewardship of the earth and needs recognition as such.

In describing the interaction of geologists and engineers, Legget (1962) points out

> ... how closely the science and the art are related and how dependent civil engineering work generally must be upon geology. It is, indeed, no mere figure of speech to say that the science of geology stands in relation to the art of the civil engineer in just the same way as do physics, chemistry, and mathematics
>
> Thus arises the need for cooperation between the civil engineer and the geologist, the practical builder and the man of science. ... This partnership is, in some ways, a union of opposites, for even the approach of the two to the same problem is psychologically different. The geologist analyzes conditions as he finds them; the engineer considers how he can change existing conditions so that they will suit his plans (pp. 2–3).

There was a time when engineering geology meant only the application of geology to such civil engineering works as roads, dams, tunnels, etc. Recent writings and trends, however, are now claiming that it embraces such subjects as soil and rock mechanics, power siting, and even landuse management. The growth of the discipline is indicated

> ... by the increasing number of citations per year under the heading "engineering geology" in the Geological Society of America's Bibliography & index of geology: 1969 (588), 1970 (586), 1971 (1,065), 1972 (2,063), 1973 (2,500 estimated) (Lee, 1974, p. 19).

The areas of rapid increase

> ... could be pinpointed in several areas, including land-use and zoning controls, resource management, nuclear-reactor siting, underground construction, subsidence due to mining and fluid extraction, construction in seismic areas, and a wide variety of costly slope-stability problems.

And in 1975, Lee (1975) announced that

> Energy materials exploration and development were the dominant issues involving
> engineering geologists in 1974. Various extraction procedures will change the
> surface and subsurface environment. These changes will be significant; some may
> be tolerable, and others intolerable and dangerous (p. 28).

In his review article on what constituted important elements in engineering geology for
1975, Throckmorton (1976) devotes most of his analysis to landuse planning and when
speaking of the third world nations states that

> We have a golden opportunity [U.S.A.] to help prevent the mistakes made when the
> United States was making similar spectacular strides in the use of land and the
> development of natural resources (p. 18).

Another approach to understanding how engineering geologists perceive their discipline is to review the types of studies that constitute the 10 case-history volumes that the
Engineering Geology Division has prepared for publication by the Geological Society of
America. Four of the volumes are organized around a central theme; these include legal
aspects, rock mechanics, rapid excavation, and *Geologic Mapping for Environmental
Purposes,* the most recent (Ferguson, 1974). The other six volumes each contain a variety
of topics and include the geology of tunnels, bridges, dams, highway embankments,
artificial recharge, aggregate, subsidence, landslides, till, rubble sources, radioactive
wastes, relief wells, rock removal, sensitive sediments, swelling of rock, reservoir loading
and waterflooding in relation to earthquakes, tsunamis, nuclear explosions, disposal wells,
etc.

Thus, it is becoming increasingly clear that engineering geologists now conceive of
their role as greatly enlarged and more multifaceted. It has been expanded to embrace
everything from the original kinship with civil engineering to analysis and decision making
in nuclear siting, landuse policy, and even economic geology. Because the discipline is
currently threatened with the danger of running in all directions, of being spread very
thin, and may have reached a critical mass (such as in New York City, which has such a
vast array of problems as to be nearly unmanageable), it is time to review goals, purposes,
and content to see whether alternative strategies may more effectively use the talents of
the engineering geologist.

In recent years, the engineering profession has reviewed its mission, and increasing
numbers of universities are broadening their base into the environmental sciences. For
example, by 1975 at least 43 major universities had instituted graduate degree programs
in departments, divisions, or schools of "environmental engineering" (Hufschmidt, 1975,
p. 46). Typical names that are now employed include Civil and Environmental Engineering, Environmental Systems Engineering, and Environmental Sciences and Engineering.
Again, it becomes apparent that, with such an all-encompassing eclecticism, service to
science and society will be enhanced by a redefinition of some of the elements that
constitute this growing morass.

Scope of Geomorphic Engineering

The education and training of a geomorphic engineer needs perceptive and balanced
treatment. Basic philosophical differences may occur in the extreme views of geo-

morphologists and engineers. The geological background of the geomorphologist has trained him to think in vast time dimensions (except for hazard and such catastrophic events as floods, hurricanes, landslides, earthquakes, and volcanic activity), whereas the engineer is generally summoned to solve rather immediate or short-range problems that are measured in days, months, or a few tens of years. Another difference concerns the scale of operations. Engineers are taught to solve individual problems at specific localities. Although many exceptions occur, construction activities generally are not regionally designed. The immediate problem may be solved as far as the local contractor, or even city, is concerned, but the construction may prove damaging to contiguous areas. The location of groins may save the individual property that it was designed for, but the natural feedback mechanism of the coastal regime may cause severe erosion in downdrift properties. Thus, the total land—water ecosystem must be understood, evaluated, considered, and managed, instead of using a piecemeal approach. Such analysis would heed the truth behind the saying, "The operation was a success but the patient died." Geomorphology and engineering are united under the EIS clauses (Environmental Impact Statement) of the National Environmental Policy Act of 1969 (NEPA). Here it is mandated that major industrial and governmental construction activities must consider alternative actions and sites and must predict the environmental changes of the construction. Thus, the complete evaluation of all terrain forms and processes is necessary for legal compliance with the NEPA.

Books on environmental awareness and the need for more earth science input in the decision-making process, such as those by Flawn (1970) and Coates (1971), are of rather recent vintage in the geological literature. Actually, the call for action and need for interdisciplinary terms in the management of terrain systems received much of its impetus from landscape architects, such as Whyte (1968) and McHarg (1969), and many elements they describe constitute some of the content of *geomorphic engineering*.

Geomorphic engineering combines the talents of the geomorphology and engineering disciplines. It differs from environmental geomorphology, wherein man is studied as one of the typical surface processes that change the landscape, and instead brings knowledge of physical systems to bear on problems that may require construction for their solution. The geomorphic engineer is interested in maintaining (and working toward the accomplishment of) the maximum integrity and balance of the total land—water ecosystem as it relates to landforms, surface materials, and processes. This approach also differs from that of the engineering geologist, who supplies the data base for construction in terms of strength, distribution, and structure of rock types to be encountered in foundations and excavations. A useful model that embodies the type of unity that should be inherent in geomorphic engineering work is that provided by such groups as CERC. This *coastal engineering* subdiscipline combines process and sediment-oriented specialists with design and planning personnel. Thus, a strong geomorphic component is brought into full orchestration with those who deal with the problems related to decision making on and management of beaches, coastlines, and harbors. In similar fashion, *river engineering* (see Part II and Chapter 16) would include those scientists with special skills in morphology of river systems, channels, and hydrogeology. *Resource engineering* includes not only the extraction and planning that accompanies mineral mining, but total management of the system in site development, delivery systems, and rehabilitation procedures.

There is a continuous spectrum of viewpoints of man and his relation to the environment. Preservationists would represent one end of the series, whereby man's inheritance of the planet mandates him to leave everything alone without disturbance. Complete adherence to such a policy would revert civilization back and beyond the horse and buggy stage. The other end of the spectrum would be represented by the money dedicated entrepreneur-type capitalist (in Marxian dogma) who is only interested in the quickest and most profits to himself. While working in Arizona for three years with the U.S. Geological Survey (1951–1954), I unfortunately encountered more than a few such land barons and absentee landlords, who instructed their tenants and workers to grow as much cotton as rapidly as possible, without regard for the amount of water required in such operations or for that loss of soil nutrients or salt buildup. Therefore, it is hoped that a certain morality might be associated with geomorphic engineering decisions, such as practiced by the Amish in their land and soil husbandry philosophy. They know their farms will be handed down to their children and to other generations, so they build a bit of immortality into their unusual care and management of the earth and its products. Thus, the geomorphic engineer should take every study and precaution to assure that (1) the structures that are built are necessary and will accomplish their intended purposes, (2) construction is located at the optimum site that will cause minimum environmental disturbance, and (3) planning and management have accounted for environmental feedback on contiguous lands and waters.

One place that requires the interaction of many disciplines is the design of "open space." Here priorities must be assessed, natural processes maintained as much as possible, and multiple landuse employed whenever feasible. Thus, geomorphology, engineering, economic geology, and landscape architecture must join forces for stewardship of the cityscape.

WATER RESOURCES

We shall now turn our attention to a brief description and analysis of those subfields where the expertise of the geomorphic engineer is an important ingredient in their use and management. Water can be considered as the basic building block in civilization.

> You could write the story of man's growth in terms of his epic concerns with water. . . . The habits of men and the forms of their social organizations have been influenced more by their close association with water than with the land by which they earned their bread (Frank, 1955, pp. 1–2).

> According to our present information, it would seem that, prior to the commercial and industrial revolution, the majority of all human beings lived within the orbit of hydraulic civilization (Wittfogel, 1956, p. 161).

Wittfogel's thesis subscribes to the old adage that "necessity is the mother of invention." He traced the development of civilization in many of the early great empires to man's requirements for water in the agrarian economy. To transport, handle, monitor, and use water, science and engineering had to combine with a strong governmental structure based on firm legal foundations for the total management of the irrigation water system.

> Irrigation demands a treatment of soil and water that is not customary in rainfall farming. The typical irrigation peasant has (1) to dig and re-dig ditches and

furrows; (2) to terrace the land if it is uneven; (3) to raise the moisture if the level of the water supply is below the surface of the fields; and (4) to regulate the flow of the water from the source to the goal, directing its ultimate application to the crop (Wittfogel, 1956, p. 157).

The use of water as a resource involves many geomorphic considerations, both in surface water and groundwater planning and management, and the knowledge of the engineer in constructing, for example, dams to impound water for storage in reservoirs, canals to transport water from surplus to deficit areas, and wells to transfer water from below ground to the surface. Each of these areas has its own special problems and constraints that need a geomorphic engineering viewpoint.

The reciprocal linkage between terrain and construction knowledge is well illustrated by the irrigation works of the hydraulic engineers who designed the canal networks in India during the nineteenth century. In irrigation agriculture it is vital to design channel shape and gradient with utmost precision; if improperly engineered, erosion will occur where gradients are too steep, and siltation where slopes are too shallow. Thus, it was no accident that some of the equations for water transport in open systems were developed using the empirical methods of the trained field observer. Using other techniques in natural stream systems, Leopold and Maddock (1953) developed the laws of hydraulic geometry, which have wide applicability to stream design characteristics when channelization is required (see Chapter 7).

Use of water in irrigation can have a double-barreled reaction: siltation *and* salinization. Jacobsen and Adams (1958) describe some of the effects of salt encrustation of soils and the resulting loss of fertility and crop productivity.

> . . . that growing salinity played an important part in the breakup of Sumerian civilization seems beyond question (p. 1252).

They discuss how, in the Diyala region of Iraq, silt in the fields raised their level 1 m (3 ft) in 500 yr, and other authors have indicated that the cleaning of canals consumed the time and energy of more than half the labor force.

Not only do surface waters introduce salinity problems in some areas, but so can groundwater usage. The excessive withdrawal and depletion of groundwater reservoirs can lead to still another problem: subsidence. The Imperial Valley of California suffers from the twin problems of salinization and subsidence. Millions of dollars are spent yearly to retile and relevel the fields. To avoid waterlogging and salt encrustation, underdrain tiles, which previously were often spaced at intervals of 200 ft and more, are now being spaced at intervals as short as 50 ft. The subsiding areas need yearly maintenance and resetting of canals, ditches, and pipes.

LIVING RESOURCES

Engineers have played a vital role in the development of the agricultural croplands that produce the living resources so necessary for man's subsistence. Although such resources are renewable, in many parts of the world good cropland is at a premium. In such situations, it is the engineer who must undertake those terrain adjustments wherein landforms are modified and new agricultural lands are created. Land-reclamation sites include coastal zones and other wetland environments, hillside terracing, and stream channelization.

The Netherlands provides the largest scale of landscape metamorphism by man. About one third of the country has been changed or created by engineering projects. Van Veen (1962) has reviewed the history of Dutch farming, and calculated that up until 1860, with the use of only hand and animal labor, the following had been accomplished: (1) moved 100 million cubic yards (yd^3) of earth into small hills, (2) moved 200 million yd^3 to build 1,750 mi of working dikes and 50 million yd^3 for dikes later abandoned, (3) dug out 800 million yd^3 for draining of land, (4) dug out 200 million yd^3 for canals, and (5) removed 10 million yd^3 of peat. By comparison, only 100 million yd^3 were removed in the digging of the Suez Canal (mostly by powered machinery). Since 1918 the Dutch government has been planning and building an immense reclamation project; when completed, it will add nearly 950 square miles (mi^2) to the country by draining about two thirds of the Zuider Zee. A 20-mi-long dike that stands 25 ft above the North Sea and 300 ft wide at the crest was completed in 1932 as the first step in the project known as the Delta Works.

The draining of wetlands and channelization of streams may be engineered for a variety of reasons, but at least some of the changes are made in the name of creating new agricultural and pastoral lands. Such changes in the natural regime and balance provide especially good examples of the feedback mechanism that can endanger the entire hydrologic regime. Although the U.S. government has been channeling rivers since the 1870s, when the Corps of Engineers initiated work along the Mississippi River, it was not until the mid-1950s that large-scale alteration of waterways for agricultural purposes got seriously underway. The Soil Conservation Service (SCS) of the U.S. Department of Agriculture (largely through the Watershed Protection and Flood Prevention Act of 1954, P.L. 566) has helped farmers widen, deepen, and straighten more than 8,000 mi of streams in all states. During the same time, the Corps of Engineers has straightened another 1,500 mi of waterways. The SCS has congressional approval (although not necessarily funding) for ditching an additional 13,524 mi of streams. Emerson (1971) has described how Blackwater River in Missouri was dredged in 1910 and shortened 24 km (about 15 mi). This resulted in tributaries becoming entrenched, serious erosion of banks, and headward erosion of gullies:

> Channelization has enabled more floodplain land to be utilized in the upper reaches of the Blackwater River. This benefit must be weighed against erosional loss of farmland, cost of bridge repair, and the downstream flood damage. . . (p. 326).

Wetlands provide another arena where man is destroying the geomorphic process that operates in this closely tuned but somewhat fragile land–water ecosystem.

> Although the nation's marshes, swamps, and bogs are among the most productive landscapes in the world, these liquid assets have suffered greater destruction and abuse than any other natural habitat manipulated by man. As a result of draining, dredging, filling, and/or pollution we have in the conterminous United States reduced the nation's wetland asset to 70 million acres, slightly more than half the original acreage (an estimated 127 million acres). And the destruction is continu-

ing at an accelerated pace of 1 per cent or more per year . . . (Niering, 1968, p. 177).

This environmental setting is especially adapted for providing excellent opportunities for the geomorphic engineer to enter into the decision-making process. It is important that he show the importance for maintaining the integrity of wetlands for living resources and their vital link in the entire food-chain ecology. Wetlands have cushioning and absorptive qualities in minimizing stresses caused by climate abnormalities in flood and drought periods. They also provide important oxidation and sedimentation basins where nutrients can be trapped and used in the total ecosystem.

Another area where man has transformed the landscape for cropland is the land reclamation of hillsides by terracing. Some of the largest terrace systems have been built in China, Indonesia, Japan, and the Philippines. The Incas were early terrace experts, building banks of 20 to 30 terraces and occasionally as many as 50. The maintenance of terraces as an efficient soil and crop resource requires constant care, and those that are most expertly designed and constructed for the local environment need minimal attention. The many different terrace types can be classified into two groups: the bench-type terrace and the ridge-type terrace. The character of the soil, hillslope orientation, degree of slope, and precipitation properties should determine the design of the terrace. For example, horizontal terraces are best adapted to steep slopes and have less erosion, but are poorly adapted if soils are impervious and require more races on the slope.

In an analysis of early agricultural methods in the Negev region, Kedar (1957) describes, in what has become a controversial paper, an unusual type of soil reclamation. Most workers in this region have argued that water was the limiting constraint to landuse by the ancient inhabitants (Evenari et al., 1961). The region contains many small check dams, and on hillslopes stones have been carefully placed in piles or along stone stripes. Evenari et al. (1961) believe that the dams were built for retention of water to be used for irrigation, and that the stones were removed from the loess-type soils to increase surface runoff into the reservoirs. Kedar (1957) believes, however, that the amount of soil and level cropland were the primary agricultural constraints, and that the region was deliberately engineered to create a land–water ecosystem wherein the accelerated erosion of debris from hillslopes was channeled into settling basins. These sediments became the source of new reclaimed land, which could then be farmed.

NONRENEWABLE RESOURCES

Resources in which there is "no second crop" are known by a variety of terms, including natural resources, raw materials, mineral resources, geologic resources, etc. They differ from living resources because the geologic processes that form them operate so slowly that additional reserves will not occur in the foreseeable future. Obviously, such resources need extremely careful processing and management. Resources of interest to the geomorphologist are those mined at the surface wherein the extraction process changes and upsets the land–water ecosystem. Not only is a new man-made landscape created, but such geomorphic agents as surface water, gravity movement, and groundwater are altered in direction and magnitude. Coal resources are a concern because strip mining vastly changes terrain. Economic materials such as sand and gravel are of special import, because the geomorphologist becomes involved in original assessments concerning their genesis, and in mapping and evaluating their quantity and quality. The geomorphic engineer should be able to broaden the entire base of operations whereby the location and occurrence of nonrenewable resources are determined, the deposits are mapped and evaluated, technology for their extraction and movement to market is developed, and plans formulated for land rehabilitation when the resource is mined out.

Open-Pit and Strip Mining

Removal of ores by extraction mining methods of open and strip surface methods has received insufficient attention by the geomorphic engineer; yet they constitute a severe type of terrain distortion, such as the copper mines at Bingham, Utah, and Bisbee, Arizona. The massive land destruction from mining of the iron ores of the Lake Superior District has left a pockmarked landscape that rivals a giant cratered lunar surface. Occasionally, the landscape architect has turned his talents to such features with suggestions for their aesthetic restoration. For example, Zube (1966) describes the "moonscape" in parts of Minnesota and Wisconsin and shows how the spoil banks can be made into appealing landforms.

> The land forms can be composed on the basis of the positive qualities of each material into a unique landscape expression that minimizes the problems of dust and erosion.... They must relate in scale and form to the pit ... and tailings dumps can be utilized as elements of transition in bridging the gap between the obvious man-made qualities of the pit and the natural landscape (pp. 138–139).

The problems of energy constitute one of the great controversies of the 1970s, and nationwide battles are being waged concerning nuclear power plants, strip mining of coal, and the possible development of oil shales. In two informative publications, the U.S. Department of the Interior (1967) and the U.S. Department of Agriculture (1968) describe many aspects of surface mining, the changes it creates, and methods that can be used to rehabilitate mined out lands. By 1965, surface operations in the United States had damaged more than 3.2 million acres; two million acres have sheet erosion, 1.2 million acres have rills and gullies, with 400,000 of these acres having gullies deeper than 1 ft. In addition, 320,000 acres have been disturbed by roads supporting mining and exploration activities; fish and wildlife have been disrupted in 2 million acres; and 14,000 mi of streams have been affected, with half of them being impaired in their water-bearing capacity. In Appalachia, 98% of the surface-mined land has inadequate control of storm runoff to prevent erosion, sediment production, or flooding. Studies in Kentucky indicate that under natural conditions 27 tons $(T)/mi^2$ of sediment is eroded from hillsides, whereas in strip-mined areas erosion amounted to 27,000 T/mi^2. The catalog of principles that can be used to prevent erosion, sedimentation, and pollution from mining activities fall within the interest and competencies of geomorphic engineers, such as (1) stabilization of slopes during mining, (2) planning for surface runoff, (3) water storage and sediment settling pond development, (4) location of operations to have minimum impact on the terrain, and (5) control of all drainages from sites and haul roads to deter contamination of adjacent areas.

Although the debate has not been decided whether the United will tap her immense oil shale reserves in such states as Utah, Colorado, and Wyoming, the extraction and environmental problems are under study by a variety of governmental and university groups. All agree that massive terrain dislocations would occur. If mined, several plans call for the rock spoil to be shoved into landforms resembling the present terrain. However, this would be unwise, according to Everett et al. (1974), because existing topography is already undergoing extensive erosion. Instead, a more appropriate geomorphic engineering solution would be to construct broader and flatter mesa-type topography capped by a man-made rubble pavement. Carefully designed man-made pediments would produce less erosion than is presently occurring in the natural landscape. Another suggested possibility is to place waste materials into unimportant tributary canyons and to engineer the crest with gentler slopes so that it rises to or above the level of existing terrain. Such solutions challenge the adage that "nature knows best." It is highly probable that, under the prevailing climatic conditions in the West, man can create topography that is more erosion-resistant and will produce less sedimentation by-products than the existing natural conditions.

Gold Mining

Economic geologists have been viewed as the culprits and perpetrators of the exploitation of gold and its resulting terrain disfiguration on the western slopes of the Sierra Nevada Mountains. Here entire hillsides and terraces have been washed away by hydraulic mining techniques or by bucket dredging machines. The waste products have been washed downstream, where additional destruction occurs. Gilbert (1917) was the first scientist to describe completely the havoc of these placer-mining operations. He showed how the tremendous amount of sediment clogged the streams, entirely changing their flow regimes and leading to increased flooding.

> The interests that suffered most acutely and consciously by reason of the great wave of mining debris from the Sierra were those of lowland farm lands and lowland towns. Some farms were buried, and for others the cost of protection from inundation was increased. Towns had to levee against sands and rising floods and were deprived of the advantages of river transportation (p. 104).

In addition, massive ecological changes occurred in wetlands of bays, such as in the San Francisco area. Gilbert calculated the total sediment addition to the bays to be in excess of 1.14 billion yd^3, causing extension of shoaled areas. The total amount of sediment produced by mining up to 1909 was eight times the material moved during the digging of the Panama Canal. Beatty (1966) analyzed some of the derelict lands, especially those produced during bucket dredging mining, and provides insight on their restoration.

> Mined land rehabilitation is a complex process involving a whole gamut of environmental sciences concerning soils, water, horticulture, engineering, and others depending on the ultimate land use (p. 128).

Sand and Gravel

Although sand and gravel deposits may not have the glamour or appeal of some of the more exotic materials, in such states as New York they are the most important and economically valuable mineral resource. There are many ramifications to their occurrence, mining, and transport (see Chapter 11). Their somewhat ubiquitous location in many valleys of New York State is a real headache for the attorney general. Numerous lawsuits are instituted by property owners when their lands are condemned and taken by "eminent domain" for construction of new roads. I have provided testimony in more than 30 such cases (Coates, 1971). In such litigation, the consulting geomorphic engineer provides a report and court witnessing to the location, quality, quantity, accessibility, and market conditions of the sand and gravel products.

The New York State Department of Transportation (NYSDOT) enacted regulations in 1972 that require a Geologic Source Report for all sand and gravel operators in the State that sell their materials for use on State roads. The reports must be written by an approved and certified geologist from a listing supplied by NYSDOT. Before the product can be used and licensed, the report must be approved by NYSDOT. I have written 12 of these reports, which are in the open file of NYSDOT, and from these and the tens of others written by scientists acting in their capacity as geomorphic engineers, the agency is compiling an important record of types of deposits and their quantity, quality, and reserve potentials throughout the State.

SERVICE ENGINEERING

This area of engineering covers those construction activities that maintain the free flow of goods and services and preserve the health, safety, and welfare of the people. Dams may be single or multipurpose for water supply, hydroelectric power, flood prevention, and recreation. Roads serve as transportation arteries for business, commerce, and recreation. Levees, dikes, seawalls, groins, jetties, etc., are built to protect investment and safety in homesites, industry, harbors, and navigation. Such features have in common a very complex problem, because by their very design they change the natural system to fit man's purposes. Thus, dams and groins impede and restrict normal flow patterns, upset equilibrium, and cause the second law of thermodynamics to operate. Such action results in a feedback reaction through which the amount and distribution of energy impacts throughout the system. For example, dams act as settling basins for sediment. Downstream from the dam, the river erodes because it no longer is burdened with sediment; still farther downstream, deposition of the eroded sediments occurs. Leopold et al. (1964) showed that 35,000 acre-ft of erosion occurred in a 100-mi stretch downstream from Denison Dam of the Red River, and the streambed was lowered 5 to 7 ft in the first 10 mi. In the case of Boulder Dam, erosion occurred immediately downstream, which led to siltation at Needles, California, where the Colorado River channel became so clogged that flooding occurs in a town that experienced no flooding prior to construction of the dam.

Highway design and planning also require a wide range of expertise because of the many different types of environmental disruptions that can occur. Parizek (1971) presents an exceptionally fine overview of changes that can occur through faulty construction and positioning of roads. The following impacts *may* occur: (1) beheading of aquifers, (2) development of groundwater drains, (3) damage and pollution to water supplies, (4) changes in ground and surface water divides and drainage areas, (5) reduction of induced streambed infiltration rates, (6) siltation of channels, causing flooding, erosion, and reduction in recharge areas, (7) obstruction of groundwater flow by abutments, (8) changes in runoff and recharge characteristics, and (9) upsets in slope stability.

The management of the coastal zone is closely linked to geomorphology and engineering considerations, and we shall single out this area for discussion as an example of involvement by geomorphic engineers in the field of service engineering.

Coastal-Zone Management

The coastal corridor has become a favorite real-estate commodity, and it is continuing to increase in importance. Shorelines are recognized as being extremely fragile areas, so extreme care must be exercised in their management. The delicate nature of beach morphology, when coupled to the feedback mechanisms inherent in coastal processes, causes man many problems when he attempts to alter the normal system. Those who urge the need to maintain beaches by man-made methods present two principal arguments: (1) the protection of property investments, and (2) the preservation of sand as a resource. Shoreline protection schemes can be placed into three main categories: (1) those that inhibit direct wave attack, which include seawalls, bulkheads, revetments, and breakwaters, (2) those designed to inhibit the transport of sand by currents, which include jetties built at bays and inlets, and groins, and (3) those that change the beach-zone topography, such as artificial beach nourishment and construction of sand dunes and dikes.

As Coates (1973) has pointed out, a paradox can occur when man attempts to manipulate coastal areas in the name of conservation. His attempts to preserve the environment often result in even greater environmental damage. As Dolan (1973) and Godfrey and Godfrey (1973) have shown, man's attitude of considering nature as an antagonist, thereby making it necessary to fight and control her, led to very sad results in the Outer Banks of North Carolina. Here there is a striking difference between the natural beaches and those that man tried to "protect" with an extensive artificial sand dune barrier system. The unaltered beaches are wider and show less shoreward movement than the controlled beaches. The beach narrowing process, where artificial systems were constructed, amplifies high wave energy, which is concentrated in an increasingly restricted run-up area, resulting in steeper beach profiles and increased turbulence with accelerated sediment attrition.

> The important point is that dunes, overwash, wind, grasses, inlets, and storms are all part of a very flexible, changing environment and successful management must likewise become flexible. The lesson which emerges, however, is clear; man, not the sea, is the worst enemy of barrier islands today (Godfrey and Godfrey, 1973, p. 257).

Groin fields and jetties are notorious among conservationists, because in case after case it is demonstrated that they distort the natural balance of coastal systems, and cause increased erosion rates in downdrift areas.

> The construction of harbor jetties is an even more important cause of beach erosion. If the longshore current is predominantly in one direction, sand will build out on the upcurrent side of the jetties, often making the beach on that side too wide, as at Santa Barbara. On the downcurrent side, the beach is cut away by winter storms, but the ordinary replenishment that would come in summer is not possible because the sand is stored on the other side of the jetties. As a result, many downcurrent beaches have been destroyed. This is particularly true in

Southern California where most harbors require jetties. Examples of erosion on the East Coast due to jetty building are found at Cape May, New Jersey, and at Ocean City, Maryland. South Cape May has virtually disappeared during the past 50 years due to the jetties to the northeast and to the southerly current. Assateague Island has been eroded at least 1,500 feet because of the Ocean City jetties (Shepard and Wanless, 1971, p. 548).

Groins produce much the same type of effect. For example, groin fields at Westhampton Beach and Ocean Beach on Long Island, New York, have caused adjacent property damage, and have contributed to a fivefold rate of accelerated erosion in downdrift areas. The extensive groin fields of Miami Beach, Florida, were started after the 1926 hurricane but have not served their purpose of beach stabilization or of prevention of sand movement by littoral drift.

Man's use of other types of construction has been equally damaging to the coastal environment. A pass was cut through the Bolivar Peninsula in Texas. The presence of the pass changed the water circulation pattern in Galveston Bay and East Bay so that erosion was accelerated by the new configuration, and Hurricane Carla in 1961 was able to erode more than 900 ft from the peninsula that otherwise would have been naturally protected. Breakwaters can also cause abnormal upsets and damages such as at Redondo and Santa Barbara, California.

> In 1929, against the advice of the Corps of Engineers, Santa Barbara, California, erected a breakwater to shelter small craft and attract tourist income. The basis of the Corps' concern immediately became evident. Above the breakwater beach frontage ballooned out some seven hundred feet. Within the breakwater sand began shoaling up at the rate of eight hundred cubic yards a day. Below the breakwater the shore receded as much as two hundred forty-five feet. Ten miles downstream a row of beach homes slumped into the sea. The breakwater that cost less than a million dollars to build caused twice that amount in property damage (Marx, 1967, p. 36).

Inman and Brush (1973) have also discussed man-made beach changes. Silver Strand Beach in Southern California has no natural inland sources of sediment since Rodriques Dam was completed in 1937. The beach has had to be artificially maintained by replacing 22×10^6 m^3 of sand in the period from 1941 to 1967. In some areas, the mining of beach sand is an important industry. During 1970, 21.3×10^6 T of sand and gravel were mined from beaches, river beds, and coastal dunes in California. The same year, 112×10^6 T were mined in Great Britain, of which 13.2×10^6 T were mined from offshore banks, many of which provide essential protection to the coastline from wave erosion. In a similar setting, when 500,000 T of gravel were mined offshore near the English village of Hallsands in 1894, it suffered severe damage from waves whose energy was no longer dampened by the shallow topography.

Prior to 1930, the federal government only performed coastal works for protection of federal property or improvement for navigation. In 1930, the Beach Erosion Board was authorized by Congress (Public Law 520, 71st) to undertake studies of beach erosion problems when so requested by other governmental levels. By the time of establishment of the CERC (U.S. Army) in 1963, the Corps of Engineers had been given much broader mandates by Congress, which included study and protection of public beaches (in 1946) and of private beaches if instrumental to the safety of public shores or for public benefits (in 1956). Between the years of 1946 to 1964, the Corps undertook more than 100 projects for prevention of beach erosion that cost $180 million. In 1968, a study was initiated for evaluation of erosion problems on all U.S. coasts and resulted in a series of publications called the National Shoreline Study (such as *Shore Management Guidelines, Regional Inventory Reports,* and *Shore Protection Guidelines,* 1971). These reports showed that 20,500 mi of national and territorial shores, one fourth of the total coastline, are experiencing significant erosion. Of these parts, 2,700 mi are critical and would cost $1.8 billion in initial costs to attempt stabilization and $73 million each year to maintain the protective structures. Many people are now saying it would be cheaper to buy back into the public domain the privately owned and precariously developed beaches than it would be to subsidize their continuing protection. Officials of the National Park Service (U.S. Department of the Interior) have recently stated a policy that may discontinue maintenance funds for rehabilitation of structures damaged by ocean processes in their system of National Seashores.

Beach-nourishment projects have become one of the preferred procedures for man-induced modification of the shoreline. Advocates of this procedure point out that (1) the beaches become suitable for recreation purposes, (2) the erosion effect is checked, but if sand is lost, it supplies other beaches, (3) the method is more economical than other methods, (4) it does not require a long-term management commitment, and (5) it does not disturb as many elements in the system as do other devices, such as groins. The Corps of Engineers has had wide experience in beach-nourishment projects and point to the success of three typical sites in Connecticut. At Seaside Park, Bridgeport, Connecticut, they widened 8,300 ft of shore to a 125-ft width above mean high water. Cost of the project was $480,000. A total of 690,000 yd^3 of borrow was obtained from an area 1,200 ft offshore in 1957, and the effectiveness of the project was measured 5 yr later.

> The data indicate that of the total fill placed in 1957, the loss of 71,500 cubic yards from the zones above mean low water represents only slightly more than a 10 percent movement of material from its zone of placement. Actually, only a little more than 1 percent (8,400 cubic yards) of the initial fill has been lost from the project area. Undoubtedly, quantities of material much greater than those indicated by the available survey data have been moved and redistributed within the study area by littoral forces (Vesper, 1965, p. 15).

The study concluded that the beach can be maintained and stabilized by 12,000 yd^3/yr at an annual cost of $3.35/linear ft/yr. A second project was done at Prospect Beach, Connecticut. Before nourishment, the shore ranged from 0 to 40 ft from seawalls, revetments, and eroding bluffs (Vesper, 1961). The Corps decided to build a 6,470-ft-long beach to a uniform 100-ft width. A total of 442,960 yd^3 was dredged from a borrow area 6,470 ft from shore at an initial cost of $348,576. During a 3-yr study, the Corps concluded that the beach fill had provided the necessary protection. Average annual loss was 13,000 yd^3 by littoral forces, and the desired stability can be maintained at an annual cost of $3.80/linear ft of shore. A third project by the Corps was done at Sherwood Island State Park, Westport, Connecticut (Vesper, 1967). Here a 6,000-ft beach was widened 150 ft in 1957 at a total cost of $559,200 for the imported 557,200 yd^3 of sand, one groin, and a training wall. By 1962, erosion had removed 30 ft of beach but never reached the former shore position. The net loss of material in the project area amounted to 1,000 yd^3/yr and to 19,500 yd^3/yr in the tidal zone. Cost of beach maintenance is \$8/linear ft for a 50-yr amortization period.

Sand is a diminishing resource in coastal areas, and programs for its conservation are becoming more widespread. Loss of this valuable resource is very great in the New York–New Jersey region where the west-flowing littoral drift along the south shore of Long Island (transporting about 600,000 yd^3 annually) merges with north-flowing longshore currents from New Jersey, and all sand is lost as they converge and disappear down the Hudson Canyon. There is great need to intercept these resources by some type of sand-bypassing project, and to place sand back into the system so it can be recycled. Similar losses are occurring in the submarine canyons of California.

> Sand is a rapidly diminishing natural resource. Although once carried to our shores in abundant supply by streams, rivers, and glaciers, cultural development in the watershed areas has progressed to a stage where large areas of our coast now receive little or no sand through natural geological processes. Continued cultural development by man in inland areas tends to further reduce erosion of the upland with resulting reduction in sand supply to the shore. It thus becomes apparent that sand must be conserved. This does not mean local hoarding of beach sand at the expense of adjoining areas, but rather the elimination of wasteful practices and the prevention of losses from the shore zone whenever feasible.
>
> Mechanical bypassing of sand at coastal inlets is one means of conservation which will come into increasing practice. Mining of beach sand for commercial purposes, formerly a common procedure, is rapidly being reduced as coastal communities learn the need for regulating this practice. Modern hopper dredges, used for channel maintenance in coastal inlets, are being equipped with pump-out capability so that their loads can be discharged on the shore instead of being dumped at sea, and it is expected that this source of loss will ultimately be eliminated (Corps of Engineers, 1971, p. 58).

The past 5 years have witnessed an exceptional national surge in the United States to cope more effectively and efficiently with coastal problems. For example, the CERC (U.S. Army) is currently spending more than \$4 million on research, five times the amount in 1965. In 1968, the Ninetieth Congress authorized a national appraisal of shore erosion and shore-protection needs. To satisfy the purposes of the authorizing legislation, a series of 12 reports was published by the U.S. Army Corp of Engineers, including regional inventory reports, shore-protection guidelines, and shore-management guidelines.

> Shore management is defined here as a process of (1) evaluating needs for preserving and enhancing the shore, (2) examining techniques to satisfy the needs, (3) formulating a plan, and (4) implementing the plan. Preservation is seen as maintaining the shore essentially in its current condition. Enhancement is seen as modifying the shore in a way that society judge to be desirable. Both preservation and enhancement may serve society, or ecological balance, or both in a symbiotic relationship (Corps of Engineers, 1971, p. 4).

Governmental planning and action are becoming increasingly significant. They can aid to modify the unfortunate precedent that was set in England in the 1820s when a landowner went to court to obtain relief from erosional destruction of his property caused by a nearby town-constructed groin. In speaking for the court, Lord Tentreden made the decision that "Each landowner may erect such defenses for the land under their care as the necessity of the case requires, leaving it to others, in like manner, to protect themselves against the common enemy." Such a "common-enemy" philosophy precipitated many of the frantic groin races that have occurred in such places as Florida and New Jersey.

Shore objectives can often be satisfied by directly controlling use of both public and private, lands through such regulatory devices as zoning, ordinances, subdivision regulation, building codes, and permits. In Massachusetts, orders have proved effective in regulation of wetlands. In Oregon, courts have upheld the right of government to prevent backshore property owners from fencing private property between mean high water and the line of vegetation. Similar rulings have been made in Texas. Although Michigan has no zoning powers related to its shoreland, other than in high-risk areas, the state does have authority to review all subdivisions along Great Lakes shores and to impose restrictions for buyer protection against inundation. Excellent building codes are in effect at Pompano Beach, Florida, and Wrightsville Beach, North Carolina. In Wisconsin, communities are authorized to zone floodplains, including the Great Lakes shore, and if they fail to do so the state will.

Since 1970, a number of states and the federal government have passed specific legislation that deals with the management of coasts. The Delaware Coastal Zone Act of 1971, the first of its kind, bars heavy industry for a 2-mi zone along the state's 115-mi coastline, and states:

> The coastal areas of Delaware are the most critical areas for the future of the state in terms of the quality of life. It is, therefore, the declared public policy of the state of Delaware to control the location, extent, and type of industrial development in Delaware's coastal areas. In so doing, the state can better protect the natural environment of its bay and coastal areas and safeguard their use primarily for recreation and tourism.

In the election of November 7, 1972, California voted for the protection of the 1,072-mi shoreline from uncontrolled commercial development. The California coastal law sets up a state coastal-zone conservation commission along with six regional commissions to guide future coastal development. When combined with the Environmental Quality Act of 1970, it controls all development within 1,000 ft of the shore and requires both state agencies and private developers to publish detailed reports on the environmental impact of their projects. The state of Washington voted that management of shorelines would be placed in the hands of state officials. Also, in 1972 the federal government passed the Coastal Zone Management Act:

> The Coastal Zone Management Act is the first piece of land control legislation passed by Congress; it includes water control as well. It carries with it encouragement that state government begin to exercise its constitutional authority regarding the control of land and water uses in coastal areas. It provides significant incentives, not only in potential funding, but in permitting states to begin to exert control over federal activities along the coastlines. This represents a considerable challenge to the traditional concepts of local autonomy in these matters (Gardner, 1973, p. 21).

The management of coastal areas involves matters of a regional scale, and consideration must be made of the total environment, which includes such diverse entities as coastal physical processes, recreation and aesthetic appreciation, living-resource extraction, non-living-resource appraisal, waste disposal, transportation, residential—commercial—industrial considerations, and ecological use. For example, on Long Island, New York, The Nassau—Suffolk Regional Planning Board has instituted a Regional Marine Resources Council that is evaluating six major problem areas of coastlines, including wetland management, shore erosion and stabilization, water use, and resource development. Even in Italy, after 27 months of heated debate, the parliament passed a law that states in part, "The safeguarding of Venice and of its lagoon is declared to be a problem of pre-eminent national interest." As a result, $510 million have been appropriated over the next 5 yr to bring freshwater from inland rivers to preserve the lagoon and its surrounding wetlands, to build Venice's first sewage system, and to stop air pollution and restore city housing and monuments.

Coastal management involves setting priorities and, if we want to save our physical environment, Inman and Brush (1973) point out:

Coastal communities are presently in the curious position of rapidly acquiring and improving beach frontage while at the same time they lack criteria for evaluating that the beach will still be in existence in 10 or 20 years. Certainly the future of any coastal man-made structure placed in the path of the longshore movement in a littoral cell is questionable, and great reservations should accompany any commitment to build such a structure.
It is imperative that we develop the means to preserve the beaches and harbors that we now have and that we develop practical techniques for creating new beaches and nearshore structures that are less damaging to the environment. From an environmental standpoint, there are three fundamental steps necessary for the good design of coastal structures: (1) identification of the important processes operative in an environment, (2) understanding of their relative importance and their mutual interactions, and (3) the correct analysis of their interaction with the contemplated design (p. 30).

Finally, Strahler and Strahler (1973) conclude:

Flooding of the fenlands and polders [England] by winter storms is a longstanding environmental problem bringing together processes and principles of many sciences, including meteorology, oceanography, geomorphology, civil engineering, and soil science. The subject makes a good example of the interdisciplinary nature of environmental science. The case we have described shows that Man, in his drive for more food to meet the needs of an increasing population, has radically altered the natural environment and ecosystem of large areas and, at the same time, has aggravated the natural hazards of coastal flooding by his land-use practices (pp. 406–407).

SUMMARY

This missionary message is necessary because the fields of geology, engineering, and environmental science are becoming increasing complex and intertwined. To designate a specific area as *geomorphic engineering* may at first glance appear to be a heretical act of nomenclatural proliferation. However, the viewpoint extolled here claims that such a field of specialization is vital. The reason for its importance is the linkage of two somewhat independent disciplines into a joint area of mutual concern, man and his works. Society will be the poorer if this union is not consummated.

The geomorphologist can aid in providing a broader perspective for the time element, the regional setting, and the nature of feedback systems that act throughout the surficial processes. He must become involved in the tools of engineering, because if construction causes irreparable damage to the land—water ecosystem due to lack of geomorphic input, the earth scientist cannot be absolved of blame. Thus, it is imperative that the geomorphic engineer be involved in the decision-making processes that plan and manage the environment. Only when there is a concerted effort by those who can blend wisdom across disparate fields will a fuller measure of "design with nature" be achieved.

REFERENCES

Beatty, R. A. 1966. The inert becomes ert: *Landscape Arch.,* Jan., p. 125–128.

Coates, D. R. 1971. Legal and environmental case studies in applied geomorphology: in D. R. Coates, ed., *Environmental Geomorphology,* State University of New York, Binghamton, N.Y., p. 223–242.

____, ed. 1972. *Environmental Geomorphology and Landscape Conservation:* Vol. I, *Prior to 1900:* Dowden, Hutchinson & Ross, Inc., Stroudsburg, Pa., 485 p.

____. 1973. *Environmental Geomorphology and Landscape Conservation:* Vol. III, *Non-urban Regions:* Dowden, Hutchinson & Ross, Inc., Stroudsburg, Pa., 483 p.

Corps of Engineers. 1971. Shore protection guidelines: National Shoreline Study, Department of the Army Corps of Engineers, Government Printing Office, Washington, D.C., var. pages in several volumes.

Dolan, R. 1973. Barrier islands: natural and controlled: in D. R. Coates, ed., *Coastal Geomorphology,* State University of New York, Binghamton, N.Y., p. 263–278.

Emerson, J. W. 1971. Channelization: a case study: *Science,* v. 173, p. 325–326.

Evenari, M., Shanan, L., Tadmor, N., and Aharoni, Y. 1961. Ancient agriculture in the Negev: *Science,* v. 133, p. 979–996.

Everett, A. G., Anderson, J. J., Peckham, A. E., and MacMillion, L. 1974. Engineering stability in the future spoil and waste piles of the western United States: *Geol. Soc. America Abstr. with Programs,* v. 6, no. 7, p. 727–728.

Ferguson, N. F., ed. 1974. Geologic mapping for environmental purposes: *Geol. Soc. America Eng. Geol. Case Histories 10,* 40 p.

Flawn, P. T. 1970. *Environmental Geology:* Harper & Row, New York, 313 p.

Frank, B. 1955. The story of water as the story of man: *Yearbook of Agriculture,* U.S. Department of Agriculture, Washington, D.C., p. 1–8.

Gardner, R. R. 1973. Policy alternatives: in *Managing Our Coastal Zone,* proceedings of conference, New York State Sea Grant Program, p. 21–24.

Gilbert, G. K. 1917. Hydraulic mining debris in the Sierra Nevada: *U.S. Geol. Survey Prof. Paper 105,* 154 p.

Godfrey, P. J., and Godfrey, M. M. 1973. Comparison of ecological and geomorphic interactions between altered and unaltered barrier island systems in North Carolina: in D. R. Coates, ed., *Coastal Geomorphology,* State University of New York, Binghamton, N.Y., p. 239–258.

Hufschmidt, M. M. 1975. The planning-environmentally oriented model of water resources: in *The Challenge of Water Resources Education,* University Council on Water Resources, Proc. Ann. Mtg., p. 39–56.

Inman, D. L., and Brush, B. M. 1973. The coastal challenge: *Science,* v. 181, p. 20–32.

Jacobsen, T., and Adams, R. M. 1958. Salt and silt in ancient Mesopotamian culture: *Science,* v. 128, p. 1251–1258.

Kedar, Y. 1957. Water and soil from the desert: some ancient agricultural achievements in the central Negev: *Geog. Jour.,* v. 123, p. 179–187.

Lee, F. T. 1974. Engineering geology: *Geotimes,* v. 19, no. 1, p. 19–20.

_____. 1975. Engineering geology: *Geotimes,* v. 20, no. 1, p. 28–29.

Legget, R. L. 1962. *Geology and Engineering,* 2nd ed.: McGraw-Hill, New York, 884 p.

Leopold, L. B., and Maddock, T., Jr. 1953. The hydraulic geometry of stream channels and some physiographic implications; *U.S. Geol. Survey Prof. Paper 252,* 40 p.

_____, Wolman, M. G., and Miller, J. P. 1964. *Fluvial Processes in Geomorphology:* W. H. Freeman, San Francisco, 522 p.

Marx, W. 1967. *The Frail Ocean:* Ballantine Books, New York, 274 p.

McHarg, I. L. 1969. *Design with Nature:* Doubleday, Garden City, N.Y., 197 p.

Niering, W. A. 1968. The ecology of wetlands in urban areas: *Garden Jour.,* v. 18, no. 6, p. 177–183.

Parizek, R. R. 1971. Impact of highways on the hydrogeologic environment: in D. R. Coates, ed., *Environmental Geomorphology,* State University of New York, Binghamton, N.Y., p. 151–199.

Shepard, F. P., and Wanless, H. R. 1971. *Our Changing Shorelines:* McGraw-Hill, New York, 579 p.

Strahler, A. N., and Strahler, A. H. 1973. *Environmental Geoscience: Interaction Between Natural Systems and Man:* Hamilton Publishing Co., Santa Barbara, Calif., 511 p.

Throckmorton, R. T., Jr. 1976. Engineering geology: *Geotimes,* v. 21, n. 1, p. 17–18.

U.S. Department of Agriculture. 1968. Restoring surface-mined land: *U.S. Dept. Agr. Misc. Publ. 1082,* 17 p.

U.S. Department of the Interior. 1967. *Surface Mining and Our Environment:* Government Printing Office, Washington, D.C., 124 p.

Van Veen, J. 1962. *Dredge, Drain, Reclaim,* 5th ed.: Martinus Nijhoff, The Hague, Netherlands, 200 p.

Vesper, W. H. 1961. Behavior of beach fill and borrow area of Prospect Beach West Haven, Connecticut: *Corps of Engineers Beach Erosion Board Tech. Memo. 127,* 29 p.

_____. 1965. Behavior of beach fill and borrow area at Seaside Park, Bridgeport, Connecticut: *U.S. Army Coastal Eng. Res. Center Tech. Memo. 11,* 24 p.

_____. 1967. Behavior of beach fill and borrow area at Sherwood Island State Park, Westport, Connecticut: *U.S. Army Coastal Eng. Res. Center Tech. Memo. 20,* 25 p.

Whyte, W. H. 1968. *The Last Landscape:* Doubleday, Garden City, N.Y., 376 p.

Wittfogel, K. A. 1956. Hydraulic civilizations: in W. L. Thomas, ed., *Man's Role in Changing the Face of the Earth,* University of Chicago Press, Chicago, p. 152–164.

Zube, E. 1966. A new technology for taconite badlands: *Landscape Arch.,* Jan., p. 136–140.

2

LANDUSE CONTRIBUTIONS
OF SOIL SURVEY WITH
GEOMORPHOLOGY AND ENGINEERING

Gerald W. Olson

INTRODUCTION

More than 1,000 soil scientists are currently mapping about 10,000 soil series and 100,000 soil map units in geomorphic landscapes of the United States, and interpreting those soils for many engineering and developmental applications. Increasingly, many disciplines, agencies, and institutions are working together in cooperative projects to improve the usefulness of soil information and maps. My purpose in this chapter is to outline some of the opportunities for interdisciplinary work relating to soil surveys, and to share some of my experiences in working with soil-survey interpretations over the past two decades. Recent advances in the pedological aspects of soil science have combined with geomorphology and engineering in interpreting detailed soil maps for specific uses. Geomorphology has contributed to a basic understanding of soil landscape units, and the use of engineering data has helped to enable quantification of soil properties and performances. Examples will be given of how geomorphology, soils, and engineering are combined to optimize environmental planning and landscape management; specific problems and benefits will be discussed. Most important, increased interdisciplinary coordination can be improved in the future to the benefit of each individual discipline involved and to their services to society as a whole. Impending demands for environmental improvement and landuse controls offer many immediate possibilities for these mutual contributions.

SOIL SURVEYS

This discussion is centered around the kind of cooperative detailed soil surveys currently being conducted in the United States. These surveys are generally conducted in the field with augers, shovels, and other digging tools to depths of 5 ft or so. Soil boundaries are delineated on aerial photographs at a scale of about 4 in. to 1 mi. In a day of hard work, about 300 acres can be traversed on foot and mapped by an experienced soil scientist; contrasting soil areas down to several acres in size are shown on the map. Smaller soil areas are sometimes located by spot symbols for wetness, bedrock outcrops, depressions, eroded knobs, etc. Detailed descriptions, laboratory analyses, and interpretations for use supplement the soil maps and the fieldwork. In the United States, more than 1,000 soil scientists are currently employed full-time in soil mapping; about 40 of these are busy mapping soils in New York State. The Cooperative Soil Survey is a joint effort of the Soil Conservation Service of the U.S. Department of Agriculture, the Agricultural Experiment Station, and other agencies in each state. Increasingly local governments and many diverse state and federal agencies are supporting soil surveys, because they need the surveys for their planning, development, and regulation activities.

Agencies and institutions involved in making soil surveys are in themselves good examples of interdisciplinary efforts among soil scientists, geologists, and engineers. Because soil surveys provide the base upon which practically all land-improvement practices of the Soil Conservation Service are planned and designed, the surveys are used to a large extent by geologists, engineers, planners, and many others with which the agency has contact. The Soil Conservation Service employs geologists and engineers to design dams, terraces, diversions, drainageways, and other structures relating to soil and water use and management. Soil scientists of the Cooperative Soil Survey at the Cornell University Agricultural Experiment Station have close contact with geologists and engineers at that institution and at other institutions, with their colleagues in the Soil Conservation Service at county, state, regional, and national levels, and with counterpart agencies and institutions on an international level. In New York State, the Department of Transportation has for many years employed geologists and engineers to work closely with soil scientists in sampling soils for engineering tests for highway construction. The resulting engineering data are now routinely published and interpreted as a part of the extensive engineering section in practically all soil-survey reports. Increasingly, town, county, and state governments are providing funding for soil surveys; increasingly, also, agencies like the Department of Environmental Conservation in New York State and the Environmental Protection Agency of the federal government are requiring soil surveys and special soil studies for environmental impact statements and other ecological evaluations.

Many people other than geologists and engineers are involved in the use of soil surveys. For example, people in numerous disciplines have been involved for many years in working with the Department of Transportation in New York State in sampling, analyzing, and interpreting soils for engineering uses and in route locations for highways. Agriculturalists with the Department of Agriculture and Markets have been involved in promoting highway locations that would minimize the loss of the best lands for farming. Agricultural districts have been established by political processes to preserve the best soil areas for food production. The New York State Board of Equalization, through local assessors, uses soil surveys as one of the bases for taxing properties. Planners at all levels make extensive use of soil surveys for many purposes.

Recently, I served as a consultant to archaeologists working in the corridor for the Genesee Expressway to link Dansville and Rochester. The archaeologists hired geologists also as consultants, and were interested in using soil information to help locate archaeological sites and to identify evidences of past activities of prehistoric peoples in the soils. Significantly, the same soil maps in this highway corridor area were being used by geologists, engineers, archaeologists, assessors, farmers, extension workers, planners, politicians, developers, and many others.

Sometimes, also, the same soil surveys are used by different people to help resolve conflicts and make difficult decisions. In the Schoharie Valley southwest of Albany, geologists and engineers used soil maps in feasibility studies for a dam site for the New York City water supply. Agriculturalists, on the other hand, are using soil maps in the same area to try to prevent or move the location of the dam construction, arguing that the soil areas to be flooded are of more value to society for food production into the future. Whatever the outcome of the negotiations, better decisions should be made because of the factual data on the characteristics of the soils provided by the soil survey.

Soil Mapping

The detailed soil map is the most basic part of a soil survey, and is published on an aerial photographic base in most of the current soil reports in the United States. In fact, the soil report is the only published source of aerial photographs for most areas. The soil survey is, in effect, an on-site evaluation of soil conditions of different areas, and is reliable to a point of 80 to 90% accuracy for many soil map units. Appendix A is an illustration of soil map sheet number 52 from the soil survey of Broome County (Giddings et al., 1971), showing the soils in part of the area occupied by Binghamton. The ChA areas are Chenango and Howard soils formed in gravelly glacial outwash deposits on 0 to 5% slopes; the Cv map units are cut and fill areas of those same soils. Areas labeled VoA, B, C, and D are Volusia soils developed on compact glacial till with fragipans on different slopes. Map units LdC and D are Lordstown soils on steep slopes with sandstone and shale at a depth of 20 to 40 in. Other map units show locations of soils of different drainage classes formed in different geomorphic materials, including alluvium as well as outwash and till. Many geologic studies have assisted soil scientists in understanding and mapping soils in this landscape (Coates, 1966).

Soil Classification

Each soil in a survey area is described in detail and classified in a basic system (Soil Survey Staff, 1973); about 10,000 soils have been described and classified into this system in the United States (Soil Survey Staff, 1972). Appendix B illustrates how each soil is described, identified, mapped, and classified in the United States; Appendix B is the official soil series description for the Cecil soils developed in weathered acid igneous and metamorphic rocks in uplands in North Carolina. The Appendix B information, typical of the format for all the 10,000 soils in the United States, identifies the Cecil soils for mapping and enables those soils to be separated from all other soils. Separations are based on soil properties mappable in the geomorphic landscapes, supported by laboratory data; each of the 10,000 soils is unique with respect to use and management of the areas it occupies in the landscape. This Cecil series was established in Maryland in 1899, but many refinements have been made in the descriptions and data of the soil since that early date. The clayey, kaolinitic, thermic, Typic Hapludult classification gives information about the soil texture, mineralogy, temperature, and general profile characteristics. Criteria used for making and interpreting a soil profile description are being made more widely available in a publication compiled by Olson (in press).

Soil Interpretation

At present, data and interpretations for uses of the soils are being entered into a computer system in the entry format illustrated in Appendix C for the Cecil series. The front page of the form contains a brief description of the soil and a considerable amount of engineering data. Interpretations are made for uses of the soils for sanitary facilities, community development, water management, and as a source material for construction activities. On the other side of the form, interpretations are made for uses of that soil for recreation, farming, woodland, wildlife, and range (grazing), considering a variety of use factors. These interpretations for all the soils in the United States are the result of at least a decade of effort and refinements. In New York State in 1972, a prototype format was used for publication of 543 pages of interpretations of soils of the state (New York Soil Survey Staff, 1972); all these data are currently being revised into the standardized national computer format illustrated in Appendix C.

Appendix C gives the computer input form used for the 10,000 soils in the United States; Appendix D is an example of the computer output format for the Valois series, which is mapped in New York State. Valois soils have formed on glacial till and are good for septic tanks, houses, roads, lawns, recreation areas, crops, woods, and most wildlife habitat where mapped on the more gentle slopes. Valois soils, however, are poor for sewage lagoons, sanitary landfills, ponds, and are poor sources of sand and gravel. Typically, these soils occupy complex slopes of kame terraces or end or lateral moraines and other geomorphic landscape positions where soils have formed on glacial till in association with outwash deposits; the till is not as compact and dense as are the fragipans and tills in most nearby soils in these landscapes. The information illustrated in Appendix D is available for all the soils in the United States; the computer printouts can be used even for areas with unpublished soil maps. Soil limitations and suitabilities for the different uses can be evaluated as soon as an area has been mapped by a soil scientist, and as soon as the soil names are identified for the different geomorphic landscape positions.

Personnel

At the present time in New York State, the Soil Conservation Service employs seven soil specialists who do intensive soils studies of a type not commonly done by soil-survey field parties in routine mapping. These men all have more than a decade of soil-survey field experience. They work primarily in making special on-site evaluations of soils for developments and projects on large-scale maps, but they also work with planners and others in making more general maps of larger areas. In many cases these soil specialists make soil interpretations in greater detail than those illustrated in Appendix D. These men often work with geologists, engineers, and many others in helping people use soils information. Currently, the offices of the soil specialists in New York State are located in Albany, Corning, Kingston, Norwich, Randolph, Rochester, and Syracuse (New York Soil Survey Staff, 1975). These soil specialists are helpful in assisting to solve many landuse, geology, and engineering problems in the areas in which they are working. Similar soil specialist positions are also maintained by the Soil Conservation Service in other states. Cooperative efforts, of course, are of mutual benefit to all participants. The work of soil specialists can be considerably enhanced by contributions of geologists and engineers, especially on special-project evaluations, feasibility studies, and environmental impact statements.

PRACTICAL APPLICATIONS

Some specific examples will illustrate the utility of soil surveys and geological and engineering contributions in landuse planning and development. In Erie County (near Buffalo) and in other places, soils and substrata are influenced by unstable silty and clayey lacustrine sediments. Recently, deep geological and engineering boreholes (drilled in conjunction with the soil survey) confirmed that some of the sediments in Erie County are more than 100 ft thick. These cooperative investigations are especially significant because some of the soil areas are sliding at the present time, and houses, roads, and other buildings are being damaged by soil instability. In the future, these areas should be avoided for building, and the soils should be put to more suitable uses like woodlands, wildlife, or recreation.

The Sleepy Hollow development south of Albany provides another example of problems caused by unstable soils. Figure 1 is a view of a lookout point overlooking the dam impounding a lake of more than 200 acres. Several miles of roads, numerous homesites, and other developments have already been constructed in the area. Soils investigations by specialists indicated hazards of landslides and erosion if the proposed subdivisions were built at the designated plat locations around the large artificial lake. The area has much scenic and historical potential, however, and could be safely and efficiently developed if careful consideration is given to the hazardous soil conditions. In this project area, numerous soil scientists, geologists, and engineers have been involved in soils investigations and in designing the dam, roads, streets, houses, and other constructions. The original proposals for the development, however, did not adequately consider the unique soil problems, and would have resulted in accelerated landslides and soil erosion, which could have cost large sums of money in damages.

The Radisson new community financed by the New York State Urban Development Corporation near Baldwinsville northwest of Syracuse is another example of hazardous soil conditions. Sandy, silty, and clayey soils at this site are wet in many places, and many of them have serious erosion limitations. Developers at the site, however, have been very much aware of the erosion potentials of the soils and have been very careful to try to maintain slopes at a proper grade and to vegetate those slopes as soon as possible after exposing them. In fact, the high water table in the wet soils has been used to good advantage in the creation of an artificial lake in a beautiful wooded setting around which

FIGURE 1 *View of lookout point overlooking dam at Sleepy Hollow development south of Albany.*

FIGURE 2 *Lake in urban setting at Radisson new community near Baldwinsville.*

houses are clustered. Figure 2 is a view of that lake in a well-designed urban setting, with the natural vegetation preserved and the clusters of houses set back from the lakeshore.

In almost every daily newspaper of every community in the United States one can read of soil problems that earth scientists could help to solve. *The Ithaca Journal* in central New York, for example, has reported for many years of the soils shallow to bedrock plaguing developmental costs on South Hill at the south end of Cayuga Lake. If soil scientists, geologists, and engineers had been able to inform all the people fully of the hazards of these conditions, millions of dollars in construction costs and human suffering could have been saved. Even at the present time, planning and constructions are underway in the area without full knowledge of the soil and ecological conditions. These examples of soil misuse (Klingebiel, 1966) show the tremendous possibilities for benefits to be realized from increased investments in resource inventories and educational programs about the uses of these data.

Computers offer many opportunities to soil scientists, geologists, and engineers in coordinating and combining their information. The work of the many scientists associated with the Department of Environmental Conservation in studying the Canadarago Lake watershed southeast of Utica is an effort that can be used for illustration. Computers (Kling and Olson, 1975) enabled erosion rates and pathways of soil movement to be predicted under present and proposed landuses. Monitoring of streams enabled checking and correction of the predictions. Modelings of soil losses to geomorphic processes and engineering constructions to slow down and control those processes offer many opportunities for interdisciplinary cooperations. Soil scientists, geologists, and engineers can mutually benefit from the contributions of each discipline in comprehensive watershed studies of these types.

Benefits to be realized from cooperative land improvement work are not confined to New York State or to the United States. Every country in the world has considerable numbers of soil scientists, geologists, and engineers working in resource inventory and developmental projects. Recently I have been teaching and consulting on soil-survey interpretations in Latin America (Olson, 1973, 1974b, 1975a, 1975b, 1975c). There are many places in Latin America that offer large potential benefits for landuse improvements. Erosion and geomorphic processes in the Andes, for example, have a profound effect on present and future landuses (Food and Agriculture Organization, 1974) at the higher elevations, and sediment and runoff water also have great effects upon the engineering structures being built to improve the productivity of the lowlands (Food and Agriculture Organization, 1973). Reducing erosion and sedimentation and improving drainage and channelization of streams will help to reduce flooding and land degradation. Many opportunities exist for soil scientists, geologists, and engineers to work together on projects for landuse improvements in many parts of the world; numerous agencies, like the Ministry of Public Works in Venezuela, are making good progress in interdisciplinary approaches to feasibility studies and project improvement implementations.

Opportunities for interactions of soils, geomorphology, and engineering in helping to solve problems are excellently outlined by Eckholm (1975) in discussions of the deterioration of mountain environments. He points out how ecological stress in the highlands takes a mounting social toll, and how "history has repeatedly shown that when ecological changes take place in the highlands, changes soon follow in the valleys and the plains." Figure 3 is an illustration of geomorphic erosion processes in the Andes mountains (Olson, 1975a), accelerated greatly by abuse of the soils. Figure 4 is a view of the resultant sediments in the Maracaibo lowlands, choking drainage channels and decreasing the productivity of the land. Solution to the problems of erosion and sedimentation, of course, lies in soil conservation and adequate engineering constructions like terraces,

FIGURE 3 *Accelerated soil erosion in the Andes mountains near Apartadero, northwestern Venezuela. Horizontal markings on the hill indicate that the steep slopes were intensively farmed, and that those human activities likely triggered the gullying that is degrading the landscape.*

FIGURE 4 *Sedimentation in drainage channel in Maracaibo basin near El Vigia, northwestern Venezuela. Sediments from erosion in the Andes mountains damage also the lowlands, reducing the flow of drainage waters and degrading the environment.*

diversions, and dams in the uplands, and in dredging (Fig. 5), channeling, and drainage of the lowlands. Figure 6 illustrates some of the tremendous productivity potentials of lands properly protected from erosion and sedimentation, and provided with drainage and good management techniques.

Bulletin 208 of the Kansas Geological Survey is a good example of interdisciplinary cooperation among soil scientists, geologists, and engineers to provide information and

FIGURE 5 *Dredging swampy areas in the Maracaibo basin near El Vigia, northwestern Venezuela. Dams, levees, channelization, drainage works, and other engineering structures help to control water and sediment, and to open the area to more productive landuses.*

FIGURE 6 *Banana plantation near El Vigia, northwestern Venezuela. These wet soils, provided with drainage and protected from flooding and sedimentation, can be extremely productive under careful management.*

FIGURE 7 *Maya terraces and other structures at the recently discovered archaeological site at La Canteada, near Copan, northwestern Honduras. The terraces and plazas were paved with cobblestones, beneath which were strata of debris from centuries of occupations. Beneath the artifactual materials were reddish slackwater clayey deposits, underlain by riverine sandy strata. Soils formed in these structures are being dated by radiocarbon analyses of charcoal discovered beneath the cobblestone pavement.*

recommendations for using the soils of Kansas for waste disposal (Olson, 1974a). The bulletin is based on the soil interpretations for sanitary facilities given on forms like those in Appendixes C and D for soils of Kansas. Additional information was added from research on waste disposal by soil scientists, geologists, and engineers. Consultations with officials of the Kansas State Department of Health made the standard soil interpretations relevant to the engineering recommendations for waste disposal in soils of Kansas. Soil information was contributed by officials of the Soil Conservation Service in the state and Kansas Agricultural Experiment Station. Most important, the work was supported and published by the Kansas Geological Survey. Such an effort demonstrates that people in the disciplines can benefit from working together, and that the composite information is of more value to people than the individual pieces produced by each person or agency working separately.

Finally, the increasing importance of the human element in the geomorphic processes should continually be emphasized. Soils, geomorphology, and engineering should function to provide information to help people live better lives and to improve the human environment. Lessons must be learned from the past if mistakes are to be avoided in the future. Figure 7 illustrates the impact that even the ancient Maya had upon their environment; all the landforms in the lower and north (left) part of the photograph are entirely artificial; these terraces, pyramids, and other structures were built by the Maya about 1,000 yr ago. These Maya modifications of the landscape had many effects on the uplands in promoting erosion and sedimentation. In the lowlands, evidences of these geomorphic disruptions can be readily observed even today, 1,000 yr later, by pedologists. Figure 8 shows a buried soil with Maya artifacts in it, underneath alluvial sediments more than 1 m thick deposited less than 1,000 yr ago, in which a modern soil has formed. The Maya soil also has accumulations of phosphorus, which apparently are at least partly artifactual from the ancient Maya occupation. With modern overgrazing and erosion in the uplands, and without proper soil-conservation and engineering procedures to control erosion and sedimentation, contemporary archaeologists are tempted to hypothesize that at a point in time about 1,000 yr in the future archaeologists will find another soil profile superimposed over those shown in Fig. 8, with the second buried profile containing bits of plastic and other artifacts of the twentieth-century occupation.

CONCLUSIONS

This chapter has provided examples from New York State, the United States, and other countries to illustrate the problems and potentials in soil-survey interpretations. Increased applications and combinations of soil, geologic, and engineering data in the future can help considerably to improve the various environments and the lives of the people living and working there.

Soil-survey interpretations are being appreciably improved through quantification and applications of geologic and engineering data. Soils are being defined more precisely than in the past, and experience is rapidly accumulating to help make design modifications for soil improvements in the future. With computers and increased cooperations among pedologists, geologists, engineers, and others into the future, we can expect to have the capabilities to prevent the land deterioration that has affected past civilizations in a detrimental manner. Future successes in land management and landuse controls will be largely determined by the effectiveness of institutions in the application and coordination

FIGURE 8 *Buried soil profile with Maya artifacts in alluvium stream-bank cut in Quebrada La Guasma, Valle de Naco, near San Pedro Sula, northwestern Honduras. Land abuse by the Maya about 1,000 yr ago likely accelerated soil erosion and deposition, contributing at least partly to the decline of that civilization and covering the soil at this site with sediment. Current overgrazing of uplands appears to be repeating the cycle, resulting in increased erosion, sedimentation, and environmental deterioration.*

of existing knowledge, and in the promotion of the development and gathering together of new techniques and innovations. Without due regard for geomorphic processes and the engineering design of structures, pedologists and archaeologists can predict repetition of the cycles of environmental degradation (like erosion and sedimentation) that have occurred in the past. Landuse patterns will degrade the environment or improve the soil productivity in accord with the effectiveness of the application of soil-survey interpretations.

ACKNOWLEDGMENTS

Many colleagues have reviewed this material, and offered suggestions and assistance during its preparation. I am grateful for all the aid, and especially for the cooperation of those in disciplines other than soil science, including geology and engineering.

REFERENCES

Coates, D. R. 1966. Glaciated Appalachian plateau: till shadows on hills: *Science*, v. 152, p. 1617–1619.

Eckholm, E. P. 1975. The deterioration of mountain environments: *Science*, v. 189, p. 764–770.

Food and Agriculture Organization. 1973. Soil survey interpretation for engineering purposes: *Soils Bulletin 19,* Food and Agriculture Organization of the United Nations, Rome, 24 p.

_____. 1974. Approaches to land classification: *Soils Bulletin 22,* Food and Agriculture Organization of the United Nations, Rome, 120 p.

Giddings, E. B., et al. 1971. Soil survey of Broome County, New York: Government Printing Office, Washington, D.C., 95 p. and 89 soil map sheets on aerial photographs at 1:15,840 scale.

Kling, G. F., and Olson, G. W. 1975. Role of computers in land use planning: *Information Bulletin 88,* New York State College of Agriculture and Life Sciences, Cornell University, Ithaca, N.Y., 12 p.

Klingebiel, A. A. 1966. Costs and returns of soil surveys: *Soil Conservation,* v. 32, p. 3–6.

New York Soil Survey Staff. 1972. Soil survey interpretations of soils in New York State: guide for selected uses of soils for resource material, engineering, community development, cropland, recreation, woodland, and wildlife: Mimeo 72-4, Department of Agronomy, Cornell University, Ithaca, N.Y., 543 p.

_____. 1975. Soil specialist: Brochure published by Soil Conservation Service, U.S. Department of Agriculture, Syracuse, N.Y., 8 folded parts on a single long sheet.

Olson, G. W. 1973. Improving uses of soils in Latin America: *Geoderma,* v. 9, p. 257–267.

_____. 1974a. Using soils of Kansas for waste disposal: *Bulletin 208,* Kansas Geological Survey, University of Kansas, Lawrence, Kans., 51 p. and map of soils of Kansas at 1:1,125,000 scale.

_____. 1974b. Field report on soils sampled around San Antonio in northern Belize (British Honduras): Mimeo 74-23, Department of Agronomy, Cornell University, Ithaca, N.Y., 11 p.

_____. 1975a. Views of land uses and opportunities for soil survey interpretations in South America (Venezuela and Brazil): Mimeo 75-11, Department of Agronomy, Cornell University, Ithaca, N.Y., and Agency for International Development, Washington, D.C., 43 p.

_____. 1975b. Training key people in soil survey interpretations in Latin America: Mimeo 75-8, Department of Agronomy, Cornell University, Ithaca, N.Y.; Centro Interamericano de Desarrollo Integral de Aguas y Tierras, Merida, Venezuela; and Ministerio de Obras Publicas, Caracas, Venezuela, 6 p.

_____. 1975c. Study of soils in Valle de Naco (near San Pedro Sula) and La Canteada (near Copan) Honduras: implications to the Maya mounds and other ruins: Mimeo 75-19, Department of Agronomy, Cornell University, Ithaca, N.Y., 74 p.

_____. In press. Criteria for making and interpreting a soil profile description: a compilation of the official United States Department of Agriculture procedure and nomenclature for describing soils: *Bulletin 212,* Kansas Geological Survey, University of Kansas, Lawrence, Kans.

Soil Survey Staff. 1972. *Soil Series of the United States, Puerto Rico, and the Virgin Islands: Their Taxonomic Classification:* Soil Conservation Service, U.S. Department of Agriculture, Washington, D.C.

_____. 1973 (preliminary abridged page proof). *Soil Taxonomy: A Basic System of Soil Classification for Making and Interpreting Soil Surveys:* Soil Conservation Service, U.S. Department of Agriculture, Washington, D.C.

APPENDIX A *Soil map sheet number 52 from the soil survey of Broome County.*

Established Series
Rev. FS:JBW:REH
7/71

CECIL SERIES

The Cecil series is a member of the clayey, kaolinitic, thermic family of Typic Hapludults. These soils have dark grayish brown or brown sandy loam or loam A horizons and red clay Bt horizons.

Typifying Pedon: Cecil sandy loam - forested.
                  (Colors are for moist soil unless otherwise stated.)

01   --   2-0" -- Very dark grayish brown (2.5Y 3/2) partially decayed leaves and twigs.
            (0 to 3 inches thick)

A1   --   0-2" -- Dark grayish brown (10YR 4/2) sandy loam; weak medium granular structure;
            very friable; many fine roots; strongly acid; clear wavy boundary. (2 to 5
            inches thick)

A2   --   2-7" -- Brown (7.5YR 5/4) sandy loam; weak medium granular structure; very friable;
            many fine and medium roots; few pieces of quartz gravel; strongly acid; clear
            smooth boundary. (3 to 13 inches thick)

B1   --   7-11" -- Yellowish red (5YR 4/8) sandy clay loam; weak fine subangular blocky
            structure; friable; few fine and medium roots; strongly acid; clear smooth
            boundary. (4 to 8 inches thick)

B21t --   11-28" -- Red (2.5YR 4/8) clay; moderate medium subangular blocky structure; firm;
            sticky; thick nearly continuous clay films on faces of peds; few fine mica
            flakes; few small pieces of quartz gravel; strongly acid; gradual smooth
            boundary. (14 to 24 inches thick)

B22t --   28-40" -- Red (2.5YR 4/8) clay; moderate and weak medium subangular blocky structure;
            firm; thin clay films on most faces of peds; few to common fine mica flakes but
            not sufficient to impart a greasy feel; strongly acid; gradual smooth boundary.
            (10 to 20 inches thick)

B3   --   40-50" -- Red (2.5YR 5/8) clay loam; common medium distinct strong brown mottles;
            weak medium subangular blocky structure; friable; few clay films on vertical
            faces; few to common fine mica flakes; strongly acid; gradual smooth boundary.
            (7 to 20 inches thick)

C   --   50-75" -- Mottled red (2.5YR 5/8) strong brown (7.5YR 5/8), and pale brown (10YR
            6/3) disintegrated gneiss mixed with clay loam; massive; friable; common fine
            mica flakes; strongly acid.

Type Location: Catawba County, North Carolina; 5-1/4 miles southeast of Newton, North
Carolina, on NC 16; then south on Maiden road, highway 1810, for 3/4 mile - wooded area -
shown on photo 2P-122.

Range in Characteristics: Thickness of the solum commonly is about 48 inches with a range of
40 inches to 60 inches or more. Depth to hard rock is greater than 5 feet, ranging to 40 feet
or more. The soil is strongly acid or very strongly acid unless limed. The A or Ap horizons
are dark grayish brown (10YR 4/2), very dark grayish brown (10YR 3/2), brown (10YR 4/3; 7.5YR
4/4, 5/4), yellowish brown (10YR 5/4), reddish brown (5YR 4/3, 4/4), yellowish red (5YR 4/6) or
red (2.5YR 4/6). Texture of the A horizons is sandy loam, fine sandy loam or loam. Gravel,
cobbles or stones may be present. Eroded pedons frequently have clay loam or sandy clay loam
Ap horizons. Some pedons have dark grayish brown (10YR 4/2) brown (10YR 4/3; 7.5YR 4/4, 5/4)
yellowish brown (10YR 5/4) or yellowish red (5YR 4/6) A2 horizons. These are sandy loam or
sandy clay loam. Some pedons have yellowish red (5YR 4/6, 4/8) or red (2.5YR 4/6, 4/8) B1
horizons. These are sandy clay loam or clay loam. The B2t horizons are red (2.5YR 4/6, 5/6,
4/8, 5/8; 10R 4/6, 5/6, 4/8, 5/8). The clay content of the B2t horizons ranges from 35 to 70
percent and silt from 15 to 30 percent. The B3 horizons are similar in color to the B2t horizons. Texture is clay loam or sandy clay loam. Brownish mottles are common in some pedons.
The C horizon is multicolored, soft felsic rock material that crushes readily and is loam or
clay loam.

APPENDIX B *Official soil series description for the Cecil series.*

Cecil Series   2

Competing Series and their Differentiae: These include Appling, Braddock, Cataula, Georgeville,
Hayesville, Herndon, Lockhart, Madison, Pacolet and Tatum. Appling soils are less red, with
Bt horizons in hues of 5YR and yellower, commonly mottled in the lower part, and commonly
have more sand in the A and upper B horizons. Braddock soils are members of a clayey mixed,
mesic family. Cataula soils have a fragipan at 18 to 36 inches. Georgeville and Herndon
soils have less sand and more silt throughout the solum, more than 30 percent silt, or more than
40 percent silt and very fine sand. Hayesville soils have mean annual soil temperatures from
47° to 59° F. Lockhart soils have less than 35 percent clay in the Bt horizons and many feldspar
crystals throughout the solum. Madison soils commonly have thinner sola, and a high content of
mica. Pacolet soils have sola less than 40 inches thick. Tatum soils contain more than 30
percent silt or more than 40 percent silt and very fine sand.

Setting: The Cecil soils are on nearly level to steep Piedmont uplands. Slope gradients are 0
to 25 percent, most commonly between 2 and 15 percent. These soils have developed on weathered
acid igneous and metamorphic rocks. Average annual precipitation is about 48 inches. Mean
annual soil temperature is about 59° F.

Principal Associated Soils: In addition to competing soils, these include Louisburg, Durham,
Louisa and Worsham soils. Louisburg and Louisa soils lack the uniform Bt horizons of Cecil
soils. Durham soils are less red and have less than 35 percent clay in the Bt horizons.
Worsham soils are poorly drained.

Drainage and Permeability: Well drained; medium to rapid runoff; medium internal drainage;
moderate permeability.

Use and Vegetation: About half of the total acreage is in cultivation, with the remainder in
pasture and forest. Common crops are small grains, corn, cotton, and tobacco.

Distribution and Extent: Piedmont area of Virginia, North Carolina, South Carolina, Georgia
and Alabama. The series is of large extent, with an area of more than 10,000,000 acres.

Series Established: Cecil County, Maryland; 1899.

Remarks: Cecil soils were formerly classified in the Red-Yellow Podzolic great soil group.

Additional Data: (1) McCracken, R. J., editor: Southern Cooperative Series Bulletin 61,
                         issued January 1959, Virginia Agricultural Experiment Station,
                         Blacksburg, Virginia.

                    (2) Engineering data in:

                         Soil Survey of Iredell County, North Carolina, issued 1964.
                         Soil Survey of Yadkin County, North Carolina, issued 1962.

                                   National Cooperative Soil Survey
                                            U. S. A.

SOIL SURVEY INTERPRETATIONS

MLRA(S) ___
STATE NORTH CAROLINA   RECORD NO. 41   AUTHOR(S) CLM   DATE 4-77   KIND OF UNIT SERIES   UNIT NAME CECIL
REVISED: UNIT MODIFIER ___
CLASSIFICATION AND BRIEF SOIL DESCRIPTION

THE CECIL SERIES CONSISTS OF WELL DRAINED NEARLY LEVEL TO STEEP SOILS ON PIEDMONT UPLANDS. TYPICALLY THESE SOILS HAVE A SANDY LOAM SURFACE LAYER ABOUT 7 INCHES THICK. THE SUBSOIL IS RED FIRM CLAY WHICH EXTENDS FROM 40 TO 60 INCHES BELOW THE SURFACE. THE SUBSTRATUM IS COMMONLY CLAY LOAM OR LOAM. THE SOIL DEVELOPED IN RESIDUUM FROM CRYSTALLINE ROCKS. SLOPES RANGE FROM 0 TO 25 PERCENT.

ESTIMATED SOIL PROPERTIES

DEPTH (IN.)	USDA TEXTURE	UNIFIED	AASHO	FRACT. >3 IN. (PCT)	PERCENT OF MATERIAL LESS THAN 3 IN. PASSING SIEVE 4	10	40	200	LIQUID LIMIT	PLASTICITY INDEX
0-7	SL, FSL, L	SM	A-2, A-4	0	94-100	90-100	67-90	26-42	NP-30	NP-6
0-7	GR-SL	GM, GP	A-2, A-1	0	40-75	35-72	25-60	18-28	NP-22	NP-4
0-50	SCL, CL	SM, SC, CL	A-4	0	74-100	72-100	68-95	32-81	21-28	3-10
7-50	C	MH, CH, CL	A-6, A-7	0	97-100	92-100	72-99	55-95	40-80	16-37
50-75	CL, L	ML, MH, SC	A-7	0	100	97-100	72-100	44-73	47-59	NP-16

DEPTH (IN.)	PERMEABILITY (IN/HR)	AVAILABLE WATER CAPACITY (IN/IN)	SOIL REACTION (pH)	SALINITY (MMHOS/CM)	SHRINK-SWELL POTENTIAL	CORROSIVITY STEEL	CONCRETE	EROSION FACTORS K	T	WIND EROD. GROUP
SAME DEPTH AS ABOVE	2.0-6.0	.12-.14	4.5-6.0	—	LOW	LOW	MODERATE	.32	4	3
	2.0-6.0	.07-.09	4.5-6.0	—	LOW	LOW		.24	4	3
	0.6-2.0	.13-.15	4.5-6.0	—	MODERATE	MODERATE		.32	3	7
	0.6-2.0	.13-.15	4.5-5.5	—	HIGH			.28		
	0.6-2.0	.13-.15	4.5-5.5	—	LOW			.32		

FLOODING FREQUENCY	DURATION	MONTHS	HIGH WATER TABLE DEPTH (FT)	KIND	MONTHS	CEMENTED PAN DEPTH (IN)	HARDNESS	BEDROCK DEPTH (IN)	HARDNESS	SUBSIDENCE INITIAL (IN)	TOTAL (IN)	HYD GRP	POTENTIAL FROST ACTION
NONE			>6					>60		—	—	B	—

	SANITARY FACILITIES		FOOTNOTES	SOURCE MATERIAL
SEPTIC TANK ABSORPTION FIELDS	0-15%: MODERATE - PERCS SLOWLY 15+%: SEVERE - SLOPE	ROADFILL	FAIR - SHRINK-SWELL, LOW STRENGTH
SEWAGE LAGOONS	0-7%: MODERATE - PERCS RAPIDLY 7+%: SEVERE - SLOPE	SAND	UNSUITED
SANITARY LANDFILL (TRENCH)	0-15%: SLIGHT 15+%: MODERATE - SLOPE	GRAVEL	UNSUITED
SANITARY LANDFILL (AREA)	0-2%: SLIGHT 8-15%: MODERATE - SLOPE 15+%: SEVERE - SLOPE	TOPSOIL	0-15%: FAIR-TOO CLAYEY 15+%: POOR-TOO STEEP
DAILY COVER FOR LANDFILL	0-15%: FAIR - TOO CLAYEY 15+%: POOR - SLOPE		

	SANITARY FACILITIES / COMMUNITY DEVELOPMENT		FOOTNOTES	WATER MANAGEMENT
		POND RESERVOIR AREA	MODERATE - PERCS RAPIDLY
SHALLOW EXCAVATIONS	0-15%: MODERATE - TOO CLAYEY 15+%: SEVERE - SLOPE	EMBANKMENTS DIKES AND LEVEES	SEVERE - COMPRESSIBLE
DWELLINGS WITHOUT BASEMENTS	0-15%: MODERATE - SHRINK-SWELL 15+%: SEVERE - SLOPE	EXCAVATED PONDS AQUIFER FED	SEVERE - NO WATER
DWELLINGS WITH BASEMENTS	0-15%: MODERATE - SHRINK-SWELL 15+%: SEVERE - SLOPE	DRAINAGE	NOT NEEDED
SMALL COMMERCIAL BUILDINGS	0-8%: MODERATE - SHRINK-SWELL 8+%: SEVERE - SLOPE	IRRIGATION	COMPLEX SLOPE
LOCAL ROADS AND STREETS	0-15%: MODERATE - LOW STRENGTH 15+%: SEVERE - SLOPE	TERRACES AND DIVERSIONS	COMPLEX SLOPE
	REGIONAL INTERPRETATIONS	GRASSED WATERWAYS	COMPLEX SLOPE

APPENDIX C Computer entry filled out for interpretations of the Cecil series.

UNIT NAME: CECIL (2)
UNIT MODIFIER: _____

RECREATION

CAMPS 30	CAMP AREAS	FOOTNOTE 0-8% SL, FSL, L, GRSL: SLIGHT 2-8% SCL, CL: MODERATE-TOO CLAYEY 8-15% SL, FSL, L, GRSL: MODERATE-SLOPE 8-15% SCL, CL: MODERATE-SLOPE, TOO CLAYEY 15+%: SEVERE-SLOPE	KEYING ONLY PLAYGD 32	PLAYGROUNDS	FOOTNOTE 0-2%: SLIGHT 2-6%SL, FSL, L, GRSL: MODERATE-SLOPE 2-6% SCL, CL: MODERATE-SLOPE, TOO CLAYEY 6+%: SEVERE-SLOPE
PICNIC 31	PICNIC-AREAS	0-8% SCL, FSL, L, GRSL: SLIGHT 0-8% SCL, CL: MODERATE-TOO CLAYEY 8-15% SL, FSL, L, GR-SL: MODERATE-SLOPE 8-15% SCL, CL: MODERATE-SLOPE, TOO CLAYEY 15+%: SEVERE-SLOPE	PATHS 33	PATHS AND TRAILS	0-15% SL, FSL, L, GR-SL: SLIGHT 0-15% SCL, CL: MODERATE-TOO CLAYEY 15+%: MODERATE-SLOPE

CAPABILITY AND PREDICTED YIELDS - CROPS AND PASTURE (HIGH LEVEL MANAGEMENT)

FOOTNOTE

CLASS-DETERMINING PHASE	CAPABILITY NIRR	IRR	CORN (BU) NIRR	IRR	COTTON (LBS LINT) NIRR	IRR	OATS (BU) NIRR	IRR	PEACHES (BU) NIRR	IRR	TOBACCO (LBS) NIRR	IRR	HAY (TONS) NIRR	IRR	GRASS-LEGUME PASTURE (AUM) NIRR	IRR
CROPS 34																
2-6% SL, FSL, L, GR-SL	2E		95		750		90		500		2100		3.2		6.5	
6-10% SL, FSL, L, GR-SL	3E		90		700		85				2000		3.0		6.5	
10-15% SL, FSL, L, GR-SL	4E		80		600		75				1900		2.8		6.0	
15+% SL, FSL, L, GR-SL	6E		70		500		60				1900		2.6		5.5	
2-6% SCL, CL, ERODED	3E		70		500		70				—		2.4		5.5	
6-10% SCL, CL, ERODED	6E		60		—		60		—		—		2.2		5.0	
10+% SCL, CL, ERODED	6E		—		—		—		—		—		1.8		4.5	

WOODLAND SUITABILITY

FOOTNOTE

CLASS-DETERMINING PHASE	ORD SYM	MANAGEMENT PROBLEMS EROSION HAZARD	EQUIP. LIMIT	SEEDLING MORTY.	WINDTH. HAZARD	PLANT COMPET.	POTENTIAL PRODUCTIVITY IMPORTANT TREES	SITE INDEX	TREES TO PLANT
WOODS 36									
0-15% SL, FSL, L, GR-SL	30	SLIGHT	SLIGHT	SLIGHT	SLIGHT	SLIGHT	EASTERN WHITE PINE	80	EASTERN WHITE PINE 4/
15+% SL, FSL, L, GR-SL	3R	MODERATE	MODERATE	↓	↓	↓	LOBLOLLY PINE	80*	LOBLOLLY PINE
							SHORTLEAF PINE	65*	SLASH PINE 5/
							VIRGINIA PINE	73*	YELLOW POPLAR
							BLACK OAK	66	
							NORTHERN RED OAK	82*	
							POST OAK	66*	
							SCARLET OAK	70*	
0-5% SCL, CL, ERODED	4C	MODERATE	MODERATE	MODERATE	SLIGHT	SLIGHT	LOBLOLLY PINE	72*	LOBLOLLY PINE
5-15% SCL, CL, ERODED	4C	SEVERE	SEVERE	↓	↓	↓	SHORTLEAF PINE	66*	SLASH PINE
							VIRGINIA PINE	65	VIRGINIA PINE

WIND BREAKS

FOOTNOTE

CLASS-DETERMINING PHASE	SPECIES	HT	SPECIES	HT	SPECIES	HT	SPECIES	HT
WINDBK 38	NONE							

WILDLIFE HABITAT SUITABILITY

FOOTNOTE

CLASS-DETERMINING PHASE	POTENTIAL FOR HABITAT ELEMENTS GRAIN & SEED	GRASS & LEGUME	WILD HERB.	HARDWD TREES	CONIFER PLANTS	SHRUBS	WETLAND PLANTS	SHALLOW WATER	POTENTIAL AS HABITAT FOR: OPENLAND WILDLIFE	WOODLAND WILDLIFE	WETLAND WILDLIFE	RANGELAND WILDLIFE
WILDLF 39												
0-5% SL, FSL, L, GR-SL	GOOD	GOOD	GOOD	GOOD	GOOD	—	V. POOR	V. POOR	GOOD	GOOD	V. POOR	—
5-15% SL, FSL, L, GR-SL	FAIR	↓	↓	↓	↓	—	↓	↓	↓	↓	—	—
15+% SL, FSL, L, GR-SL	POOR			↓		—		FAIR			—	—
2-15% SCL, CL	FAIR	↓	FAIR			—		↓			—	—
15+% SCL, CL	POOR	FAIR	↓			—					—	—

POTENTIAL NATIVE PLANT COMMUNITY (RANGELAND OR FOREST UNDERSTORY VEGETATION)

FOOTNOTE

COMMON PLANT NAME	PLANT SYMBOL (NLSPN)	PERCENTAGE COMPOSITION (DRY WEIGHT) BY CLASS-DETERMINING PHASE
PHASE 40		
PLANT 41		
WITCH HAZEL	HAVI4	
FLOWERING DOGWOOD	COFL2	
AMERICAN HOLLY	ILOP	
PAW PAW	ASTR	
FARKLEBERRY	VAAR	
MUSCADINE GRAPE	VIRO3	
SUMAC	RHUS	
RED MULBERRY	MORU	
BLACKBERRY BRIAR	RUBUS	
PERSIMMON	DIVI5	
SOURWOOD	OXAR	
BLACK CHERRY	PRSE2	
HOPHORNBEAN	OSVI	
AMERICAN HORNBEAM	CARO	
TRUMPET CREEPER	CARA2	

POTENTIAL PRODUCTION (LBS/AC. DRY WT):

	FAVORABLE YEARS	NORMAL YEARS	UNFAVORABLE YEARS
PRODUC 43			

FOOTNOTES

SYM	
NOTES 44	1 BASED ON TEST DATA FROM 11 PROFILES IN NORTH CAROLINA, GEORGIA, AND SOUTH CAROLINA
	2 RATED IN SOUTH REGION ONLY
	3 DATA FROM "SOIL SURVEY INTERPRETATIONS FOR WOODLANDS", PROGRESS REPORT W-13, SEPTEMBER
	4 UPPER PIEDMONT OF NORTH CAROLINA ONLY
	5 EXCEPT IN NORTH CAROLINA PIEDMONT.
	* BASED ON SITE INDEX MEASUREMENTS

PLRA(S): 101, 140, 144   VALOIS SERIES
JUN. 3-73
TYPIC DYSTROCHREPTS, COARSE-LOAMY, MIXED, MESIC

THE VALOIS SERIES CONSISTS OF DEEP, WELL DRAINED SOILS ON UPLANDS. THEY FORMED IN GLACIAL TILL. TYPICALLY THESE SOILS HAVE A BROWN GRAVELLY LOAM SURFACE LAYER 7 INCHES THICK. THE SUBSOIL FROM 7 TO 30 INCHES IS STRONG BROWN GRAVELLY LOAM AND FROM 30 TO 47 INCHES IS DARK BROWN GRAVELLY SILT LOAM. THE SUBSTRATUM FROM 47 TO 60 INCHES IS DARK GRAYISH BROWN VERY GRAVELLY FINE SANDY LOAM. SLOPES RANGE FROM 0 TO 60 PERCENT.

ESTIMATED SOIL PROPERTIES (A)

DEPTH (IN.)	USDA TEXTURE	UNIFIED	AASHO	FRACT >3 IN (PCT)	PERCENT OF MATERIAL LESS THAN 3" PASSING SIEVE NO.				LIQUID LIMIT	PLAS- TICITY INDEX
					4	10	40	200		
0-7	L, SIL, FSL	ML, SM	A-4, A-2	0-2	80-95	75-90	50-90	30-80	NP-35	NP-5
0-7	GR-L, GR-SIL, GR-FSL	ML, GM, SM	A-4, A-2, A-1	0-2	60-80	55-70	35-70	20-65	NP-25	NP-5
7-47	GR-L, GR-SIL, GR-FSL	GM, ML, SM	A-4, A-2, A-1	0-10	60-75	55-70	40-70	20-60	NP-25	NP-5
47-60	GRV-FSL, GRV-L	GM, GW-GM	A-1, A-2	0-15	30-60	20-55	15-45	10-35	NP-25	NP-7

DEPTH (IN.)	PERMEABILITY (IN/HR)	AVAILABLE WATER CAPACITY (IN/IN)	SOIL REACTION (PH)	SALINITY (MMHCS/CM)	SHRINK-SWELL POTENTIAL	CORROSIVITY		EROSION FACTORS		WIND EROD. GROUP
						STEEL	CONCRETE	K	T	
0-7	0.6-2.0	0.12-0.21	4.5-5.5	-	LOW	LOW	HIGH	.24	3	-
0-7	0.6-2.0	0.08-0.16	4.5-5.5	-	LOW	LOW	HIGH	.17	3	
7-47	0.6-2.0	0.07-0.14	5.1-6.0	-	LOW	LOW	MODERATE	.43		
47-60	0.6-6.0	0.03-0.09	5.6-7.3	-	LOW	LOW	MODERATE	.28		

FLOODING			HIGH WATER TABLE			CEMENTED PAN		BEDROCK		SUBSIDENCE		MYC (POTENT)
FREQUENCY	DURATION	MONTHS	DEPTH (FT)	KIND	MONTHS	DEPTH (IN)	HARDNESS	DEPTH (IN)	HARDNESS	INIT. (IN)	TOTAL (IN)	GRP FROST ACTION
NONE			3-6			-		>60		-	-	B LOW

SANITARY FACILITIES (B)

SEPTIC TANK ABSORPTION FIELDS	0-8%: SLIGHT 8-15%: MODERATE-SLOPE 15+%: SEVERE-SLOPE		ROADFILL
SEWAGE LAGOONS	0-7%: SEVERE-PERCS RAPIDLY 7+%: SEVERE-PERCS RAPIDLY,SLOPE		SAND
SANITARY LANDFILL (TRENCH)	0-25%: SEVERE-PERCS RAPIDLY 25+%: SEVERE-PERCS RAPIDLY,SLOPE		GRAVEL
SANITARY LANDFILL (AREA)	0-15%: SEVERE-PERCS RAPIDLY 15+%: SEVERE-PERCS RAPIDLY,SLOPE		TOPSOIL
DAILY COVER FOR LANDFILL	0-8%: GOOD 8-15%: FAIR-SLOPE 15+%: POOR-SLOPE		

SOURCE MATERIAL (B)

ROADFILL	0-15%: GOOD 15-25%: FAIR-SLOPE 25+%: POOR-SLOPE
SAND	UNSUITED
GRAVEL	POOR-EXCESS FINES
TOPSOIL	0-15%: FAIR-SMALL STONES 15+%: POOR-SLOPE

WATER MANAGEMENT (B)

POND RESERVOIR AREA	PERCS RAPIDLY,SLOPE
EMBANKMENTS DIKES AND LEVEES	PERCS RAPIDLY
EXCAVATED PONDS AQUIFER FED	NO WATER
DRAINAGE	NOT NEEDED
IRRIGATION	PERCS RAPIDLY,SLOPE
TERRACES AND DIVERSIONS	SLOPE
GRASSED WATERWAYS	SLOPE

COMMUNITY DEVELOPMENT (B)

SHALLOW EXCAVATIONS	0-8%: SLIGHT 8-15%: MODERATE-SLOPE 15+%: SEVERE-SLOPE
DWELLINGS WITHOUT BASEMENTS	0-8%: SLIGHT 8-15%: MODERATE-SLOPE 15+%: SEVERE-SLOPE
DWELLINGS WITH BASEMENTS	0-8%: SLIGHT 8-15%: MODERATE-SLOPE 15+%: SEVERE-SLOPE
SMALL COMMERCIAL BUILDINGS	0-4%: SLIGHT 4-8%: MODERATE-SLOPE 8+%: SEVERE-SLOPE
LOCAL ROADS AND STREETS	0-8%: SLIGHT 8-15%: MODERATE-SLOPE 15+%: SEVERE-SLOPE

REGIONAL INTERPRETATIONS

LAWNS, LANDSCAPING, GOLF FAIRWAY	0-8%: SLIGHT 8-15%: MODERATE-SLOPE 15+%: SEVERE-SLOPE

APPENDIX D   *Computer output format printed out for interpretations of the Valois series.*

RECREATION (C)

CAMP AREAS	0-8%: SLIGHT 8-15%: MODERATE-SLOPE 15+%: SEVERE-SLOPE		PLAYGROUNDS	0-2%: SLIGHT 2-6%: MODERATE-SLOPE 6+%: SEVERE-SLOPE
PICNIC AREAS	0-8%: SLIGHT 8-15%: MODERATE-SLOPE 15+%: SEVERE-SLOPE		PATHS AND TRAILS	0-15%: SLIGHT 15-25%: MODERATE-SLOPE 25+%: SEVERE-SLOPE

CAPABILITY AND PREDICTED YIELDS -- CROPS AND PASTURE (HIGH LEVEL MANAGEMENT)

CLASS-DETERMINING PHASE	CAPA-BILITY	CORN (BU)		CORN SILAGE (TONS)		OATS (BU)		WHEAT (BU)		PLTATOES, IRISH (CWT)		ALFALFA HAY (TONS)		PASTURE (AUM)	
		NIRR	IRR	NIRR	IRR	NIRR	IRR	NIRR	IRR	NIRR	IRR	NIRR	IRR	NIRR	IRR
0-3%	1	100		20		75		45		300		4.0		8.0	
3-8%	2E	100		20		75		45		300		4.0		8.0	
8-15%	3E	95		19		75		45		-		4.0		6.0	
8-15% SEV. ER.	4E	-		-		-		-		-		-		-	
15-25%	4E	90		18		70		40		-		3.5		5.0	
15-25% SEV. ER.	6E	-		-		-		-		-		-		-	
25-35%	6E	-		-		-		-		-		-		-	
25-35% SEV. ER.	7E	-		-		-		-		-		-		-	
35+%	7E	-		-		-		-		-		-		-	

WOODLAND SUITABILITY (D)

CLASS-DETERMINING PHASE	ORD SYM	EROSION HAZARD	EQUIP. LIMIT	SEEDLING MORT'Y.	WINDTH. HAZARD	PLANT COMPET.	IMPORTANT TREES	SITE INDX	TREES TO PLANT
0-15%	3O	SLIGHT	SLIGHT	SLIGHT	SLIGHT		SUGAR MAPLE	61 *	EASTERN WHITE PINE
15-35%	3R	SLIGHT	MODERATE	SLIGHT	SLIGHT		NORTHERN RED OAK	70	WHITE SPRUCE
35+%	3R	MODERATE	SEVERE	SLIGHT	SLIGHT				NORWAY SPRUCE
									RED PINE
									EUROPEAN LARCH

WINDBREAKS

CLASS-DETERMINING PHASE	SPECIES	HT	SPECIES	HT	SPECIES	HT	SPECIES	HT
	NONE							

WILDLIFE HABITAT SUITABILITY (E)

CLASS-DETERMINING PHASE	POTENTIAL FOR HABITAT ELEMENTS								POTENTIAL AS HABITAT FOR			
	GRAIN & SEED	GRASS & LEGUME	WILD HERB.	HARDWD TREES	CONIFER PLANTS	SHRUBS	WETLAND PLANTS	SHALLOW WATER	OPENLD WILDLF	WOOOLD WILDLF	WETLAND WILDLF	RANGELD WILDLF
0-3%	GOOD	GOOD	GOOD	GOOD	GOOD	-	POOR	V. POOR	GOOD	GOOD	V. POOR	-
3-8%	FAIR	GOOD	GOOD	GOOD	GOOD	-	POOR	V. POOR	GOOD	GOOD	V. POOR	-
8-15%	FAIR	GOOD	GOOD	GOOD	GOOD	-	V. POOR	V. POOR	GOOD	GOOD	V. POOR	-
15-25%	POOR	FAIR	GOOD	GOOD	GOOD	-	V. POOR	V. POOR	FAIR	GOOD	V. POOR	-
25-35%	V. POOR	FAIR	GOOD	GOOD	GOOD	-	V. POOR	V. POOR	FAIR	GOOD	V. POOR	-
35+%	V. POOR	POOR	GOOD	GOOD	GOOD	-	V. POOR	V. POOR	POOR	POOR	V. POOR	-

POTENTIAL NATIVE PLANT COMMUNITY (RANGELAND OR FOREST UNDERSTORY VEGETATION)

COMMON PLANT NAME	PLANT SYMBOL (NLSPN)	PERCENTAGE COMPOSITION (DRY WEIGHT) BY CLASS DETERMINING PHASE

POTENTIAL PRODUCTION (LBS./AC. DRY WT):		
FAVORABLE YEARS		
NORMAL YEARS		
UNFAVORABLE YEARS		

FOOTNOTES

A ESTIMATES OF ENGINEERING PROPERTIES ARE BASED ON TEST DATA FROM 1 PEDON FROM CHEMUNG CO., NEW YORK.
B RATINGS BASED ON GUIDE FOR INTERPRETING ENGINEERING USES OF SOILS, NOV. 1971
1 SEASONAL WATER TABLE WITHIN 5 FT. DEPTH BUT CONSIDERED A SLIGHT LIMITATION FOR THESE USES
2 RATINGS BASED ON NORTHEAST REGIONAL CRITERIA, MARCH 1966
C RECREATION RATINGS BASED ON SOILS MEMORANDUM-69, OCT. 1968
D RATINGS BY SOILS MEMO. 26, SEPT. 1967 AND REGIONAL CRITERIA. SITE INDEX VALUES MAY RANGE + OR - 5 OR MORE
E WILDLIFE RATING BASED ON SOILS MEMORANDUM-74, JAN. 1972
* SITE INDEX IS A SUMMARY OF 5 OR MORE MEASUREMENTS ON THIS SOIL.

USDA-SCS-HYATTSVILLE, MD. 1973

MAPPING OF MOUNTAIN SOILS
WEST OF DENVER, COLORADO,
FOR LANDUSE PLANNING

Paul W. Schmidt and Kenneth L. Pierce

INTRODUCTION

The foothills of the Front Range west of Denver are currently experiencing rapid suburban growth. This area is underlain largely by Precambrian crystalline bedrock that is mantled by a discontinuous covering of surficial debris, regolith. The distribution and thickness of the regolith, or more properly the absence of regolith, are of prime importance for determining the suitability of a given area for various landuses. The absence of regolith places constraints on urbanization; it increases the costs and engineering problems encountered in excavation of roads, utility corridors, foundations, and septic systems.

We shall describe a simple classification system and reconnaissance method for mapping the distribution of regolith and hard bedrock at and near the surface, and present the information in a manner that can be easily understood and used by planners, developers, engineers, and landowners. The mapping is not intended for specific site evaluation, but rather for classification of areas of several acres or more for use in broad-scale landuse planning.

In the mountainous part of the Front Range urban corridor (Fig. 1), probably the most important geologic variable affecting urbanization is the distribution of soil (regolith) above hard bedrock. The distribution, depth, or absence of regolith places constraints on urbanization, increases the engineering problems, and affects the cost encountered in the excavation of septic systems, roads, utility corridors, and foundations, as well as increasing health problems due to ground- and surface-water contamination.

The material presented here is based primarily on the area mapped to date in the western part of Jefferson County, Colorado. Our study pertains only to the mountainous area underlain by Precambrian crystalline rock. It does not include the plains and lowlands to the east, which are underlain by sedimentary rocks and present different problems, such as swelling clays and impermeable beds. We think the procedures employed can be used in environmental geologic mapping elsewhere, particularly in hilly terrain underlain by crystalline rocks.

In the past decade, extraordinary population growth has occurred and is continuing to occur in the Front Range urban corridor (Fig. 1) near Denver, Colorado. For instance, the population in the mountainous parts of Jefferson County more than doubled in the 1960s (Fig. 2). The growth rate is expected to be about 5.4% yearly for the mid-1970s, in contrast to an annual growth rate of about 2.6% for Colorado and about 1.3% for the United States. Population in the mountainous part of Jefferson County was about 7,500

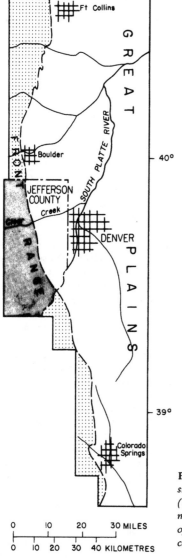

FIGURE 1 *Index map showing mountain soils study area in the Front Range urban corridor. (W. R. Hansen and E. R. Hampton, unpublished mapping 1971.) Stippled pattern indicates area of Mountain Soils project. Shaded pattern indicates area of this report.*

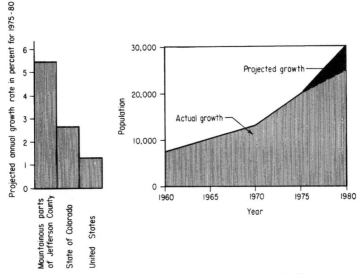

FIGURE 2 *Population growth in the mountainous part of Jefferson County. (Modified from Schalman and Blackburn, 1974.)*

people in 1960, but by the year 1980 is expected to reach 25,000 to 30,000 people (Schalman and Blackburn, 1974).

This growth has been accompanied by a need for pertinent geological information to assist state, county, and local planners and land developers in the evaluation of land for its best possible utilization and maximum value. To aid in satisfying this need, the U.S. Geological Survey began a program of geologic and hydrologic studies, including the preparation of maps of the Front Range urban corridor, at scales ranging from 1:24,000 to 1:100,000. Our study of mountain soils is part of this ongoing program.

Five map units are defined by the areal percentage of regolith more than 6 ft deep: alluvial soil, mostly soil, soil with subordinate rock, rock with subordinate soil, and mostly rock. The units have been mapped by a combination of field reconnaissance and air-photo interpretation. Depths and degree of weathering have been estimated by the study of roadcuts and other excavations and by refraction seismology. Secondary characteristics of the map units, such as percentages of outcrops in a given area, steepness of slope, and type and amount of vegetation cover, were used to draw lines separating map units.

GEOLOGIC SETTING

The mountainous part of Jefferson County has a total relief of more than 4,000 ft (1,200 m), ranging in altitude from just under 6,000 ft (1,800 m) to more than 10,000 ft (3,000 m). The landscape has a long history of uplift, erosion, and deposition. Scott (1975) has divided the geomorphic development of the southern Rockies into six stages:

(1) uplift, erosion, and deposition in Laramide time when mountains were elevated about 11,500 ft (3,600 m), streams removed the sedimentary cover and beveled Precambrian rocks, and the pre-Mesozoic surface on the Precambrian rocks was exhumed and locally is still preserved; (2) erosion of a widespread surface by late Eocene time; (3) deposition of fluvial, lacustrine, and volcanic material in Oligocene time; (4) uplift, erosion, and deposition in early Miocene through Pliocene time, when mountain blocks were displaced as much as 40,000 ft (12,000 m) vertically in relation to valley blocks (much less in Jefferson County); (5) accelerated uplift and canyon cutting in Pliocene time resulting in canyons 600 to 1,000 ft (180 to 300 m) deep in the mountain flanks; and (6) glaciation and formation of pediments and terraces in Quaternary time.

Although relics of the old erosional features are preserved locally in the map area in the form of saprolite and old gravel deposits, the erosional stripping of weathered rock during uplift and canyon cutting has left a landscape ranging from deeply weathered bedrock to nearly continuous fresh bedrock.

Different igneous and metamorphic bedrock types, each possessing different weathering characteristics, occur within the study area. Although the map units are not based on bedrock type, knowledge of the way that the different rock types weather aids in field mapping and photo interpretation. The Precambrian bedrock of the study area is comprised of granitic and metamorphic rocks. Even on relatively steep slopes, most granitic rocks tend to weather uniformly to gruss, separated by scattered outcrops and tors. In gruss areas, core stones occur locally and range in size from that of a basketball to that of a large house. Of the different granites in the area, the Pikes Peak Granite is most susceptible to gruss alteration, the Boulder Creek Granite is intermediate in susceptibility, and the Silver Plume Granite is least susceptible. The metamorphic rocks range from mica schists that weather rapidly, to quartzose gneiss and pegmatites that are quite resistant to weathering. Where different rock types occur in narrow bands, weathering may be highly variable. The amount and degree of weathering correlates relatively well with the concentration of biotite and the gentleness of the slope. Along fault zones, the thickness of the regolith is greater owing to either fractures in the bedrock or, especially in the southern part of the area, to hydrothermal alteration.

Excepting alluvium along the valley floors, most regolith is formed by in-place weathering. Where complete, this weathering profile consists of a gradational sequence from "dirt" through mostly weathered and partly weathered bedrock to hard bedrock (Fig. 3). The character of this sequence varies in depth and in degree of weathering with its topographic position and rock type. Dirt, the material between the ground surface and weathered bedrock, is an amorphous mixture of sand, silt, clay, and rock fragments, commonly containing organic matter in its upper part. It can easily be dug with hand tools and light power equipment. Mostly weathered bedrock is commonly saprolitic, i.e., highly decomposed by chemical weathering in place. It is typically yellowish brown or reddish brown in color, and retains original bedrock structures such as foliation, layering, and minor folds. It can easily be crushed by hand, and dug with hand tools and light power equipment. Partly weathered bedrock is formed by both mechanical and chemical weathering and is typically weak brown. When struck with a hammer, it gives off a thud. Fragments, grasped in both hands, can be broken. Heavy equipment is generally required for its removal. Hard bedrock rings when struck with a hammer. Blasting is generally required for its removal.

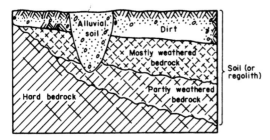

FIGURE 3 *Schematic section showing relations among different surficial materials and bedrock. (Modified from Schmidt, 1976a.)*

MAPPING STRATEGY

To rapidly and inexpensively produce a map that is useful for landuse planning, we decided on five simple soil-terrain units: alluvial soil, mostly soil, soil with subordinate rock, rock with subordinate soil, and mostly rock. The alluvial soil is singled out because of its distinctive lithologic, hydrologic, and topographic properties. The other four map units are based on the areal percentage of soil more than 6 ft (1.8 m) thick to hard bedrock (Table 1, column 1). Soil (regolith) as used here includes all unconsolidated or partly weathered material above hard bedrock.

There are probably many other different strategies for mapping regolith in the mountains west of Denver, the most complete being grain size and isopach of the different layers of regolith. However, owing to the highly irregular distribution of the regolith, many years of comprehensive study would be required for such mapping. This would involve detailed geophysical data and numerous drill holes, and the end result would probably be overly detailed for most planning purposes.

The reconnaissance maps we have made are published in black and white on 7.5-min quadrangles. To date, two maps have been released (Pierce and Schmidt, 1975a, 1975b), two are in press (Schmidt, 1976a, 1976b), and five quadrangles are in preparation. Although classification, unit size, and accuracy would permit reduction to a scale of 1:50,000, we have published at 1:24,000, which is the scale preferred by local planners.

Accompanying each map is a short text explaining the mapping procedures and uses, suggestions for coloring the map units, references to other geologic maps, and a tabular text showing the most important environmental properties of the map units (Table 1).

The procedure for construction of these maps involves field observations and aerial photograph interpretation. Soil thickness and the degree of weathering of the bedrock are measured by studying roadcuts and other excavations and, if necessary, by seismic profiling. A portable seismograph (Bison signal enhancement seismograph model 15708) was used to augment information on subsurface weathering. It was used to acquire a better control on depths to hard bedrock where there are few or no outcrops or excavations. Weathering horizons were easily approximated by velocity changes (Fig. 4), and this information has contributed to a better understanding of the nature of weathering on different rock types. Velocities recorded were 1,000 to 2,000 fps (feet per second)

TABLE 1 *Description and Environmentally Important Properties of the Map Units*

COLUMN 1	COLUMN 2
DESCRIPTION	LANDSCAPE CHARACTERISTICS
Alluvial soil 5 ft (1.5 m) to more than 10 ft (3 m) thick with variation in average grain size depending on topographic position. In small upland basins consists primarily of silt, sand, and clay which grades marginally into colluvium containing angular rock fragments. Beneath valley floors along major drainages, the upper 1-4 ft (0.3-1.2 m) generally is silty sand which overlies coarse sand and gravel, and at the base of valley slopes commonly includes colluvium with angular rock fragments. Fine grained alluvium is commonly rich in organic matter	In upland areas, gently sloping valley bottoms are cut locally by small steep-walled gullies; lush grassy vegetation. On valley floors adjacent to larger drainages, narrow sinuous flats support coniferous and deciduous trees, shrubs, and grasses
More than 70 percent of area is covered by dirt and weathered bedrock to a depth of at least 6 ft (1.8 m). The dirt at the surface averages 2-4 ft (0.6-1.2 m) in thickness. It generally overlies at least 3 ft (0.9 m) of mostly weathered bedrock, which, in turn, overlies at least 2 ft (0.6 m) of partly weathered bedrock, which grades down into hard bedrock. This map unit includes scattered outcrops of hard bedrock (pegmatite dikes, granitic corestones, quartzose metamorphic rocks, etc.)	Gently rolling to nearly flat slopes covered mostly by grass and, locally, widely spaced coniferous trees. Scattered sparse bedrock outcrops, mostly pegmatite dikes, locally protrude through the vegetation
From 40-70 percent of area is covered by dirt and weathered bedrock to a depth of at least 6 ft (1.8 m). The dirt at the surface averages 1-3 ft (0.3-0.9 m) in thickness. It generally overlies at least 4 ft (1.2 m) of mostly to partly weathered bedrock. At a depth of 6 ft (1.8 m), partly weathered bedrock is more common than mostly weathered bedrock. Hard bedrock is at or just beneath the ground surface in about 10-30 percent of the area	Gentle to moderate slopes covered by grass or open coniferous forests. Slopes locally are steep; north-facing slopes have dense coniferous forests and south-facing slopes have scattered coniferous trees and shrubs. Bedrock outcrops are commonly small and scattered
From 10-50 percent of area is covered by dirt and weathered bedrock to depths of 6 ft (1.8 m). Where soil is 6 ft (1.8 m) deep, the dirt at the surface is commonly about 1 ft (0.3 m) thick but may be as much as 4 ft (1.2 m). Much of this dirt has moved downslope and concentrated in pockets. This dirt generally overlies a few feet of partly weathered bedrock. The areas with soil as thick as 6 ft (1.8 m) commonly occur in pockets between rock outcrops. Hard bedrock is at or very near the ground surface in more than 30 percent of the area	Moderate to steep slopes covered with a dense to scattered coniferous forest. Rubbly surfaces covered by shrubs and sparse grass are common on south-facing slopes and along the mountain front, whereas very dense coniferous forests that often mask the soil and bedrock surface are common on north-facing slopes. Bedrock outcrops are generally extensive, locally with large knobs or pinnacles
Less than 10 percent of area is covered by dirt and partly weathered bedrock to a depth of 6 ft (1.8 m). Soil, where present, fills small pockets where dirt and rock rubble have been concentrated by downslope movement. More than 50 percent of the area is hard bedrock outcrop; much of the remainder consists of rock fragments underlain by hard bedrock at shallow depths. Generally occurs on ridge crests and very steep slopes	Mostly rock outcrops on ridge crests and cliffs, covered with scattered trees and shrubs growing in cracks between rocks and on pockets of soil. Rock fragments mantle the slopes between outcrops

MAP UNIT	COLUMN 3	COLUMN 4
	RELATIVE CONSTRAINTS REGARDING SEPTIC SYSTEMS WITH LEACH FIELDS	RELATIVE EASE OF EXCAVATION IN UPPER 6 FT (1.8 M) [Roads, utility corridors, foundations, etc.]
Al Alluvial soil	WATER-TABLE CONSTRAINTS--Although the unconsolidated materials in this area are easily excavated for septic systems with leach fields, the seasonal rise of the water table and local inundation of the surface make this area one of high potential constraint. Slopes are gentle. Only after careful site evaluation, especially during infiltration of the spring snowmelt, can this area be considered safe for septic systems with leach fields. Along major valleys, septic systems with leach fields may drain through highly permeable alluvial gravels and contaminate near by streams	EASY--Hard bedrock absent at or near the surface, making most shallow excavations easy. Machinery can get mired in areas of seasonally wet ground, especially in humus-rich sediment
MS Mostly soil	FEWEST CONSTRAINTS--Fewest geologic constraints for septic systems with leach fields, except for possible water-table situations where sites near local drainages may be seasonally saturated, especially following spring snowmelt. Detailed site study should be made to determine if seasonal saturation will be a problem. Constraints are fewest because hard bedrock is only locally present within 6 ft (1.8 m) of ground surface and because slopes are gentle. With proper planning, this area can accommodate greatest density of septic systems with leach fields	EASY--Hard bedrock generally absent at or near the surface, making most shallow excavations easy
SR Soil with subordinate rock	FEW CONSTRAINTS--Few geologic constraints because hard bedrock is at or near the surface in less than 40 percent of the area and because slopes are mostly gentle to moderate. With proper planning, can accommodate a moderate density of septic systems with leach fields	MODERATELY EASY--Hard bedrock at or near the surface only locally, making most shallow excavations moderately easy if such hard bedrock areas are avoided
RS Rock with subordinate soil	MANY CONSTRAINTS--Many geologic constraints because of the large amount of hard bedrock at or near the ground surface and because slopes are mostly moderate to steep. With proper site evaluation, this area can accommodate only a limited density of septic systems with leach fields	DIFFICULT--Hard bedrock common at or near the surface, making most shallow excavations difficult. Many hard bedrock area may be avoided if local conditions are carefully studied and development is limited
MR Mostly rock	MOST CONSTRAINTS--Most geologic constraints because hard bedrock or coarse rock rubble is common at or near the ground surface and because slopes are mostly steep. Septic systems with leach fields should be restricted to those few areas where acceptable amounts of unconsolidated material can be located on suitable slopes or where pervious fill can be added	VERY DIFFICULT--Hard bedrock and rock fragments are nearly continuous at or near the surface, making most shallow excavations very difficult

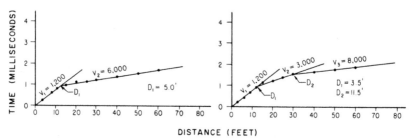

FIGURE 4 *Seismic profiles showing velocity zonation of the regolith: V, velocity as determined by the slope of the line; D, depth as determined by velocity change and distance from geophone. The velocity of 1,200 fps (365 mps) probably represents dirt and mostly weathered bedrock; 3,000 fps (915 mps), partly weathered bedrock; and 6,000 to 8,000 fps (1,830 to 2,440 mps), hard bedrock.*

(300 to 600 mps) for dirt and mostly weathered bedrock, averaging 1,200 fps (350 mps); 3,000 fps (1,000 mps) for partly weathered bedrock; and 4,000 to 12,000 fps (1,200 to 3,650 mps) for hard bedrock, with most measurements between 5,000 and 10,000 fps (1,500 and 3,000 mps).

Along with the data on the thickness of regolith, secondary landscape characteristics (Table 1, column 2, and Figs. 5 and 6) were also noted in the field and used extensively in aerial photo interpretation. Two scales of stereopaired aerial photographs were used:

FIGURE 5 *Moderately rugged landscape showing appearance of associated map units. MR, areas of mostly rock on ridges grade downslope into RS, areas of rock with subordinate soil, and some more meadowy areas of soil with subordinate rock, SR. Meadows of mostly soil, MS, in foreground.*

FIGURE 6 *Subdued landscape showing appearance of associated map units. Clear Creek (just behind foreground) has cut a canyon 1,000 ft (300 m) deep into this old landscape. Meadowy areas of soil with subordinate rock, SR, and mostly soil, MS, grade into alluvial soil, Al, which fills the lows and is cut by historic gullies.*

AMS high-altitude photos at a scale of 1:63,000, and low-altitude photos at a scale of 1:26,800. After the map unit assignments and limited drawing of provisional boundaries between units were made in the field, detailed photo interpretation was made in the office. The lines separating the map units (different from geologic "contacts") were drawn primarily on the basis of such secondary characteristics as percentage of outcrops, degree of slope, topographic (geomorphic) position, underlying bedrock type, and vegetation type and amount of cover (Fig. 7). After photo delineation was complete, the data were transferred to the base map from the annotated photos with a Kern PG2 plotter.

CONCLUSIONS

The mapping of regolith in the mountains has been, and is, an exercise in devising map units that are as simple as possible so that the environmental significances can be readily communicated to a nongeologist. The last two columns of Table 1 consider problems encountered in urbanization. The map units are ranked in terms of constraints in regard to septic systems (column 3) and ease of excavation (column 4).

Because of the dispersed population of this area, sewage disposal plants and water treatment and distribution systems are generally impractical. The result is the common use of single septic systems and wells. In a recent hydrologic study, Hofstra and Hall (1975) showed that water from 20% of the wells sampled and many local streams exceeded drinking water standards in coliform bacteria. Many factors enter into ground- and surface-water contamination. The primary cause is, however, septic systems with

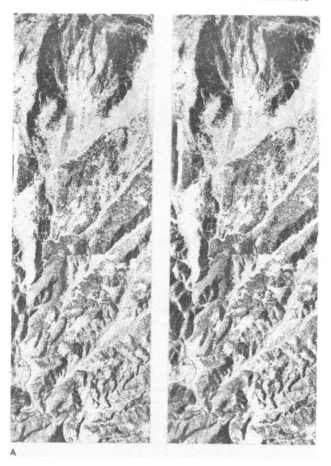

A

leach fields that have been improperly maintained, poorly designed, improperly located, or simply overcrowded.

Depending on need, there are different waste-disposal systems. The most common system by far in the mountainous part of Jefferson County is a septic system with a leach field. Although many factors enter into septic-system requirements, some are the ability of the soil to absorb the effluent, the physical composition of the soil, the depth to hard bedrock and (or) water table, the degree of slope, and the character of the underlying bedrock. Unless individually designed and approved, Jefferson County regulations require that 4 ft (1.2 m) of suitable soil be available between the bottom of the absorption system [generally about 1.5 ft (0.46 m) below the ground surface] and either the bedrock or the highest seasonal groundwater table.[1] Although our soils maps are not

---

[1] Regulations adopted by Jefferson County Board of Health governing individual sewage-disposal systems; Regulation 25-10-102, effective January 20, 1975.

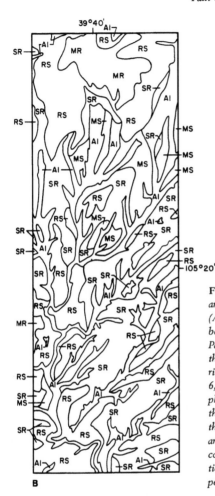

B

FIGURE 7 *Stereopair and map of an area that transects a variety of terrains. (A) Stereopair strip that, from top to bottom, goes from rocky area of Bergen Park [altitude 9,400 ft (2,900 m)] through meadowy flats and then wooded ridges to the Bear Creek Canyon [altitude 6,500 ft (2,000 m)]. These high-altitude photos show most terrain types found in the study area, but lack the resolution of the photos actually used in mapping this area. (B) Map showing units along strip covered by stereopair. Some simplification from actual mapping at 1:24,000 to permit reduction to this scale (1:63,000).*

intended to be used for individual site evaluations, they do indicate general areas that differ in the ratios of soil to hard bedrock (Table 1, columns 1 and 3). Areas mapped as mostly soil have the fewest geological constraints for septic systems owing to the high ratio of soil to hard bedrock; those areas mapped as mostly rock have the most geologic constraints. Although alluvial soils provide a thick filtration medium, they do have constraints. Some reasons for the constraints are the seasonal fluctuation of a shallow water table, the possibilities of flooding and inundation along major drainages, especially during spring snowmelt, and the possibility of contamination of local streams because of the high permeability of alluvial gravels. The degree of slope also affects the performance of septic systems. Although the degree of slope is not a primary basis for the mapping, it correlates well with the map units (Table 1, column 2).

The units mapped also indicate problems likely to be encountered in shallow excavations. Column 4 of Table 1 gives the relative ranking of the units. Shallow excavations in mostly soil, for example, obviously will be much easier than those in mostly rock.

One known use of these maps is for map data that can be programmed into the composite computer mapping system being developed for Jefferson County. This program was developed by the Department of the Interior RALI Program in cooperation with other concerned bodies (Smedes and Turner, 1975; Schalman and Blackburn, 1974). Briefly, individual maps of 20 different attributes, such as vegetation, population, soils, and other environmental factors, are prepared and programmed into the computer using a grid system of 10-acre cells. The cellular map matrix then can be quantitatively weighed, manipulated, and overlain to reflect different landuse strategies or needs and produce composite maps rapidly and with little expense.

REFERENCES

Hofstra, W. E., and Hall, D. C. 1975. Geologic control of supply and quality of water in the mountainous part of Jefferson County, Colorado: *Colorado Geol. Survey Bull. 36*, 51 p.

Pierce, K. L., and Schmidt, P. W. 1975a. Reconnaissance map showing relative amounts of soil and bedrock in the mountainous part of the Ralston Buttes quadrangle and adjoining areas to the east and west in Jefferson County, Colorado: *U.S. Geol. Survey Misc. Field Studies Map MF-689.*

_____, and Schmidt, P. W. 1975b. Reconnaissance map showing relative amounts of soil and bedrock in the mountainous part of the Eldorado Springs quadrangle, Boulder and Jefferson Counties, Colorado: *U.S. Geol. Survey Misc. Field Studies Map MF-695.*

Schalman, G., and Blackburn, B. 1974. Jefferson County mountain area research project; a study performed by faculty and students at the University of Colorado at Denver in conjunction with the Jefferson County Planning Department, Sept. 1974: Jefferson County Advanced Planning Personnel, vols. 1–5.

Schmidt, P. W. 1976a. Reconnaissance map showing relative amounts of soil and bedrock in the mountainous part of the Morrison–Evergreen quadrangles and adjoining areas to the west in Jefferson County, Colorado: *U.S. Geol. Survey Misc. Field Studies Map MF 740.*

_____. 1976b. Reconnaissance map showing relative amounts of soil and bedrock in the mountainous part of the Indian Hills quadrangle, Jefferson County, Colorado: *U.S. Geol. Survey Misc. Field Studies Map MF 741.*

Scott, G. R. 1975. Cenozoic surfaces and deposits in the southern Rocky Mountains: in M. Bruce Curtis, ed., Cenozoic history of the Southern Rocky Mountains: *Geol. Soc. America Mem. 144*, p. 227–248.

Smedes, H. W., and Turner, A. K. 1975. Cooperative development of an environmental information system for Jefferson County, Colorado: ASP–ACSM Fall Convention, Phoenix, Ariz., Oct. 26–31, 1975, 20 p.

4

PRECONSTRUCTION TERRAIN EVALUATION FOR THE TRANS-ALASKA PIPELINE PROJECT

Raymond A. Kreig and Richard D. Reger[1]

INTRODUCTION

The route of the Trans-Alaska Pipeline System (TAPS) traverses 789 mi of diverse terrain, including three major mountain ranges and 590 mi of permafrost. Before the pipeline could be designed, detailed information on soil, bedrock, groundwater, permafrost, and other environmental conditions had to be gathered along the entire route and analyzed. Field reconnaissance information and conventional airphoto interpretation were combined with computer-assisted methods in an integrated program designed to evaluate terrain conditions over large areas, where acquisition of ground-truth data is limited by high costs and difficult access.

Airphoto interpretation was used to prepare terrain unit maps on a photomosaic base at a scale of 1:12,000 to show the distribution of landforms in a 2-mi-wide strip along the project route as well as locations of soil borings. This document served as a basis for designing the pipeline and determining construction techniques suitable for each landform along the route. It has also been useful in locating materials sources and disposal sites, establishing oil-spill and erosion-control contingency plans, anticipating avalanche problems, evaluating slope stability conditions, and numerous other applications.

The construction of TAPS at a projected cost of over $6 billion is the largest and most expensive private project in history. Designing and building a 48-in. hot oil pipeline across diverse terrain, including three major mountain ranges (Fig. 1) and 590 mi of permafrost, presented a truly awesome challenge. To comply with stringent government stipulations designed to protect the environment and ensure the integrity of the pipe from hazards such as earthquakes and thawing of ice-rich permafrost, the pipeline route had to be investigated in more detail and with greater accuracy than any previous large construction project in Alaska. Detailed geotechnical information on soil, bedrock, groundwater, permafrost, and other environmental conditions had to be gathered along the entire route, most of which crossed extensive undeveloped and uninhabited areas. Very expensive field operations and difficult access problems led to the development of techniques for evaluating terrain conditions over large areas where ground-truth information is limited.

Investigation of the TAPS route began with the customary compilation of data from the literature and unpublished documents. Because of its importance in mineral investigation, bedrock geology had been mapped by the U.S. Geological Survey on a reconnaissance level or better for most of the route. However, the character of the surficial

[1] The work described in this report was performed while the authors were affiliated with R & M Consultants, Inc., P.O. Box 2630, Fairbanks, Alaska 99707.

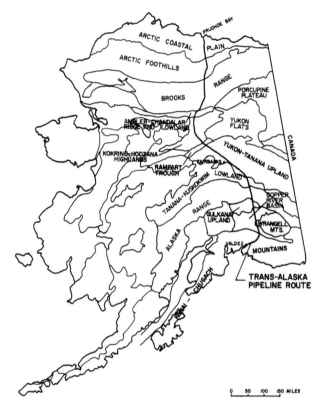

FIGURE 1 *Physiographic provinces in the vicinity of the Trans-Alaska Pipeline route. (Modified from Brew, 1974, Fig. 1.)*

deposits and the distribution of permafrost, which are of primary concern in pipeline design, were largely unmapped and poorly known.

Following a preliminary soil-boring program, in which boreholes were spaced 1 to 10 mi apart, a reconnaissance soil map of the route corridor was prepared at a scale of 1:12,000. A detailed soil-boring program was then undertaken to refine the soil map; define soil properties such as texture, moisture, density, specific gravity, and thaw-settlement characteristics; and provide more detailed information on the distribution of permafrost, groundwater, and bedrock. As of January 1974 over 3,500 soil borings had been drilled along the proposed pipeline route; 2,100 of these were on pipe centerline at an average spacing of 2,000 ft. These borings provided soil test data from over 33,700 samples.

A computer-based data bank was designed for storage and rapid retrieval of the geotechnical information from this extensive sampling and field program. From this massive volume of data, a quantitative assessment of the natural variation of critical soil properties in each landform was summarized by the computer. These summaries have been very useful in comparing conditions in different landforms, establishing exploration priorities, allocating field expenditures, and planning pipeline construction. Timely preparation of construction planning estimates would not have been possible if manual examination of all soil information had been necessary.

Similar techniques have been used to analyze terrain in Australia, Canada, Africa, and Asia.

LANDFORM CLASSIFICATION USED IN
THE TAPS PROJECT

In most large construction projects, an intensive soil-boring program is undertaken to determine specific soil properties needed for design. For example, borings are usually spaced about 500 ft apart (or less, in critical areas) along proposed highway alignments Soil properties are then correlated from one boring to the next, and a geologic cross section or series of cross sections is prepared. However, an investigation at this level of detail for the TAPS route was not practical because as many as 20,000 boreholes would have been required during the preconstruction soil investigation of the 789-mi-long alignment. Even this large amount of soil sampling would not have been adequate to delineate completely certain highly variable subsurface conditions found in permafrost terrain. For example, attempts were made during the early soil-sampling program to delineate each occurrence of massive ground ice along the proposed route so that soils potentially subject to excessive differential thaw settlement could be identified. As many as 11 holes were drilled within a 300- by 300-ft area to define a single ground ice mass. It soon became obvious that such detail was prohibitively expensive. Another approach was needed for predicting soil variations without an excessive number of borings in areas where previous knowledge was minimal or generalized, access difficult, and logistics extremely expensive.

The technique of terrain analysis that was ultimately adopted for the TAPS project consisted of identifying landforms by airphoto interpretation and subsequently defining the variation of geotechnical conditions in each landform by field observations and soil borings. The landform approach to terrain analysis is based on the premises that (1) terrain classification by landform is a reliable means of arranging and correlating borehole and soil-test properties, because each landform represents either a single geologic process or a combination of processes that commonly function together; and (2) each landform consequently has a characteristic range of soil properties, such as unified soil classification, dry density, soil moisture, and thaw settlement (Belcher, 1946, 1948). Each ground-truth observation not only provided information about a particular location, but also—when considered with all other observations in the same landform—helped develop a pattern of variation for that landform. Once the variation of properties in each landform was known, an acceptably conservative design for each route segment was developed by identifying landforms through stereoscopic examination of airphotos and the placement of a few strategically located confirmation borings. This system allowed the most efficient use of exploration funds.

A *landform* has been defined by Belcher (1946, 1948) as an element of the landscape that has a definite composition and range of physical and visual characteristics, such as topographic form, drainage pattern, and gully morphology, which occur wherever the landform is found. It should be emphasized, however, that there is no universally accepted standard definition for the term "landform." Some earth scientists and geographers prefer that this term be restricted to the description of topographic features, i.e., mountains, valleys, and basins. But this particular use of "landform" is only of limited value in a geotechnical investigation, because it does not include a consideration of such three-dimensional properties as soil characteristics and other physical and environmental conditions at the surface and at depth. Although the word "form" indicates shape only, it has become generally acceptable to use the term "landform" to describe not only surface topography, but also the deposits comprising the feature (Howard and Spock, 1940).

The landform classification developed for the TAPS project grouped landforms genetically, because similar geologic processes usually result in landforms with similar characteristics and engineering problems. Each landform is identified by letter symbols, the first letter of which is capitalized and indicates the basic genesis of the deposit (e.g., C for colluvium and F for fluvial deposits). Subsequent lowercase letters differentiate specific landforms in each genetic group. These symbols are chosen mnemonically for simplicity and as an aid to users: Fp, floodplain alluvium; Fg, granular fan.

For the specific purposes of the TAPS project, it was necessary to define two supplemental terms more precisely than the general definition of landform given above. A landform consists of one or more single components, called *landform types,* each of which usually represents a single geologic process. Where exposed at the ground surface, the landform type is a morphostratigraphic unit (Gary et al., 1972, p. 464) because it is usually identified primarily by its surface form. Where buried, it is identified by boring information. *Terrain units* are defined as the landform types expected to occur from the ground surface to a depth of about 25 ft. They are used only in map (plan) views to give three-dimensional information on landform types present near the surface of the ground; they are not used in geologic sections. A limiting depth of 25 ft was chosen because deeper soils generally have minimal effect on pipeline design and construction.

Figure 2 illustrates the relationship of terrain units to landform types. In the geologic cross section there are three landform types: floodplain alluvium (Fp), granular alluvial

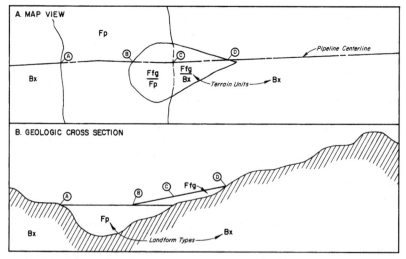

FIGURE 2 *Relationship of terrain units in map view to landform types in geologic cross section.*

TABLE 1 *Landforms Identified Along the Trans-Alaska Pipeline Route*

Symbol	Landform	Symbol	Landform
Bx	Bedrock	Fpb-c	Braided floodplain cover deposits
Bx-u	Unweathered, well-consolidated bedrock	Fpb-r	Braided floodplain riverbed deposits
Bx-w	Weathered or weakly consolidated bedrock	Fpc	Creek or small watercourse deposits
C	Colluvium	Fpm	Meander floodplain deposits
Ca	Avalanche deposits	Fpm-c	Meander floodplain cover deposits
Cg	Rock glacier		
Cl	Slide deposit	Fpm-r	Meander floodplain riverbed deposits
Cm	Mudflow		
Cs	Solifluction deposits	Fpt	Old terrace deposits
Css	Silty solifluction deposits	Fs	Retransported deposits
Ct	Talus	Fss	Retransported silt
Ctc	Talus cone		
Ctp	Protalus rampart	G	Glacial deposits
		Gt	Till sheet
E	Eolian deposits	Gg	Glacier
El	Loess		
Ell	Lowland loess	GF	Glaciofluvial deposits
Elr	Frozen complex upland silt[a]	GFo	Outwash
		GFk	Kames and eskers
Elu	Upland loess		
Elx	Frozen upland loess[b]	H	Man-made deposits
Es	Sand dune deposits	Hf	Fills and embankments
		Ht	Tailings
F	Fluvial deposits		
Fd	Delta deposits	L	Lacustrine deposits
Ff	Alluvial fan	Lt	Thaw-lake deposits
Ffg	Granular alluvial fan		
Fp	Floodplain deposits	M	Marine deposits
Fp-c	Floodplain cover deposits	Mc	Coastal and coastal-plain deposits
Fp-r	Floodplain riverbed deposits		
Fpa	Abandoned floodplain deposits	Mcb	Beach deposits
Fpa-c	Abandoned floodplain cover deposits	Mct	Tidal-flat deposits
Fpb	Braided floodplain deposits	O	Organic deposits
		Ox	Organic basin fillings[a]

[a]On the North Slope only.
[b]In the Yukon-Tanana Upland and Kokrine-Hodzana Highlands.

fan (Ffg), and bedrock (Bx). Between points B and D, a relatively thin granular alluvial fan overlies both bedrock and stream alluvium. This superposition is indicated by terrain unit symbols on the map view between the letters B and C and between C and D. Elsewhere along this geologic section, either floodplain alluvium or bedrock is exposed as a simple landform. Thus, landform types and terrain units are necessary to describe fully and spatially define landforms.

Fifty-five basic landforms were identified along the pipeline route (Table 1). Many landforms commonly occur together in complex relationships and could not be mapped separately; these landforms are represented by composite symbols. An example is the

TABLE 2 *Frequency of Landform Occurrence by Physiographic Province and for Entire Route Based on Cross-sectional Area to Depth of 50 Feet Beneath Pipeline Centerline on Landform-type Profile*

	By physiographic province							For entire route		
Physiographic province	Landform	% Area	Cumulative % area	Physiographic province	Landform	% Area	Cumulative % area	Landform	% Area	Cumulative % area
Chugach Mountains	Bx-u	41.2	41.2	Kokrine—Hodzana Highlands	Bx	43.2	43.2	Bx	13.6	13.6
	Ffg	14.5	55.7		C	19.3	62.5	Gt	10.5	24.1
	Bx	12.9	68.6		Bx-w	13.3	75.8	G + L	10.3	34.4
	Gt	11.3	79.9		C or F?	7.1	82.9	Fp-r	9.1	43.5
	Fpb-r	6.1	86.6		Fpt	3.6	86.5	Fpb-r	7.5	51.0
	Fp-r	1.2	87.8		Fs	3.4	89.9	Bx-w	4.8	55.8
Copper River Basin	G + L	89.4	89.4		Fp-r	3.3	93.2	Bx-u	3.8	59.6
	L	2.4	91.8		C? or F	1.7	94.9	C	3.2	61.8
	Bx	1.5	98.3	Brooks Range	G + GF	16.1	16.1	Ffg	3.2	65.0
Alaska Range	Gt	35.1	35.1		GF	11.9	28.0	G + GF	3.0	68.0
	GFo	14.4	49.5		Fp-r	8.9	36.9	Fss	3.0	71.0
	Bx	10.8	60.3		Bx	6.8	43.7	GFo	2.6	73.6
	Ffg	7.7	68.0		L	6.5	50.2	GF	2.3	75.9
	Fp-r	6.5	74.5		Fpb-r	5.7	55.9	Elx	1.7	77.6
	Fpb-r	6.5	81.0		Ffg	5.4	61.3	Fs	1.6	79.2
	G + L	2.7	83.7		GF or L	5.4	66.7	L	1.5	80.7
	Bx-u	1.7	85.4		Fs	4.6	71.3	Mc	1.2	81.9
Yukon–Tanana Upland	Bx	22.6	22.6		Gt	3.5	74.8	GF or L	0.9	82.8
	Fp-r	17.8	40.4		Bx-u	1.3	76.1	Es	0.7	83.5
	Fss	12.4	52.8	Arctic Slope	Fpb-r	28.3	28.3	Elr	0.7	84.2
	C	7.4	69.0		Gt	21.3	50.1	Fpm-r	0.6	84.8
	Es	2.6	78.8		Fp-r	10.8	60.9	Fp-c	0.6	85.4
	Fpm-r	2.3	81.1		Bx-w	7.9	68.8	Fpb-c	0.4	85.8
	Fs	1.5	82.6		Mc	7.0	75.8	Ht	0.4	86.2
	Fpc + Cs	1.5	84.1		Bx	5.0	80.8	Fpa-c	0.3	86.5
	Fp-c	1.5	85.6		Elr	4.0	84.8	Fpt	0.3	86.8
	Ht	1.5	87.1		GFo	2.6	87.4	Elu	0.2	87.0
					Fpb-c	1.6	89.0	Ell	0.1	87.1
					GF	1.3	90.3	Ell + Lt	0.1	87.2
					Fpa-c	1.1	91.4			

complex glaciolacustrine deposits (symbol G + L) in the Copper River Basin. Although 250 combinations of single landforms were mapped, only 29, representing 87% of the soils along the route, are of major importance (Table 2).

In our terrain analysis, some consideration was also given to different conditions at the ground surface, even though these conditions do not significantly affect the soils at depth. Terraces and dissected remnants of alluvial fans, for example, are flooded infrequently and may have distinctive vegetation and surface characteristics, but these characteristics do not affect the soil properties. Because these types of surface differences do not reflect different soils, they are not ranked at the same level as landform types or terrain units in the classification, but are treated as subordinate *surface phases* of landforms (Table 3). Surface phases are used with terrain units in map views. They are symbolized with lowercase letters in parentheses after the terrain-unit symbols describing the deposits beneath the surface. For example, Fp(ft) designates a relatively young alluvial terrace. In contrast, the alluvium in very old terraces, such as those just north of the Yukon River in the Ray River drainage, is extensively weathered. Because of its altered condition, this ancient, high-level alluvium was not mapped as a surface phase but as old terrace deposits (Fpt).

TERRAIN UNIT MAP
AND LANDFORM-TYPE PROFILE

Terrain unit maps were prepared at a scale of 1:12,000 by the interpretation of airphoto and boring information to illustrate geotechnical conditions in a 2-mi-wide zone along the route (Fig. 3). Their main features are (1) a photomosaic showing the areal extent of each terrain unit and the locations of the pipeline alignment, roads, streams, pump stations, and selected borings; and (2) a profile showing the landform types expected to a depth of 50 ft along the pipe centerline. Landform types appear on this profile because boring spacing was generally too great for meaningful correlation of soil types. Also shown on the landform-type profile are groundwater levels, borehole depths and numbers, and permafrost distribution. Other information illustrated includes surface soil classification, a topographic profile, and survey stationing along pipeline centerline.

TABLE 3   *Landform Surface Phases Identified Along the Trans-Alaska Pipeline Route*

Surface phase	Symbol	Topographic condition
Young terraces or dissected remnants	(ft)	Former floodplain or alluvial fan surfaces that are no longer actively flooded. Terrace deposits are not significantly weathered.
Permafrost-modified floodplain	(fk)	A hummocky floodplain surface modified by the formation and/or thawing of permafrost.
Moraine	(gm)	Irregular topography of discontinuous ridges, knolls, and hummocks surrounding closed depressions on till sheets.
Drumlin	(gd)	Low, linear ridges separated by broad, shallow, linear troughs formed in unconsolidated deposits by the flow of glacial ice.

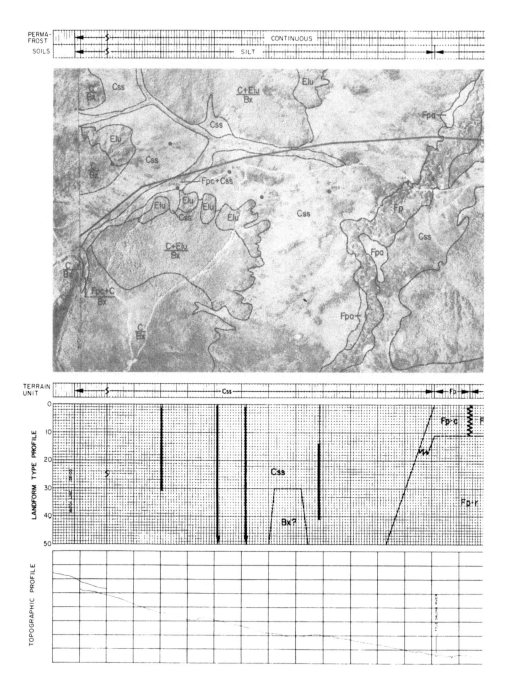

FIGURE 3 *Typical terrain unit map prepared during the preconstruction terrain evaluation of the Trans-Alaska Pipeline route.*

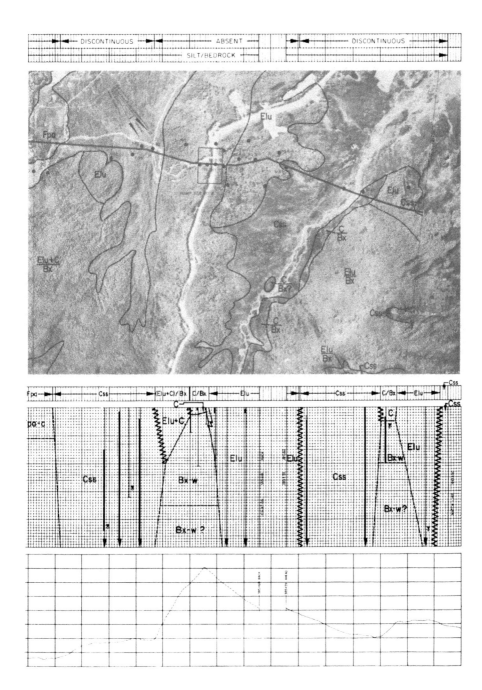

LABORATORY AND SOIL DATA BANKS

A computer-based storage system was set up for geotechnical information from all field investigations and laboratory tests for two reasons: (1) to facilitate report publishing and revision as new information became available, and (2) to facilitate data handling as studies were made of soil property variations in each landform.

Two data banks were created: (1) The *laboratory data bank* (LDB) stored boring location, permafrost conditions, water-table level, and the results of laboratory soil tests, such as gradation and hydrometer analyses, unified soil classification, organic content, specific gravity, dry density, Atterburg limits, and moisture content (Fig. 4). (2) The *soil data bank* (SDB) stored most of the soil-test information on the LDB in addition to all estimated or calculated properties derived from the laboratory results. Calculated properties included dry density, saturation, and moisture content in both frozen and thawed states, excess ice content, thaw strain, and thaw-settlement values (Fig. 5). To prepare a comprehensive thaw-settlement analysis, it was necessary to have actual or estimated soil properties to a projected depth of 99 ft for most borings along the centerline. Therefore, in situ soil properties were estimated for strata intervals that were not actually sampled or tested; for a particular boring, these estimated values are designated by an "E" suffix in the SDB.

Thus, the LDB contains only soil-test results, and the SDB contains both test results and estimated or calculated engineering properties. Both banks contain landform types for all borings and samples. From these two banks a series of summaries were produced to compare conditions in different landforms.

Landform soil property summary tabulates soil test results and estimates of five soil properties for each landform. Moisture content and dry density distributions for specific

```
STATION  1632+59/ 4                  ALYESKA PIPELINE SERVICE COMPANY                        DM-SDB-006
                                             LAB DATA BANK                                    58 -027
PARTY-BORING    9- 66                                                                         07/18/74

**BORING INFORMATION
    ALIGNMENT      OFFSET          DATE     BORING   DIA-      WATER TABLE          BORING
    SHEET                         DRILLED   DEPTH    METER     FT     FT           STATUS
    58 -R 9        E   237        8/01/70    40      2.5     NONE ENCWTRD  SOILS DATA BANK

**LANDFORM TYPES * FP-C              0.0-  4 * FPR  (0)        4 -  8 * FPR              8 - 40 *

**COBBLES & BOULDERS    XX   0.0- 7.5* SC   7.5- 40.0*

**SOIL STATE * T    0.0-  1.0* F   1.0- 40.0*

**SOIL INDEX PROPERTY DATA                                                PCT PASSING
    SAMPLE    DEPTH     FROZ-    SOIL     DRY     MOIST  PCT  SPC GR  PCT  SIEVE SIZES  TESTED
    NO      TOP BOT    THAW    CLASS  DENSITY   CONT   SAT   FINE   ORGNC  1   4  200    BY

    2       1.5-  3.0    F      ML     68.2     49.2    90   2.73          100  93     RGM
    4       5.0-  6.5    F                      34.4                                   RGM
    5       6.5-  8.0    F      SM     92.2     26.4    84   2.75          100  24     RGM
    6       8.0-  9.5    F                       8.5                                   RGM
    7      15.0- 16.5    F                      20.6                                   RGM

**SOIL CLASSIFICATION DATA
    SAMPLE    DEPTH     SOIL            PERCENT PASSING SIEVE SIZES          HYDROMETER  ATTER.  WEIGHT
    NO      TOP BOT    CLASS   3   2 3/2  1 3/4 1/2 3/8   4   10  40 200    .02  .002    LL PI   (GM)

    2       1.5- 3.0   ML                                         100  93  31.7        NV NP
    5       6.5- 8.0   SM                             100  99  98  24   2.5            NV NP


STATION  1632+59/ 4      PARTY-BORING   9- 66                                 58 -027
```

FIGURE 4 *Typical computerized listing of soil index properties stored in the laboratory data bank (LDB).*

```
FEET FROM VALDEZ 1772337            ALYESKA PIPELINE SERVICE COMPANY              PAGE  881 OF BOREDATA
ALIGNMENT SHEET  58  REV  9      LISTING OF VERTICAL PROPERTIES AT EACH BORING                DATE  05/28/74

BORING   9- 66     OFFSET E   237  WATER-TABLE          DEPTH  40       BORE-DATE  8/ 1/70     DIAMETER  2.5

SAMP  SAMPLE    STRATA      FROZEN  FROZEN  FROZEN EXCESS STRAIN  THAW   ACC   THAWED  THAWED THAWED SPEC  SOIL
 ID  INTERVAL  INTERVAL     D-DEN  MOISTURE  SAT    ICE          SETT   SETT   D-DEN  MOISTURE  SAT  GRAV  CLASS
                0.0- 0.4    50.0E   67.0E   84.4   0.00    2.7   0.01    .01   51.4   67.0   88.2  2.20E  OL
                0.4- 1.8    60.0E   58.0F   87.1  12.30   22.6   0.31    .32   77.5   37.4   87.1  2.67E  SP-SM
  2   1.8- 3.0  1.8- 3.0    68.2    49.2    89.6   5.24    9.3   0.11    .43   75.2   41.5   89.6  2.73   ML
                3.0- 4.0    75.0F   40.0E   85.8   2.75    7.2   0.07    .50   80.8   36.3   89.6  2.73E  ML
  4   5.0- 6.5  4.0- 7.5    82.0F   34.4    87.1   7.46   13.7   0.48    .98   95.0   25.2   87.1  2.73E  SM
  5   7.5- 8.0  7.5- 8.0    92.2    26.4    84.3   4.21    8.0   0.04   1.02  100.2   21.8   84.3  2.75   GM
  6   8.0- 9.5  8.0-13.0   127.0E    8.5    67.2   0.00    0.0   0.00   1.02  127.0    8.5   67.2  2.74E  GP
  7  15.0-16.5 13.0-18.0   107.0E   20.6    94.4   2.40    4.0   0.20   1.22  111.5   18.3   94.4  2.74E  GP-GM
               18.0-73.0   178.0E   10.0E   81.6   0.00    0.0   0.00   1.22  128.0   10.0   81.6  2.74E  GP
               23.0-28.0   138.0E    5.0E   57.3   0.00    0.0   0.00   1.22  138.0    5.0   57.3  2.74E  GP
               28.0-53.0   138.0E    5.0E   57.3   0.00    0.0   0.00   1.22  138.0    5.0   57.3  2.74E  GP
               53.0-99.9   127.0E   10.0E   79.1   0.91    1.8   0.86   2.08  129.3    9.2   79.1  2.74E  GP-GM

BORING - SOIL STATE
  T  1.0* F  40.0*

THAW SETTLEMENT SUMMARY
0-8=1.02 * 8-13=        * 8-18= .20 * 8-23= .20 * 8-28= .20 * 8-53= .20 * 13-53= .20*

BORING - LANDFORM TYPES
FP-C             4 * FPR  (0)        8 * FPR         40 *

BORING - COBBLES & BOULDERS
XX   7.5 * SC  40.0 *




STATION   1632+59/ 4            ALIGNMENT SHEET  58  REV  9              BORING   9- 66
```

FIGURE 5 *Typical listing of measured, calculated, and estimated soil index properties stored in the soil data bank (SDB).*

depth increments are displayed in addition to the percentage of the landform that is frozen. The percentage of samples in each unified soil classification[1] is shown, as well as the amount of massive ground ice encountered and the occurrence of cobbles and boulders (Fig. 6).

Textural triangle plots illustrate the range of soil textures that can be expected within each landform. Because soil textures are defined in terms of four particle sizes (gravel, sand, silt, and clay), and because a triangle plot can show only three variables, it was necessary to use two triangles to represent fully the range of soil samples tested (Fig. 7). The computer program used in developing these triangles extracted gradation and/or hydrometer data from the LDB. The percentages of the clay, silt, and coarse (sand and gravel) fractions were plotted in the left triangle, and the fines (clay and silt), sand, and gravel were plotted in the right triangle.

Modified textural triangle is a graphic display of the range of engineering soil types in each landform (Fig. 8). It is based on the unified soil classification and a textural triangle of the ternary system: gravel-sand-fines (clay and silt). Because the unified soil classification is based entirely on plasticity characteristics when more then 55% of the soil is fines, the upper part of the triangle was replaced by a format showing plasticity-liquid limit relationships and organic content. Data from the LDB are entered on the left side of each box; data from the SDB are entered on the right side and underlined. Soils with 5 to 55% fines are shown in the middle and lower parts of the triangle; they are classified by particle-size gradation and plasticity characteristics of the silt-clay fractions. Soils con-

---

[1] On the TAPS project, the unified classification system was modified to better define borderline soils by adding additional classifications for soils with 45 to 55% fines.

LANDFORM
G +L

ALYESKA PIPELINE SERVICE COMPANY
LANDFORM SOILS PROPERTY SUMMARY
PROPERTY STATISTICS BY VERTICAL LANDFORM

PAGE 80 OF DM-SDB-008
10/16/74

ALIGNMENT SHEET 12 TO 27 STATION 3+51/ 2 TO 3148+00/ 3 MILEFEET 340514 TO 807255

FROM LAB DATA BANK

Depth of property (feet)	Moisture content by weight less than 5	12	15	20	25	30	35	50	C	30	Total no. samp.	MC at pct less than 5	20	50	80	95
0– 3	0	1	4	11	24	40	53	67	88	97	118	15	24	33	68	274
3– 5	0	6	11	15	21	44	58	83	96	95	124	11	23	32	47	64
5– 8	1	9	13	28	45	63	78	94	99	C	325	9	17	26	35	50
8– 10	0	14	22	38	54	68	81	95	C	C	85	9	13	22	34	46
10– 13	1	10	17	33	53	71	80	92	99	C	332	9	16	24	34	62
13– 18	0	15	27	43	62	73	80	93	99	C	350	9	13	21	35	54
18– 23	1	16	33	49	64	76	83	92	99	C	305	9	12	20	33	58
23– 30	0	16	30	50	68	79	88	95	99	C	397	8	12	20	30	49
30– 50	0	12	24	45	66	80	86	94	99	C	828	10	13	21	29	53
50–100	0	13	23	46	71	83	90	95	98	C	386	9	14	20	27	48
ALL DEPTH	0	12	23	40	59	73	82	93	98	C	3250	9	14	22	33	57

Property (feet)	Less than 25	50	70	80	90	C	+10	20	30	40	No. samp.	Less than 5	20	50	80	95
0– 3	7	16	37	51	66	92	98	C	C	C	105	16	54	75	95	100
3– 5	1	4	27	40	58	84	92	97	C	C	112	54	68	84	97	114
5– 8	0	1	10	27	47	71	85	93	97	C	286	67	77	91	105	123
8– 10	0	0	5	21	34	57	70	84	95	C	77	68	79	94	115	131
10– 13	0	2	10	18	34	60	81	91	97	C	276	64	81	96	108	126
13– 18	0	2	10	18	34	54	74	87	99	C	280	64	82	98	113	126
18– 23	0	0	9	17	30	48	65	83	98	C	234	68	83	101	118	125
23– 30	0	1	5	13	28	44	62	82	95	99	304	70	86	102	118	129
30– 50	0	2	5	9	27	46	72	89	98	C	611	68	85	101	114	125
50–100	1	2	5	11	27	49	72	90	99	C	286	70	85	100	115	124
ALL DEPTH	1	2	9	18	34	55	75	89	98	C	2571	63	81	97	113	125

	TOTAL	BELOW 5 FEET	BELOW 8 FEET
SAMPLES	3278	3028	2701
PCT FROZEN	100	100	100
PCT THAWED			
PCT QUESTIONABLE			

FROM SOILS DATA BANK

Depth of property (feet)	Moisture content by weight less than 5	12	15	20	25	30	35	50	C	30	Total feet drilled	MC at pct less than 5	20	50	80	95
0– 3	0	10	15	24	36	44	50	60	97	99	327	10	16	34	67	75
3– 5	1	14	19	39	54	66	75	92	99	C	246	9	15	22	35	67
5– 8	1	13	21	46	58	69	78	94	99	C	394	10	14	20	35	52
8– 10	1	15	29	50	65	77	84	95	99	99	268	9	12	19	32	50
10– 13	1	17	31	52	66	79	86	95	98	99	399	7	12	19	30	49
13– 18	1	27	42	59	71	81	85	91	99	C	680	8	11	16	29	56
18– 23	2	23	43	61	70	82	87	91	99	99	679	7	11	16	28	57
23– 30	1	20	37	59	73	84	90	96	99	C	978	8	12	16	27	47
30– 50	0	17	30	45	66	80	88	94	99	C	1777	8	12	21	30	52
50–100	0	18	30	53	82	89	94	98	C	C	666	8	12	19	24	41
ALL DEPTH	1	18	32	51	67	79	85	92	99	C	6418	8	12	19	30	57

Property (feet)	Less than 25	50	70	80	90	C	+10	20	30	40	Feet drilled	Less than 5	20	50	80	95
0– 3	1	10	41	45	56	71	81	93	97	C	325	40	50	81	107	125
3– 5	0	4	8	22	37	56	75	91	98	99	246	50	79	97	112	124
5– 8	0	4	7	19	34	52	71	89	98	99	394	64	80	98	115	124
8– 10	1	3	6	10	23	38	63	83	99	C	268	64	88	104	119	126
10– 13	1	3	5	10	20	36	61	81	97	C	398	69	89	105	119	127
13– 18	0	1	10	11	20	35	55	72	96	C	678	65	90	107	123	129
18– 23	0	1	8	13	22	32	49	73	94	C	671	66	87	110	121	130
23– 30	0	1	4	9	21	36	54	79	92	99	975	70	89	106	120	132
30– 50	0	1	5	10	30	50	66	81	93	99	1769	68	85	100	119	132
50–100	0	0	2	9	22	51	66	79	90	C	666	76	88	99	120	132
ALL DEPTH	0	2	8	13	27	45	62	80	94	C	6355	62	85	102	119	130

	TOTAL	BELOW 5 FEET	BELOW 8 FEET
FEET DRILLED	6418	5845	5450
PCT FROZEN	100	100	100
PCT THAWED			

*NO OVERSIZE ENCOUNTERED

UNIFIED SOIL CLASSIFICATION	TOTAL	CH	MH-CH	MH	OH	PT	CL	ML-CL	OL	ML	GM-ML	GC-CL	ML-GM
SAMPLES TESTED (PCT)	1211	3.2	1.6	.5	.2	.4	36.8	20.2	.2	14.0		.1	.1
EST FT ENCOUNTERED (PCT)	6335.8	1.5	2.3	.1		.1	26.7	16.6	.6	13.1	.3		

UNIFIED SOIL CLASSIFICATION	CL-GC	GM	GC	GM-GC	GP-GM	GW-GM	GP-GC	GW-GC	GW	GP	SM-ML	SC-CL	ML-SM
SAMPLES TESTED (PCT)	.3	1.3	.1	.3	.2	.1			.4	.2	3.4	1.7	2.7
EST FT ENCOUNTERED (PCT)		3.6	.2	.2	.3				.5	1.1	6.8	1.4	2.8

UNIFIED SOIL CLASSIFICATION	CL-SC	SM	SC	SM-SC	SP-SM	SW-SM	SP-SC	SW-SC	SW	SP	ICE	BX	BX-W
SAMPLES TESTED (PCT)	.8	8.3	.3	.6	1.1	.3				.6			
EST FT ENCOUNTERED (PCT)	.1	16.1	.6	.3	1.9	.1				2.3	.4		

UNIFIED SOIL CLASSIFICATION	OTHERS
SAMPLES TESTED (PCT)	
EST FT ENCOUNTERED (PCT)	

UNIFIED SOIL CLASSIFICATION	ICE+SOIL	BX+SOIL
SAMPLES TESTED (PCT)		
EST FT ENCOUNTERED (PCT)	5.8	

LANDFORM G +L CONSISTS OF G +L *

LANDFORM - G +L

FIGURE 6 *Typical summary of soil index properties extracted from the laboratory data bank (LDB) and soil data bank (SDB).*

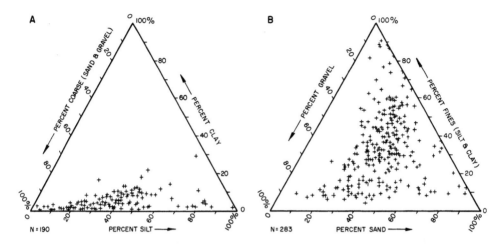

FIGURE 7 *Textural triangle plots of clay-silt-coarse (A) and fines-sand-gravel (B) fractions of glacial till (landform Gt) in the Alaska Range.*

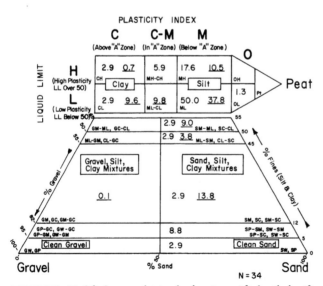

FIGURE 8 *Modified textural triangle showing unified soil classification of lacustrine deposits (landform L) in the northern Brooks Range.*

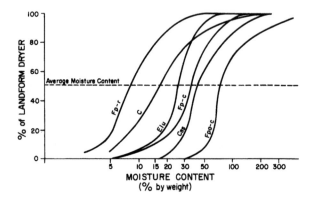

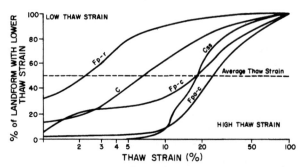

FIGURE 9 *Comparison of moisture-content (upper) and thaw-settlement (lower) curves for various landforms on a typical terrain unit map (Fig. 3).*

taining less than 5% fines are plotted along the base of the triangle and are classified solely on the basis of their size gradations.

Additional summaries of thaw-strain, soil-saturation, and grain-size curves were prepared for each landform. Figure 9 illustrates the differences in moisture-content and thaw-settlement predictions for the landforms on the sample terrain unit map (Fig. 3).

SOIL VARIATION WITHIN LANDFORMS

Landforms may be homogeneous or heterogeneous, depending on the nature of the processes forming them. Homogeneous landforms, such as sand dunes (Es), are usually the result of eolian, fluvial, or lacustrine processes that deposit well-sorted materials. When identified, they define a fairly narrow range of geotechnical characteristics. Heterogeneous landforms contain poorly sorted deposits usually formed by colluvial or glacial processes. The range of soil properties encountered in these landforms often varies considerably. A till sheet (Gt) can contain not only material deposited directly from the melting glacial ice, but also minor amounts of alluvium deposited by streams flowing on or in the glacier, as well as lacustrine deposits laid down in ponds occupying depressions in the stagnant ice.

Variation of landform is best considered within the framework of the physiographic province (Fig. 1). Not only does the pattern of landforms differ significantly in each physiographic province, but our investigation of the TAPS route also demonstrates that the geotechnical properties of landforms may vary from one province to the next because of differences in climate, weathering rates and processes, and predominant bedrock type.

Principles of Landform Variation Analysis

The variation of soil properties within a landform can best be evaluated using data derived from field observations and soil tests of that landform. However, unweighted averaging of such data can be very misleading because of biases incurred through different drilling and sampling methods. When soil properties in a particular landform are studied during an alignment investigation, data should be considered not only from borings drilled along centerline, but also from borings drilled off-line in the same landform. Figure 10 is a hypothetical terrain unit map showing the pipeline route crossing several landforms on a hillside including bedrock (Bx) at the hill crest, colluvium over bedrock [$\frac{C}{Bx}$] on the upper slope, colluvium (C) on the lower slope, and retransported silt (Fss) in the valley bottoms. In this example, the alignment traverses short segments of these landforms, which were sampled by only one or two borings. Within landform Fss there are only two boreholes along centerline, neither of which encountered ice-rich soils. By just considering the results of these two boreholes, one would erroneously conclude that the segment is free of massive ice. On the other hand, if all the borings within landform Fss are considered, whether on or off-line, four out of ten borings encountered ice-rich soils, indicating that about 40% of the retransported silt contains significant ground ice. There is a strong probability that both borings drilled in the short segment of retrans-

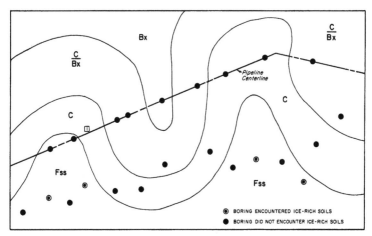

FIGURE 10 *Hypothetical terrain unit map showing distribution of on-line and off-line boreholes encountering different ice conditions in various landforms.*

ported silt could miss the ice-rich sediments. Thus, examination of all available data from local as well as other borings in the same landform provides the most accurate range of expected soil properties.

The use of all available information from borings drilled in a particular landform also provides a reliable basis for predicting soil characteristics in segments of the pipeline route where there are no borings. For example, at locality 1 in Fig. 10, the centerline crosses a short, undrilled colluvial deposit (C). Six borings were drilled in landform C; four are off-line and two are located in a nearby section of the alignment. Soil data from all these borings can provide a reliable basis for predicting the range of characteristics that might be found at locality 1.

Use of Landform Types and Terrain Units in Soil Property Studies

Soil properties and test data generally should only be grouped by landform type, not terrain unit. In Fig. 11, although none of the test pits are sufficiently deep to penetrate the thin glacial till (Gt) and encounter the underlying bedrock (Bx), the terrain unit was classified glacial till over bedrock $[^{Gt}_{Bx}]$ on the basis of information derived from airphoto interpretation and fieldwork. If the soil properties of the deposits encountered in the test pits in this terrain unit were to be averaged without regard to landform type, the importance of the organic deposits (O) would be overestimated and the presence of bedrock missed completely. These erroneous conclusions would not be reached if a more valid sample of all the landforms present were obtained by deeper drilling. A better method of summing soil properties is by landform type: only the five samples of glacial till should be used to predict properties for landform type Gt, and only the samples from the organic material should be used to predict properties of the organics. When soil properties are summarized by terrain unit rather than by individual landform type, mislead-

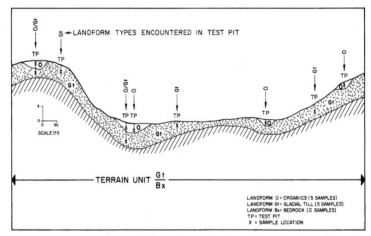

FIGURE 11 *Hypothetical cross section showing relationship of test pits to complex soil conditions in a glaciated terrain.*

ing averages commonly result from the mingling of sample data from dissimilar landforms.

EVALUATION OF METHODS AND DATA

Prediction of Soil Properties

The prediction of subsurface soil properties is generally difficult because these soils are masked by vegetation and surface deposits; furthermore, economic considerations limit the number of borings that can be made to sample them. The number of borings required to evaluate subsurface soil conditions can be considerably reduced if variations in data from subsurface samples can be correlated with visible surface features. Once this correlation is established, reliable estimates of subsurface soil conditions can be made by studying surface patterns.

What is to be done, however, when conditions encountered in boreholes cannot be reliably correlated with surface features or landforms? An example is the erratic and unpredictable occurrence of massive ground ice without distinctive surface expression—a common situation in interior Alaska. The landforms in which massive ice can occur are readily recognized, but the distribution of large ground-ice bodies within these landforms cannot be determined without intensive soil sampling. Another common example is the variability of thaw-settlement values between boreholes without obvious correlation with recognizable surface or geologic patterns. In these situations it is best to consider the *probability* of occurrence within landforms until further research demonstrates a discernible pattern or reason for property variations. Such probabilities should be determined from a random sample population of boreholes or field observations of a recognizable landform.

Weighting Sample Data

The soil-test data used to prepare the landform soil property summaries were generally not collected in a statistically random manner. Biases were introduced in selecting boring locations and in testing the different strata encountered in individual borings; because of these biases, certain soils are often emphasized. A representative determination of soil properties from nonrandom data requires a series of weighting procedures.

Figure 12 illustrates a hypothetical, although typical, situation in a segment of the TAPS route across perennially frozen, retransported silt. Inclusion of all sample data from boring cluster "A," which was drilled to delineate an ice mass, would overemphasize their importance in a landform soil property summary if their significance was not weighted in some fashion. An excellent, although tedious, technique for weighting nonrandom data was used by Thiessen (1911) to evaluate data from irregularly spaced sample locations. Use of this technique considers each boring as representing an area around it defined by lines equidistant between it and surrounding borings (Fig. 13). By weighting boreholes in this manner, the effect of each boring in "A" on the landform soil property summary is minimized.

Another, simpler, technique for minimizing the emphasis of a cluster of borings is to select a single representative boring from the cluster and discard the remaining data. The landform soil property summaries used this method for information derived from the

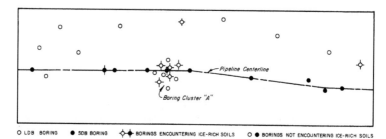

O LDB BORING   ● SDB BORING   ◇ ◆ BORINGS ENCOUNTERING ICE-RICH SOILS   O ● BORINGS NOT ENCOUNTERING ICE-RICH SOILS

FIGURE 12 *Hypothetical distribution of soil borings in perennially frozen, retransported silt (landform Fss).*

SDB, since only selected representative borings along centerline were included on the SDB.

Sampling bias within an individual test hole can result because in situ samples are difficult to obtain from unfrozen coarse-grained soils. Use of standard penetration samplers in these materials is hindered by frequent refusal and poor recovery; samples for density tests are particularly difficult to obtain. A considerable number of the borings in frozen soils along the TAPS route were drilled with diamond-set core barrels utilizing cooled fluids (Hvorslev and Goode, 1966). This technique is an ideal sampling method because it permits the retrieval of undisturbed cores through all frozen materials. Therefore, when summarizing soil property data in coarse-grained landforms, it is necessary to compensate for variations in drilling and sampling methods.

The SDB contains estimated properties for soil strata not sampled or tested in borings—in addition to strata that were sampled. These properties are weighted according to strata thickness to reduce biases, and appear in the SDB section of the landform soil property summaries (Fig. 6). Soil property information from the LDB is not weighted. The differences between weighted (SDB) and unweighted (LDB) data are illustrated in the density distributions of three landforms (Fig. 14). Dry densities from the two data banks differ significantly only for partly frozen granular alluvium (Fp-r) in the Yukon-Tanana Upland, which was primarily sampled with auger borings. Almost all sampling in frozen granular floodplain alluvium (Fp-r) on the Arctic Slope was accomplished using refriger-

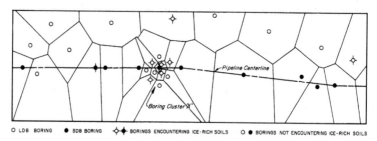

O LDB BORING   ● SDB BORING   ◇ ◆ BORINGS ENCOUNTERING ICE-RICH SOILS   O ● BORINGS NOT ENCOUNTERING ICE-RICH SOILS

FIGURE 13 *Technique for establishing areas represented by data from individual soil borings.*

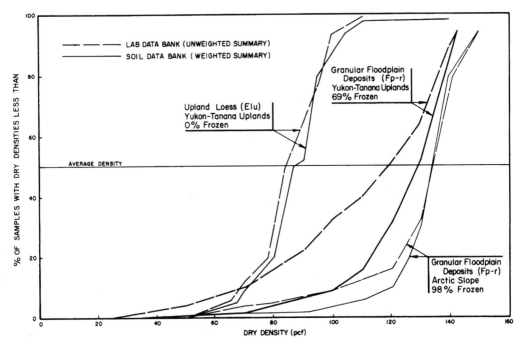

FIGURE 14 *Comparison of weighted (SDB) and unweighted (LDB) dry densities of various landforms in selected physiographic provinces along the trans-Alaska Pipeline route.*

ated coring, and there is little difference between density distributions derived from weighted and unweighted data. The situation is similar in unfrozen upland loess (Elu), where the testing of standard penetration samples provided representative data. Differences between SDB and LDB unified soil classification information are also illustrated by the example of lacustrine deposits (L) in the northern Brooks Range (Fig. 8); these results are due to nonrandom sampling. Most of the borings in this landform were drilled with compressed air methods, and relatively few samples were obtained.

APPLICATIONS OF LANDFORM ANALYSIS

The terrain unit map has served as a design and planning document for construction of TAPS. It was used for evaluating reroute possibilities, locating materials sources and disposal sites, establishing erosion-control and oil-spill contingency plans, anticipating avalanche problems, evaluating slope stability, conducting resistivity studies for establishing cathodic protection procedures, determining work-pad thicknesses, and many other purposes where geotechnical input was required. It was also distributed to contractors for bidding purposes and to government agencies and consultants reviewing the project.

One of the most important applications of the landform approach to the TAPS project was its use in computerized construction planning where input of soil conditions was required. For example, as an aid in materials management, the volume of earthwork was computed for all cuts and fills in each landform. Using soil-texture characteristics and moisture-content data, an estimate was made of excavated material suitable for use as embankments in each landform along the entire route. The landform soil property summaries were very useful for comparing conditions in different landforms, allocating exploration funding and efforts, and estimating ditching and pile-drilling rates so that construction activities could be effectively scheduled and equipment ordered. Timely preparation of these construction planning estimates would not have been possible if manual examination of all soil-boring logs and field data had been required.

COMPARISON WITH
OTHER TERRAIN ANALYSIS SYSTEMS

Several other landform classification systems and terrain evaluation methods have been developed to assess terrain over large areas where ground truth is limited or acquisition of data is difficult.

The land resources surveys by CSIRO[1] in Australia (Christian and Stewart, 1968) and the land system atlases published by the MEXE[2] group and Cambridge and Oxford Universities in Great Britain (Beckett et al., 1972) were developed for agricultural land utilization and general-purpose terrain classification. The MEXE system and its derivatives are used in central and southern Africa, Malaysia, and India, and the similar CSIRO system is used in Australia and eastern New Guinea. These systems are based on the recognition of local landform associations, called *land systems,* which are named after a locality in the same fashion that soil series are named by the USDA.[3] However, land systems differ from soil series in that they are in a higher rank of terrain classification generally corresponding to the soil association of the USDA. Unlike the USDA soil series and soil associations, which are defined almost entirely on the basis of pedologic soil characteristics, land systems are defined in terms of all terrain parameters, such as geology, climate, vegetation, and surface morphology, in addition to pedologic soils. The land systems, once defined, are divided into facets or land types that in many cases correspond to individual landforms, such as floodplains or moraines, or minor subdivi-for agricultural reconnaissance and landuse purposes, was also developed for military uses Because the land system units are named after geographic localities, the classification does not relate units to one another by genesis. The MEXE system, in addition to being used for agricultural reconnaissance and land-use purposes, was also developed for military uses such as trafficability and engineering construction problems. It was specifically set up for the storage of terrain information in a data-bank system.

In Australia, CSIRO has also developed the PUCE[4] program of terrain evaluation for engineering purposes (Grant, 1973, 1974). This system is based on parent material and geologic age. It has four ranks of subdivisions that allow the classification of terrain down

[1] Commonwealth Scientific and Industrial Research Organization, Melbourne, Australia.
[2] Military Engineering Experimental Establishment, Christchurch, U.K.
[3] U.S. Department of Agriculture, Soil Conservation Service.
[4] Pattern Unit Component Evaluation.

to very minute components. The units, however, are distinguished by number and not by name. This designation is somewhat inconvenient and confusing to use on a map; however, it is superbly adapted to use with computers. The PUCE system is apparently unrelated to the CSIRO land-system classification.

A terrain classification based on the genesis of landforms is being used in studies of the Mackenzie River valley in Canada (Zoltai and Pettapiece, 1973). Their units are symbolized with letters keyed to geologic processes (such as eolian and fluvial). Other letter symbols indicate landform morphology and surface soil texture. This system is very similar to the system developed for the TAPS project.

All the above systems were designed to classify surface soils for mapping purposes. They are not used in the construction of cross sections or in the grouping of soil data from borings deep enough to encounter buried deposits of different genesis. The landform classification developed for use during the TAPS project required this capability.

SUMMARY

The preconstruction geotechnical investigation of the TAPS route utilized airphoto analysis and landform classification as an aid in correlating geotechnical information from over 3,500 boreholes and numerous field observations. Soil properties in each landform were summarized on two computerized data banks and used for many engineering purposes where geotechnical input was required. The landform approach allowed the timely preparation of construction planning estimates, which, because of the magnitude of the project, would not have been possible using manual procedures.

Several two-dimensional terrain analysis techniques utilizing the landform approach have been developed for different purposes in areas where ground-truth data are scanty and access is difficult. The system of terrain evaluation developed for the TAPS project introduces a three-dimensional concept.

ACKNOWLEDGMENTS

We would like to thank Alyeska Pipeline Service Company for permission to publish this report. Many employees of R & M Consultants, Inc., have contributed to various phases of the terrain evaluation. Although their number is too great to acknowledge them individually, several deserve special recognition, including Robert L. Schraeder, Wendy L. Kline, Lois D. Sala, and John H. Patterson. Comments and criticism of this manuscript by Robert L. Schraeder, Walter T. Phillips, and Wendy L. Kline have been very helpful. The support of Alyeska Pipeline Service Company and R & M Consultants, Inc., is gratefully appreciated.

REFERENCES

Beckett, P. H. T., Webster, R., McNeil, G. M., and Mitchell, C. W. 1972. Terrain evaluation by means of a data bank: *Geog. Jour.*, v. 138, p. 430–456.

Belcher, D. J. 1946. Engineering applications of aerial reconnaissance: *Geol. Soc. America Bull.*, v. 57, p. 727–734.

____. 1948. The engineering significance of landforms: *Highway Res. Board Bull. 13*, p. 929.

Brew, D. A. 1974. Environmental impact analysis: the example of the proposed Trans-Alaska Pipeline: *U.S. Geol. Survey Circ. 695,* 16 p.

Christian, C. S., and Stewart, G. A. 1968. Methodology of integrated surveys: in *Aerial Surveys and Integrated Studies:* Proceedings, UNESCO Conference on Aerial Surveys and Integrated Studies, p. 233–280, UNESCO, Paris.

Gary, M., McAfee, R., Jr., and Wolf, C. L. 1972. *Glossary of Geology:* American Geological Institute, Washington, D.C., 805 p.

Grant, K. 1973. The PUCE programme for terrain evaluation for engineering purposes. I. Principles: *Tech. Paper 15,* Div. Appl. Geomechanics, Commonwealth Sci. and Indust. Res. Organ. of Australia, 32 p.

———. 1974. The PUCE programme for terrain evaluation for engineering purposes. II. Procedures for terrain classification: *Tech. Paper 19,* Div. Appl. Geomechanics, Commonwealth Sci. and Indust. Res. Organ. of Australia, 68 p.

Howard, A. D., and Spock, L. E. 1940. A classification of landforms: *Jour. Geomorphology,* v. 3, p. 332–345.

Hvorslev, M., and Goode, T. B. 1966. Core drilling in frozen soils: Proceedings, First International Permafrost Conference, *Natl. Acad. Sci–Natl. Res. Council Publ. 1287,* p. 364–371.

Thiessen, A. H. 1911. Precipitation for large areas: *Monthly Weather Rev.,* v. 39, p. 1082–1084.

Zoltai, S. C., and Pettapiece, W. W. 1973. Terrain, vegetation and permafrost relationships in the northern part of the Mackenzie Valley and northern Yukon: *Canada Environmental Social Program for Northern Pipelines, Task Force on Northern Oil Development Rept. 73-4,* 105 p.

II
RIVER
ENGINEERING

5

THE MISSISSIPPI RIVER FLOOD
OF 1973

Charles C. Noble[1]

FEDERAL FLOOD CONTROL PROGRAM
IN THE LOWER MISSISSIPPI VALLEY

Introduction

The amount of water that pours down the Lower Mississippi River during a major flood is enormous. The river drains 41% of the 48 contiguous states, an area that includes all or parts of 31 states and two Canadian provinces. The basin, draining 1,245,000 mi^2 in the heart of the continent, is wide at the top and very narrow at the bottom. It roughly resembles a funnel with its upper part spread across the nation and its spout at the Gulf of Mexico. When too much water pours into the funnel, more water accumulates than the spout can carry away. Then the Lower Mississippi begins to spill out of its banks, flooding unprotected lowlands along its length, and creeping into the low-lying backwater areas of its tributaries (Fig. 1).

Under the Mississippi River and Tributaries Flood Control Project (MR&T Project), the long-held concept of protecting the valley from flood by means of levees alone was considered untenable and was abandoned. It was recognized that to control the great volume of water associated with Mississippi River floods, a plan consisting of additional flood-control features would be required. The MR&T Project is designed to control the "project flood," which is the result of combining some of the most severe storms of record that have occurred in the valley and placing them in a pattern to produce the greatest flood having a reasonable probability of occurrence. The project flood would produce an estimated 3 million ft^3/s flow at the latitude of Old River, half of which would pass down the leveed Mississippi River Channel to the gulf, the other half down the Atchafalaya Basin. Figure 2 shows the source and distribution of one particular combination of tributary flows that would produce the project flood downstream from St. Louis, Missouri.

The four major elements of the plan are levees, floodways, channel improvements, and major tributary improvements.

Levees

The main levee system (Fig. 3) includes about 2,200 mi of levees and floodwalls. Main-line lower Mississippi River levees are formidable earthen embankments that average 30 ft in height. These main-line levees of the Lower Mississippi begin just below Cape Girardeau, Missouri, on the west bank of the river, and stretch to the Gulf of Mexico. On

---

[1] General Noble was President of the Mississippi River Commission and Division Engineer for the Lower Mississippi Valley Division prior to his retirement from the U.S. Army in August 1974.

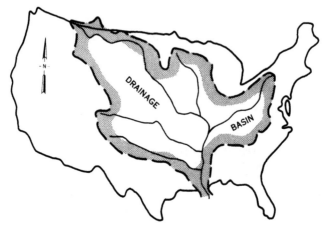

FIGURE 1   *Mississippi River drainage basin.*

the east bank between Cairo, Illinois, and the gulf, the levees alternate with natural bluffs, providing protection only where it is needed. There are intentional gaps in the system where tributaries enter the Lower Mississippi, and during a flood the river often over-powers its tributaries and spreads out over the natural backwater areas.

Floodways

Four floodways have been built in the MR&T Project, one in Missouri and three in Louisiana (Fig. 3). These are located at strategic positions along the river to siphon off excess flood flows. Since the completion of the floodways, some as early as the mid-1930s, there has been a great deal of development in some of the floodway rights-of-way by individuals. But the U.S. Government has flowage rights in the flood-ways, and they are always available for use if needed.

Fortunately, great floods that required operation of the floodways have been rela-tively rare. The Birds Point–New Madrid Floodway, for example, has been operated only once since its completion during the 1937 flood. The Bonnet Carre Spillway above New Orleans was operated in 1937, 1945, 1950, 1973, and 1975. The Morganza Spillway south of Old River was completed in 1954 and was opened for the first time during the 1973 flood. The West Atchafalaya Floodway, which is expected to be the last one to function in the event of a project flood, has never been used.

Channel Improvements

The MR&T Project includes channel improvements to stabilize the alignment of the channel and improve its flood-carrying capacity. One element of the channel improve-ment program consists of cutoffs to improve the capacity of the channel. By 1942, a total of 16 cutoffs and two major chutes had been developed. One effect of these improve-ments was to lower project-flood river stages about 16 ft at Arkansas City, Arkansas, and 10 ft at Vicksburg, Mississippi.

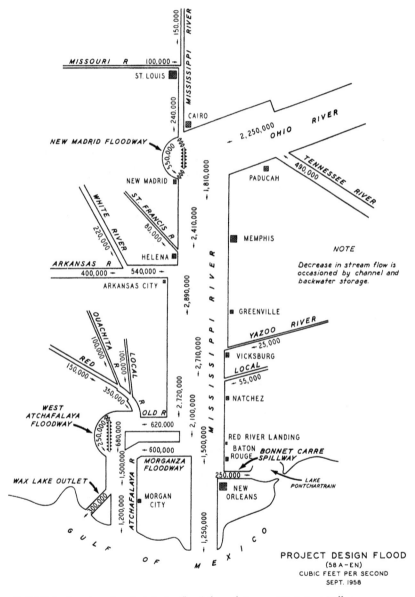

FIGURE 2 *Routing of project design flood through Lower Mississippi Valley.*

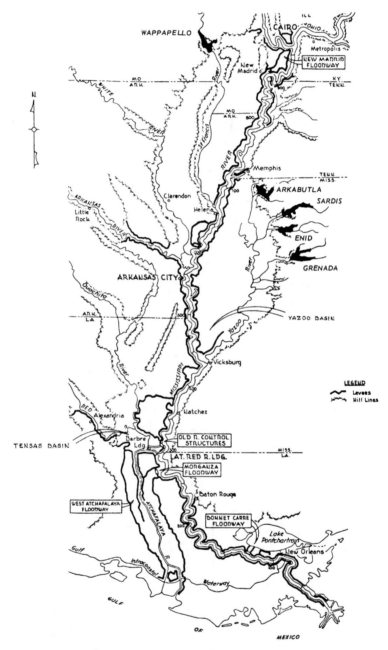

FIGURE 3 *Levees, floodways, and reservoirs forming an integral part of the Mississippi River and Tributaries Flood Control Project (MR&T Project).*

FIGURE 4 *Revetment or landing mat being laid from specially constructed barge near Vicksburg. Revetment consists of articulated concrete blocks about 2 by 4 by 0.5 ft, held together with copper wire. The mat is anchored to the bank, and as the barge moves toward the center of the river, the continuous mat is laid to or slightly beyond the thalweg of the stream.*

The channel-improvement program includes bank revetments, which are placed in the river by plants (Fig. 4) that are unique to the Lower River. These were developed by the Mississippi River Commission and are probably the only such units in the world. The revetments consist of articulated concrete mattress extending from the water's edge to the deepest part of the channel. They protect the riverbank from the cutting action of river currents.

Dikes are used to direct the flow of the river into desired alignments. These features consist of stone structures that restrict low-water flows to the designated channel. A dredging program, to maintain authorized channel depths, is the final part of the channel-improvement program.

It is not feasible to protect all the banks on the Mississippi. Priorities are established based upon which reaches are critical for the protection of the levees or maintenance of the navigable channel.

Figure 5 shows a reach of the river where some revetment is in place but where a particularly troublesome split-flow condition exists. The revetment, dikes, and dredging that will be used to correct this situation are shown. The upper set of dikes will be constructed to shut off the secondary flow and force the water into the left or main channel. Corrective dredging will be performed to widen and deepen the main channel. The lower set of dikes on the left bank is designed to induce the channel to cross to the right bank at the proper point, and the revetment on the right bank will be constructed to

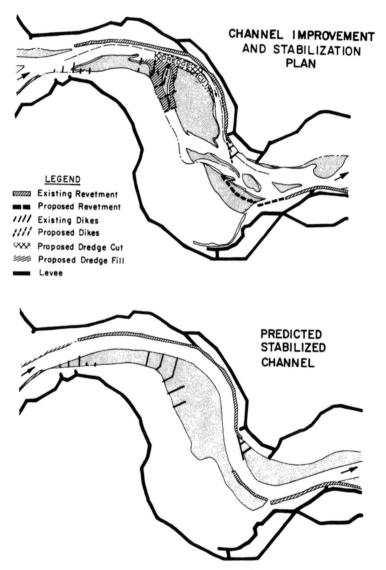

FIGURE 5 *Example of corrective works used to alleviate a troublesome divided-flow condition along a selected Mississippi River reach.*

prevent caving that could be caused by the direct impingement of the main channel current against the bank.

The lower half of Fig. 5 shows how the reach is expected to look after the proposed work is all in place. It should be noted that this is a simplified example in which relatively little additional work is required. In many cases this is not so, and it takes many years to do all the work required to stabilize a particular reach because changes often occur before all work is in place. The river must be continually surveyed and studied to evaluate the changes and make necessary adjustments in plans for future work.

Unusually high flows, such as those experienced in 1973, can damage some of the bank protective works. Big holes were scoured behind some revetments, and some dikes were flanked, with the river taking out the landside end of the dike and chewing into the dike field.

Dredging work on the river has been unusually demanding in the last several dredging seasons because the floodwaters of 1973 and the abnormally high waters experienced in 1974 carried a tremendous sediment load. Wherever these sediments have settled out, excessive shoaling occurred, greatly increasing maintenance dredging requirements.

Tributary Improvements

A fourth element of the MR&T Project involves major tributary basin drainage and flood-control improvements. It includes five reservoirs (Fig. 3) in the hills that border the valley. Those in the Yazoo Basin (Enid, Arkabutla, Sardis, and Grenada) played particularly important roles during the flood of 1973.

The MR&T Project as a whole is 41% complete. Much remains to be done before the valley will be secure against a project flood. This incompleteness was a continual problem during the 1973 flood fight.

THE FLOOD OF 1973

Autumn and Winter (1972–1973)

Over a continuous period of eight months, beginning in October 1972, the Mississippi Valley experienced heavy precipitation, often torrential in nature, over a wide area. Almost all the tributary basins were involved. The low-water period was missed entirely in 1972. At Vicksburg, Mississippi, hydrologists of the Army Corps of Engineers noted an unhappy parallel between the autumn of 1926 and the autumn of 1972. A very wet autumn in 1926 had brought the Lower Mississippi up to abnormally high stages by the end of November of that year. A rainy winter had followed, and the spring rains had brought the great flood of 1927.

Hydrographs comparing the October–November period of 1926 with the same period in 1972 (Fig. 6) revealed all too clearly the relationship between the preflood pattern of 1926 and the pattern that was developing in 1972.

As a result of incessant precipitation during the winter months, flood-control reservoirs along tributary streams began to fill. The ground became saturated, ponding areas were filled, and continuing rains could no longer be absorbed. Most of what fell became runoff. St. Louis, Cairo, and particularly Greenwood, Mississippi, felt the effect of these rains of early winter.

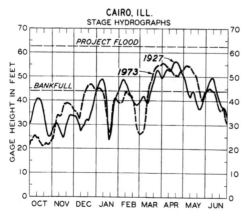

FIGURE 6 *Cairo hydrograph showing similarity between the stages recorded by the flood of 1973 and the disastrous flood of 1927.*

Late Winter (1973)

The Mississippi reached flood stage in the St. Louis area on March 11, 1973. In succeeding weeks, several hundred non-Federal levees in the middle reach of the Mississippi were breached or overtopped, inundating thousands of acres. The higher stages on the Middle Mississippi were quickly reflected by the Cairo gage, which recorded a rise of more than 21 ft during the first 15 days of March. On March 13, the river spilled out of its banks again at Cairo, and tributary flooding was wide-spread in Kansas, Missouri, Iowa, parts of Oklahoma, Arkansas, and Mississippi.

The Missouri River was at flood stage from Kansas City to its junction with the Middle Mississippi. The Arkansas and St. Francis Rivers were out of their banks. The White River was flooding. The Lower Mississippi was at flood stage from Cairo to Caruthersville, Missouri.

Current velocities were so swift around all the river bridges and in the bends that towboat operators were beginning to have trouble. Some boats lacked the power to get their barges upstream through difficult passages, and others were having difficulty controlling their tows when going downstream.

The storm that hit the Lower Mississippi Valley on the night of March 15 was one of the most severe on record. Within the next 30 hr northern Mississippi received up to 11 in. of rain. The reservoirs on the Yazoo River headwater project rose to record levels. Sardis and Grenada emergency spillways began spilling waters on March 15, and the spillway at Enid began a few days later. It was the first time that water had ever run over the emergency spillways on these three Yazoo Basin reservoirs. Later, Arkabutla would also have water running over its spillway.

Flash floods hit many towns and cities in the Yazoo Basin. An estimated 600 homes were flooded (Figs. 7, 8, and 9). In the St. Louis District, hundreds of private levees had

FIGURE 7 *Flooded residential areas in southeast Greenwood, Miss., between U.S. Highway 49E and U.S. Highway 82E.*

broken and hundreds of families were forced to flee. Subdivisions, industries, and houses built outside the protective works were under water (Fig. 10).

Spring (1973)

By March 24, the river had risen to 54.5 ft at Cairo, a stage that was uncomfortably close to the point at which a decision would have to be made about using the Birds Point–New Madrid Floodway. The Lower Mississippi rose almost 4 ft at Vicksburg during

FIGURE 8 *U.S. Highway 49E near Blaine, Miss. Highway was overtopped by water in numerous places in lower Yazoo Basin.*

FIGURE 9   *Agricultural aircraft in flooded cotton field near Valley Park, Miss.*

FIGURE 10   *Flooded River Des Pere area of south St. Louis, at Alabama Street.*

the week in which it reached flood stage, and every inch in stage the river gained meant more misery for residents of the backwater areas. Backwater was spreading over the Red River Basin as well, and evacuations were underway. Nothing could stop backwater flooding in the unprotected areas.

By April 8, the floodwaters were exceeding the capacity of the main stem of the river, making it necessary to open Bonnet Carre Spillway to keep the levees protecting New Orleans and points downstream from being overtopped. But as bad as the flood was by mid-April 1973, it threatened to get worse. Several times during that spring, it appeared that the flood might rise to project proportions. On April 17, the Morganza Spillway below Old River was opened to alleviate flood flows downstream from that point and to aid in combatting a critical situation that had developed at the Old River Control Structure.

As the river rose to unprecedented heights, records were broken from one end of the St. Louis District to the other. It crested at St. Louis on April 28–29 at a historic high of 43.2 ft. On the main stem of the Lower Mississippi, seepage and sand boils were growing steadily worse. In the backwater areas around Vicksburg, residents were told that the new crest would raise the backwater level 2 to 3 ft. Hastily constructed dirt levees around homes and buildings were doomed unless they could be raised 3 to 4 ft. Most of them were already surrounded by water, and there was no dirt available. In the Lower Mississippi Valley Division, more than 10 million acres of land were flooded. Almost 30,000 people had been forced to leave their homes.

The first day of May brought the Mississippi Valley another bad storm accompanied by high winds. Seepage, sand boils, and wave wash continued to be serious problems everywhere. The long-awaited crest at Cairo arrived on May 4 at 54.7 ft. On May 8, the Mississippi crested at Memphis, surpassing the highest stage recorded in the flood of 1950. At Vicksburg, the river had been rising steadily since early in March. It had gone above the flood stage of 43 ft on March 10 and had been over 50 ft on the Vicksburg gage since April 8. On May 13, the river crested at Vicksburg at 53.1 ft, exceeding the 1950 flood by 6.4 ft. On May 14, the river also reached its crest at New Orleans. On May 15, the river was falling slowly at St. Louis, Cairo, Memphis, Vicksburg, and New Orleans. The river's fall was a slow and lingering one, but by June 20 it was back in its banks at Vicksburg. The flood of 1973 was on its way out into the gulf.

SELECTED PROBLEM AREAS

The damages to the MR&T Project itself were considerable. Siltation in the Lower Mississippi and at the Passes in the gulf was extremely heavy. Revetments and dikes were damaged. However, four problem areas for which much data have been collected and much detailed analyses must still be made stand out sharply in the aftermath of the flood: (1) underseepage beneath the levees, (2) bank failures in the lowermost reach of the river, (3) scour at the Old River Control Structure, and (4) stage–discharge relationships and channel deterioration.

Underseepage

Most of our Mississippi River levees between Cairo, Illinois, and Port Allen, Lousiana, are founded on permeable point bar soils. Even where levees have sufficient foreshore to secure them from bank failures, these point bar soils cause problems in underseepage beneath the levee during periods of flood. Geologists and soils experts work together closely on this problem by mapping the areal extent of the deposits and designing landside seepage berms to control underseepage and prevent dangerous boils.

During high-water periods such as occurred during the 1973 flood, the development of underseepage and sand boils on the landside of the levees became a major concern. A sand boil (see Chapter 6) is a small crater that develops on the landside of a levee subject to water pressure from the river over a long period of time. The water forces its way through the ground beneath the earthen levee and finds a weak spot beneath the surface of the ground on the landside of the levee, where it rises and flows out, looking very much like a spring.

As long as the water coming from a sand boil remains clear, it is not a real threat to the levee, however ominous it may appear. If it begins to throw out particles of soil, however, it must quickly be brought under control, or it can carve a channel that will enlarge until it undermines the levee and causes it to collapse.

When a levee is subject to a differential hydrostatic head of water as a result of river stages being higher than the ground surface landward of the levee, seepage entering the pervious substratum sands through the bed of the river and the riverside top stratum creates a flow of seepage beneath and landward of the levee, and the development of hydrostatic uplift pressures in the previous sand below the more impervious top-stratum materials. If the hydrostatic pressure in the pervious substratum sands landward of the levee becomes greater than the weight of the top stratum, the upward pressure will cause heaving of the top stratum, and it may rupture at thin or weak spots, with a resulting concentration of seepage flow in the form of sand boils. When sand boils develop, they may cause erosion or piping of subsurface materials, which can result in progressive collapse of the levee foundation and ultimate undermining of the levee.

The control of underseepage and piping is normally accomplished by constructing properly designed landside seepage berms to increase the weight of resisting top stratum, or by installing relief wells landside of the levee to reduce the uplift pressures in the sand.

Bank Failures

During the 1973 flood, seven Mississippi River bank failures occurred in Louisiana. All were quickly and safely contained; however, their potential for undermining adjacent levees was real and serious. Bank failures normally occur following a sharp fall in the river. When river stages drop quickly, water trapped in the soaked banks tends to trigger bank failures in certain soils due to liquefaction and unbalanced hydrostatic pressure. The hydrograph (Fig. 11) explains the first bank failure at Montz, Lousiana, near the Bonnet Carre Floodway. As shown on this hydrograph, the river stage remained between 20 and 25 ft for approximately 2 months. The stage then dropped sharply 7 ft in 9 days, and on March 9, at the lowest stage since November 20, 1972, the bank at Montz failed.

Six hundred feet of bank slid into the river, threatening the main line levee (Fig. 12). A levee setback was urgently required, and the work was accomplished on emergency, around-the-clock basis.

The Montz bank failure is a good example of the relation of many of our river engineering problems to the geologic setting of the Lower Mississippi Valley. It was a typical flow failure caused by liquefaction of fine sands in point bar materials comprising the bank. Eyewitness accounts indicate that flow slides develop without warning and progress to completion in a matter of hours.

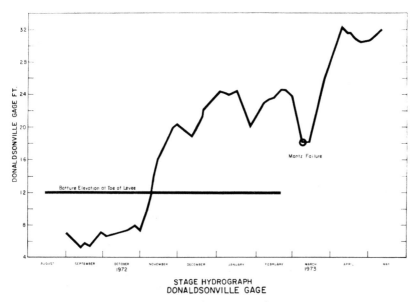

FIGURE 11 *Donaldsville, La., hydrograph (upstream from Montz) showing pronounced drop in river level prior to the Montz, La., riverbank failure of March 9.*

FIGURE 12 *Failure in point bar deposits forming Mississippi River bend at Montz, La., on March 9 after rapid fall in river level. Photograph taken after completion of levee setback.*

Flow failures occur only in point bar deposits. These deposits are usually structured with a somewhat cohesive overburden underlain by fine sands called the "upper sand series" and deeper coarse sands and gravels designated the "lower sand series." The flow failures are confined to the cohesive overburden and the fine sands.

Unfortunately, because of intensive developments behind the levees, it is not practicable everywhere to locate the levees sufficiently landward of the riverbank to avoid threats to their continued stability and safety from failing banks.

Scour at Old River

One critical problem surfaced about mid-April when a guide wall at the Old River Low Sill Structure was undermined by raging floodwaters, setting in motion a disturbed flow pattern at the entrance to the structure. The guide wall was not essential to the structure itself, but there was a definite danger that the eddy pattern set up by the disturbed flow pattern of the water at the entrance to the structure would undermine the levee abutment or the Old River Structure itself (Fig. 13).

To reduce the punishing velocities, the adjacent overbank structure was opened. This raised the tailwater below the structure and lowered the flow velocities through the gates. To further help the Low Sill Structure and to protect downriver levees, the Morganza Floodway was partially opened. This lowered the headwater at the Low Sill Structure. By raising the tailwater and lowering the headwater, the discharges and velocities of flow through the structure were reduced. Because of the eddy pattern, the powerful currents dug a deep scour hole in the silty river bed just upstream of the structure. To stop this attack, a rock dike was constructed and rock was dumped into the scour hole. The eddying waters also chewed out a substantial cavity beneath the structure.

FIGURE 13   *Old River Control Structure during flood of 1973. Mississippi River enters the structure from lower right. Note the absence of the curved guidewall at lower left, a structure that failed during the flood. A scour hole formed near this point and was filled with large rocks.*

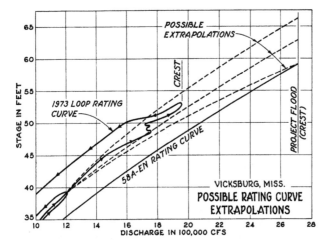

FIGURE 14 *Projections of rating curves (dashed lines) of stage–discharge relationships at Vicksburg based on actual 1973 rating curve (line with arrows). Extrapolations of stages at project-flood crest vary from 59 to 66 ft.*

As floodwaters subsided, the Corps set about to determine just how seriously the Low Sill Structure had been damaged and to what extent large voids may have been created under the structure. Holes were drilled through the base slab to determine their extent. This exploration confirmed the Engineers' concern. The eddies had created a void completely under the structure, extending beneath the downstream stilling basin. The hole in front of the structure had been filled with 100,000 tons of rock during the flood. The void found beneath the structure was filled with more than 32,000 yd³ of a special grouting mixture. A rock dike, designed with the aid of model studies, was built to replace the function of the lost wing wall.

The problem at Old River arose partly from the surprisingly high stages for the discharges experienced during the flood. These higher-than-expected stages caused very high river velocities.

Stage–Discharge Relationships

Early in the flood, it was noticed that the river stages were higher for given discharges than rating curves indicated they should be. This was a very serious development, since it meant that the available rating curves could not be reliably used to predict stages as the river continued to rise. As an example, on the rating curve for Vicksburg shown on Fig. 14, it was impossible to extrapolate with any assurance in late March to determine stages for higher discharge predictions. The curve could have followed any path within the outer dashed lines, either continuing its drift away from the existing rating curve, or closing back on the old rating curve. A spread of about 8 ft in stage prediction was involved. This was far beyond the limit of acceptable error, since freeboard on the levees is considerably less than that.

The line with arrows is the record of plotted river stages recorded for the discharges experienced as the flood developed. This line indicates that the stage–discharge relationship is unstable, that stages are higher for any given discharges than indicated by the 58 A-EN rating curve, and that at various times during the flood there were different stages for the same discharges.

An unstable stage–discharge relationship is characteristic of alluvial rivers because sand waves, bank migrations, and other bed movements constantly change channel dimensions during great floods. The higher-than-expected stages indicated a reduction in channel efficiency, and the multiple stage values are the result of the "hydraulic loop" effect. The flood of 1973 was characterized by a substantial loop effect.

It is an axiom of river hydraulics that a rising river will pass a given discharge at a lower stage than will a falling river. Figure 15 is a typical pattern for a rising river and the beginning of the fall. The dashed line is the pattern for a hypothetical continued fall. However, if a new rise begins before the loop closes on itself, as shown in the figure, the new pattern will have higher stage values than before for the same discharges.

In an actual situation, the river is traced through its erratic stage discharge pattern at Vicksburg on Fig. 16. The river passed 900,000 ft^3/s on January 2 at a stage of 32 ft. It subsequently passed the same amount at stages of 33 and 34 ft. The river wavered again in mid-April, with the result that there were three different stages for a flow of 1.75 million ft^3/s, stages ranging from about 48 to 50.4 ft. This 2-ft difference resulted in much additional flooding in the Yazoo backwater area. Some floods, like that of 1937,

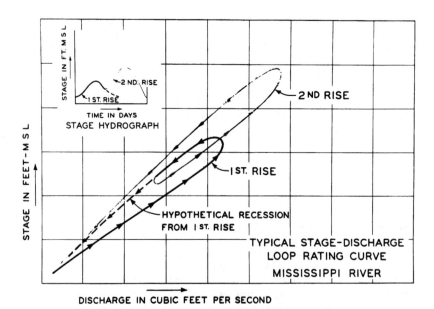

FIGURE 15 Stage–discharge loop rating curve showing typical increase in stage that accompanies a rise, a subsequent moderate fall, and another rise in stage.

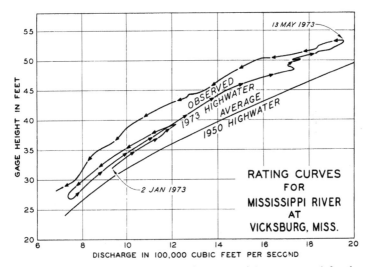

FIGURE 16 *Stage–discharge relationships at Vicksburg compared for the 1973 and 1950 floods, suggesting amount of channel deterioration that has occurred during intervening years.*

had a fairly smooth rise and fall; 1973 was a very long flood with a number of successive rises. This greatly compounded the loop effect.

The loss in channel efficiency is a different matter. The 58A-EN project design flood flowline had been established using data from the 1945 and 1950 floods (Figs. 2 and 14). These floods followed the cutoff program very closely, so the stages experienced showed a greatly improved channel capacity. Since 1950, the river has been losing channel capacity gradually as the river attempted to regain its original length. The channel-stabilization program was designed to develop and hold the channel at its peak efficiency; but because of the magnitude of the undertaking, it was not practicable to pursue it fast enough and some capacity has been lost. Stage–discharge measurements made during the 1973 flood showed that a serious reduction in channel capacity had taken place in the middle reach of the Lower Mississippi River since 1950. The channel is still more efficient than before the cutoffs were constructed, but not as efficient as it was in 1950.

Because of the urgency during the flood to determine the effect of channel deterioration on the stage for the 58A-EN project design flood discharge, it was necessary to develop an expeditious method to extrapolate from observed data to higher flows. Because of the unreliable nature of the graphically extrapolated data, a computational method was utilized. The revised stage was computed by adjusting 1950 channel characteristics to reflect the changes in bankfull channel efficiency that were identified from the 1973 flood data, with the overbank channel characteristics held constant. Overbank conditions were assumed to have remained essentially constant since 1950; therefore, the entire effect on project-flood stages arising from channel deterioration was assigned to the channel deterioration between the riverbanks. The net effect of the reduced channel capacity therefore diminishes as stages rise above bankfull. Utilizing these revised channel characteristics, it was possible to expeditiously compute the order of magnitude increase in flowline attributable to channel deterioration (Fig. 17). As shown on the chart for Vicksburg, the increase of 4.5 ft in stage observed at bank-full capacity translated by computational methods to be 2.8 ft at project-flood level. This computational approach was verified at the huge Mississippi River Basin model at Clinton, Mississippi.

It was then necessary to determine the effect of the loop on project flood stages. The loop effect had not previously been taken into account in establishing the project design flood flowline. As a matter of fact, in the interest of economy of levee construction, an average rating curve had been used in establishing the 58A-EN project design flood flowline. It is now evident, in a valley such as the Mississippi, where the consequences of levee overtopping are unacceptable, that the use of an average rating curve is not a viable option. Protection must be based on the most severe case reasonably expected to occur.

In correcting this situation, other Mississippi River floods exhibiting a multiple-loop effect were analyzed to arrive at a reasonable value to add to the 58A-EN project design flood flowline as corrected for channel deterioration. As a result, a stage increase for loop effect of 1.7 ft was added to the stage increase for channel deterioration at Vicksburg, and other increases were calculated at other points along the river.

In developing the 58A-EN project design flood flowline, the possibility of a decrease in channel efficiency was considered, but no special allowance was made for this loss. In retrospect, a substantial allowance should have been made for loss of a portion of the channel efficiency gained by the cutoff program. The cutoffs are still intact and effective;

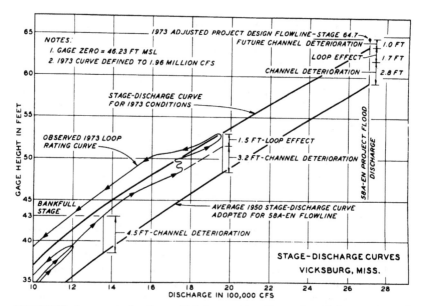

FIGURE 17 *Increase in flowline at Vicksburg attributable to channel deterioration and loop effect.*

however, channel efficiency has since diminished somewhat through the cutoff reach owing to changes in the river arising from its dynamic nature.

The high flows experienced in 1973 permitted the quantitative evaluation of the degree of deterioration in channel capacity associated with large floods. Adjustments in the project-flood stage or project flowline, taking into account past and future channel deterioration and the loop effect, add up to a total stage adjustment of 5.5 ft at Vicksburg, Using the same techniques, the 58A-EN flowline was adjusted upward throughout the affected middle reach of the lower river.

Since levee grades are based on the project-flood flowline plus freeboard, the upward adjustment of the project design flood flowline will require that many miles of MR&T Project levees be constructed to grades higher than previously planned in order to protect against the project flood.

SUMMARY

The flood of 1973 was one of the great floods in the history of the Mississippi River. It reached a flood of record between St. Louis and Cairo and was the greatest flood since 1937 south of Cairo. For 90 days the Mississippi was out of its banks at Vicksburg. The massive flood fight along more than 1,200 mi of river was centralized in the office of the president, Mississippi River Commission and Division Engineer, Lower Mississippi Valley Division, with work conducted by district offices in St. Louis, Memphis, Vicksburg, and New Orleans. The lingering flood created great problems in backwater areas. The great discharges and high velocities attacked and damaged the Low Sill Structure at Old River, and forced the use of two of the major diversion floodways to ensure the integrity of the levee system. By coming in successive crests, this flood created higher flowlines than would be expected of a flood of one crest. As a consequence of the 1973 flood, more than 800 mi of main-line Mississippi levees are being raised. The great Mississippi River and Tributaries Flood Control Project proved its value and is getting renewed emphasis and attention.

IN RETROSPECT

The flood of 1973 disrupted the lives of hundreds of thousands of people. The economic losses are estimated at about $1 billion. Few lives were lost as a direct result of the flood, but some flood-related accidents might not have occurred under other circumstances. Each of these incidents was a personal tragedy for the family of the victim, but there was no wholesale loss of life as there had been in the 1927 and earlier floods. About 69,000 people were made homeless by the 1973 floodwaters, and around 16,500,000 acres of land were inundated.

In the face of such staggering loss estimates, it may seem paradoxical to assert that the flood fight in 1973 was a magnificent success, but losses that actually occurred must be measured against losses that would have been suffered without the federal flood-control project and the flood fight that prevented its destruction. Without the MR&T Project, it is estimated that economic losses would have amounted to more than $15 billion instead of $1 billion. Without the project, an additional 14,500,000 acres of land would have been flooded. This includes just about all the major population and industrial centers in the Lower Valley. What the loss in human lives might have been had the MR&T

levees and structures failed, no one can estimate. In 1927, when the main-line levees broke, conditions were so chaotic that no accurate count could be made, but it is believed that more than 200 people died.

In retrospect, the MR&T Project was remarkably effective in containing one of the great floods in the history of the Mississippi Valley and routing it safely to the gulf. The most sobering reflection, perhaps, is the knowledge that the flood of 1973 came very close to assuming project proportions. Such a flood would have almost certainly overcome the incompleted project. Completion of the MR&T Project is an urgent necessity.

The Mississippi is a dynamic, living, changing river—in short, an alluvial stream. It cannot be unreasonably confined. It has to have a floodplain. It encounters obstacles in its meanderings, such as the stable bordering bluffs and bed deposits here and there of tough nonerodible materials. Man-made revetments and dikes also impose restrictions on natural development in channel and bed mobility. Despite man's efforts to hinder channel shifting, he has been largely unsuccessful. The Lower River is anchored securely at the head of the Alluvial Valley by the gorge in the Commerce Hills. It has been largely fixed below Old River by the works of man and is relatively stable. Between these points, it is in a stage of constant change, seeking to establish and maintain a state of equilibrium among its length, slope, and volume and velocity of discharge. Much work remains to be done to control the Mississippi River for the beneficial use of man.

REFERENCE

Mississippi River Commission. 1975. *The Flood of '73:* Lower Mississippi Valley Division, U.S. Army Corps of Engineers, Vicksburg, Miss., 61 pp.

6

GEOLOGIC CONTROL
OF SAND BOILS ALONG
MISSISSIPPI RIVER LEVEES

Charles R. Kolb[1]

INTRODUCTION

A common and potentially hazardous phenomenon associated with a flooding Mississippi River is seepage beneath the levees and the formation of sand boils. Sand boils consist of sand carried to the surface on the landward side of levees by seepage forces, which often deposit these granular materials in the form of conical mounds (Fig. 1) with water issuing from the top of the mound. Although limited underseepage and through-seepage of the levees are generally acceptable, seepage beneath levees in the form of sand boils indicates active piping and poses a threat to levee safety.

The U.S. Army Corps of Engineers has observed and recorded seepage phenomena during floods along Mississippi River levees in some detail since the early 1930s, and in the early 1950s the first comprehensive studies were made of the phenomena of underseepage and sand boils (Mansur et al., 1956). The purpose of this chapter is to review and reevaluate some of the findings and conclusions reached in this earlier study concerning the effect of geologic factors on underseepage; to discuss underseepage data collected along a randomly selected 40-mi reach of the river during the 1973 flood (USAE District, Vicksburg, 1974); and to relate such data to geologic mapping completed in the late 1950s (Kolb et al., 1958).

THE SETTING

Approximately 2,000 mi of levees flank the Mississippi River in the Lower Mississippi Alluvial Valley. These massive earth embankments are from 30 to 40 ft high and effectively confine flood flows throughout most of the Lower Valley. The distance between levees on either side of the river varies widely. Overbank flood flows range from 15 mi wide between some of these embankments to less than 2 mi wide in some reaches. Figure 2 shows a typical surface profile along an east bank levee at levee station (LS) 2860 or approximately at river mile 563. Flood heights against this levee during the floods of 1937, 1950, and 1973 are shown in the figure. The height of the project flood (the maximum flood expected in the valley) is also shown. Note that overbank flood heights contained by the levee in this reach were moderate during the 1973 flood. Underseepage was correspondingly moderate.

In fact, few sand boils of consequence were reported by the various teams who surveyed underseepage along the levees during the maximum height of the 1973 flood

[1] Former Chief of Engineering Geology Division, U.S. Army Engineer Waterways Experiment Station, Vicksburg, Mississippi.

FIGURE 1 *Sand boils rising above the water level of a sack sublevee near Friars Point, Miss., 1937 high water.*

(USAE District, Vicksburg, 1974). Many pin boils were reported. These are springs or upwellings of water on the landward side of the levee, which carry almost no material to the surface and have no buildup of sediment around their mouths. As flood heights increase and the head difference on either side of the levee increases, however, pin boils may become sand boils that pipe material to the surface.

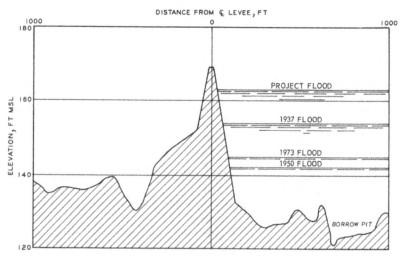

FIGURE 2 *Surface profile along a typical levee showing crest heights of three major floods and of the project flood. Profile taken on east bank of river at levee station 2860.*

Figure 3 summarizes the underseepage data collected during the 1973 flood in a 40-mi reach of the river, which includes Arkansas City, Arkansas. The river reach was more or less randomly selected. Levees are shown in this figure with an appropriate symbol. Levee stations are shown at 20,000-ft intervals. Landside areas where moderate to heavy, heavy, and very heavy seepage occurred are symbolized in Fig. 3, and notes taken by the survey parties along such seepage areas are summarized. Areas of light and moderate seepage recorded by the survey parties are not shown. An asterisk is used to indicate areas where pin boils were reported.

Every attempt was made to make the adjectival classifications of heavy, medium, and light seepage as meaningful as possible. Moist areas on the landside levee, on the berm, or in the field landside of the levee were classified as "light" seepage. The designation "heavy" seepage was reserved for areas where water was visibly flowing, often from pin boils or small sand boils. "Medium" seepage was reserved as an intermediate classification. The river crested for the first time in this reach of the river on April 18, and a field survey was scheduled to correspond with this high-river stage. However, rain (see Table 1) interfered considerably with the judgments used in classifying underseepage. Another survey was made during a subsequent crest near the middle of May. Sunny weather prevailed before and during this latter survey. Thus, the judgments of underseepage made during the May 10–13 time span were given more weight in arriving at the adjectival underseepage ratings plotted in Fig. 3.

Seepage values corresponding to the three classifications of light, moderate, and heavy are approximate at best. Mansur et al. (1956) classified "heavy" seepage as more than 10 gal/min/100 ft of levee, "medium" seepage as between 5 and 10 gal/min/100 ft of levee, and "light" as less than 5 gal/min/100 ft of levee.

DEVELOPMENT OF UNDERSEEPAGE

A convenient distinction in the Mississippi Valley, as in other alluvial valleys, is that between a more or less impervious, fine-grained top stratum and an underlying substratum of sand. Figure 4 schematically depicts the top stratum–substratum relationship at right angles to a typical levee, the irregular thickness of the top stratum, and the depth to the generally impermeable Tertiary horizon (to be discussed more fully in the following section). Note the generalized seepage pattern as flood flows rise against the levee, the zone where seepage typically occurs on the landward side of the levee, and the effect of borrow pits, which often penetrate substratum sands on the riverside of the levee and form an effective and troublesome avenue for seepage.

The first sign of underseepage is usually a dampening of the top-stratum soil at the levee landside toe, along drainage ditches landside of the levee, or up through the ubiquitous crayfish holes that often decorate low-lying areas by the tens of thousands (Fig. 5). As overbank flood flows rise against the levees, hydrostatic pressure in the pervious substratum landward of the levee becomes greater than the submerged weight of the top stratum. Pitcher pumps sunk into the substratum sands at dwellings and in cow pastures, and often miles from the river, begin to flow. Uplift pressures seeking relief along paths of least resistance carry seepage to the surface through root holes, shrinkage cracks, minute fissures, and along man-made and natural depressions and drainage channels. As underseepage increases, springs begin to flow from thousands of pin boils. Some of these eventually develop into sand boils as sand and silt are carried to the surface

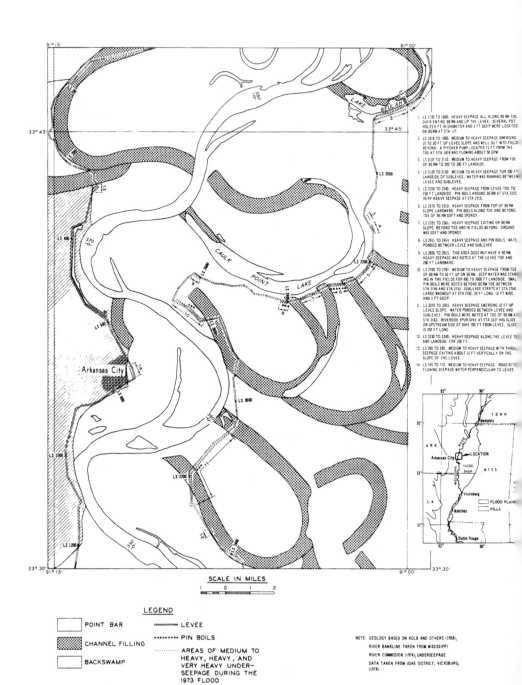

FIGURE 3 *Geologic environments of deposition along a typical reach of the Mississippi River. Areas of significant seepage and of pin-boil development during the 1973 flood are shown.*

TABLE 1 *1973 Stage and Rainfall Data Pertinent to Underseepage Inspection of Levees Shown in Figure 3 (from USAE District, Vicksburg, 1974)*

Date	Arkansas City river stage (ft)	Rainfall (in)
Apr. 16	43.9	1.65
Apr. 17	43.7	
Apr. 18	43.8	3.32
May 10	47.5	
May 11	47.6[a]	
May 12	47.6	
May 13	47.6	
May 14	47.5	
May 15	47.4	
Gage zero	96.7 msl	

[a]Maximum stage during 1973 flood. West bank levees were inspected for underseepage on Apr. 18 and on May 14 and 15. East bank levees were inspected on Apr. 17 and 18 and on May 10 and 11.

from the substratum. A common method for combatting boils is to surround the features with rings of sandbags. Impounding water within such rings to a height equal to the effective hydraulic head stops seepage and sand boil activity at a given point, but subsurface pressures continue to seek avenues for relief by welling through countless other openings. As the flood continues to rise and hydraulic pressures in the substratum increase, seepage keeps pace until the landward sides of the levees are often covered by broad sheets of water with springs and sometimes the more ominous sand boils welling up to heights slightly above the surface of the impounded water (Fig. 6). Although such impounded water makes many roads impassable in and near the levees, preempts

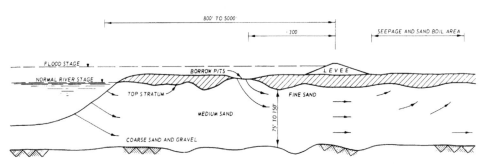

FIGURE 4 *Generalized geological cross section beneath levees in the Arkansas City area. Relatively impervious top stratum is underlain by pervious substratum, which is, in turn, underlain by Tertiary clay.*

FIGURE 5   *Clay encircling typical crayfish hole in top-stratum deposits. These holes are often paths for seepage and eventual sand-boil development.*

FIGURE 6   *Sand boils and seepage water on main-line Mississippi River levee near Greenville during 1973 flood. Notice floodwater elevation on riverside of levee. Most of the boils are ringed with sandbags.*

farming, and covers vast areas with quagmires, a serious situation arises only where sand boils form and piping beneath the levee becomes a possibility.

Top stratum landward of the levees can be classified into three categories (Mansur et al., 1956): (1) no significant top stratum, (2) top stratum of insufficient thickness to withstand the hydrostatic pressures that tend to develop, and (3) top stratum of sufficient thickness to withstand any hydrostatic pressure that may develop during the maximum design flood.

The situation in category 1 occurs only at the extreme northern part of the valley or where top stratum has been removed. Seepage under such conditions can be heavy, as uplift pressures are readily dissipated, but piping and the formation of sand boils are rare. Where large seepage volumes cause problems, drainage sumps and pumps can be used to keep critical areas reasonably dry. Other methods, such as the installation of berms riverward or landward of the levees, the installation of sublevees or cutoffs, etc., have proved effective. Such measures will be discussed more fully later.

Category 3 presents no underseepage problems except at localized spots where the landside top stratum has been removed or partially removed. An interesting case in point is where a soils boring has been made to the underlying substratum and been left open or backfilled with pervious material.

Potentially dangerous underseepage most frequently encountered along the levees is category 2. In this case, the resistance to seepage flow through the top stratum is so great in comparison with the low resistance to seepage flow through the substratum sands that appreciable artesian pressures are built up beneath the top stratum landward of the levee toe. During high water, such artesian pressures range from 25 to 75% of the net head on the levee, and may extend appreciable distances landward of the levee.

The amount of underseepage and uplift hydrostatic pressure that develops landward of the levee is related to the location of the point where seepage enters the substratum on the riverside of the levee and the configuration, thickness, and distribution of the relatively impervious top stratum on the landward side of the levee. One of the most useful tools for determining these important factors and the general distribution and configuration of the top-stratum and substratum deposits is a knowledge of the geology of the Lower Mississippi Valley and the alluvial morphology of the floodplain. The use of air-photo interpretive methods to subdivide alluvial landforms into such basic types as point bars, abandoned channel fillings, natural levees, backswamp deposits, etc., is an important first step in determining where and what kinds of underseepage should be expected along a given reach of levee.

ALLUVIAL VALLEY GEOLOGY
AND ITS EFFECT ON UNDERSEEPAGE

The Alluvial Valley of the Lower Mississippi is a broad flatland about 500 mi long and averaging 50 mi wide. It begins at the confluence of the Mississippi and the Ohio Rivers at Cairo, Illinois, and extends southward to the vicinity of Baton Rouge, Louisiana, where it merges with the Deltaic Plain. The configuration of the valley between Memphis, Tennessee, and Baton Rouge, Louisiana, is shown on the inset map in Fig. 3.

The shape of the floodplain (its outline where it joins the hill lands) is the culmination of erosional and depositional processes during waxing and waning stages of Late Wisconsin glaciation. Glacial meltwaters flowing to the gulf, then some 450 ft lower than today, during the glacial maximum, scoured an entrenched valley into underlying Tertiary and older deposits to depths 100 to 400 ft below the level of the present floodplain. As sea level began to rise about 17,000 yr ago, remnant sands and gravels within the entrenched valley were covered by additional sands and gravels and at higher levels by sand alone. As a result, a variable thickness of sand and gravel lies above an irregularly eroded and relatively impermeable basement of Tertiary and pre-Tertiary deposits (Fig. 4).

Beginning about 10,000 yr ago, a top stratum of clay, silt, and sandy mixtures of clay and silt was deposited above the sandy substratum, first in the lower part of the valley and then in the northern portions. At the southern end, deltas were built and abandoned. Northward, within the Alluvial Valley itself, the Mississippi River changed from a shallow, braided, anastomosing stream to a deep, sinuous, meandering one. Meander belts were built and courses were abandoned about as frequently as were the deltas to the south. The result of this alluvial activity is the deposition of a top-stratum sequence that is highly variable in thickness, often increasing from a superficial cover less than 2 ft thick to a massive clay 100 ft thick within a horizontal distance of 200 ft.

Point Bar Deposits

Point bar or accretion deposits underlie perhaps 60% of the Mississippi River levees. They form on the insides or the convex sides of bends as the bends meander and enlarge. Topographically, the point bar consists of low ridges of silty sand or sand with intervening arcuate lows called swales. Swales, filled with silt and clay, mark quiescent stages in growth of the bend, their directions paralleling the former active river channel. Because of downstream migration of meanders, however, successive ridges and swales tend to overlap in a complex fashion. As individual bends grow, central parts of the bend and those portions most distant from the active channel are covered with vegetation, which traps additional fine-grained soils, so that, even though the ridge-and-swale topography is preserved, the entire sequence is buried eventually beneath a thin cover of finer-grained material. The result is a soil sequence in the ridge areas that tends to grade downward from sandy silt into silty sand and eventually into the clean pervious sand of the substratum. The thickness varies with latitude but can range from inches to as much as 25 ft in the southern part of the valley. The swales, on the other hand, consist of essentially impervious materials, generally varying in depth from 10 to 50 ft. Some are unusually shallow, their depth often depending on effectiveness of scour in the swale during flood flows.

Figure 7 shows the effect of these elongate clay bodies on underseepage, particularly where they pass beneath a levee at an acute angle. Seepage is often heaviest and boil formation most marked within the acute angle. The clay body tends to concentrate seepage in the pervious ridge areas where the geometry of the levee vis-à-vis the trend of the swales resembles that shown in Fig. 7. Note that boils also tend to form adjacent to the swale within the obtuse angle formed by the swale and the levee. However, such seepage is generally less pronounced. Figure 8 illustrates the distribution of boils where swales cross beneath levees at roughly right angles. Boils still concentrate in the ridge soils next to the clay swales, but their distribution is more random than in the case shown in Fig. 7.

Point bar deposits are generally the only deposits along the river thin enough or permeable enough to pose underseepage problems beneath the levees. Note that, in the

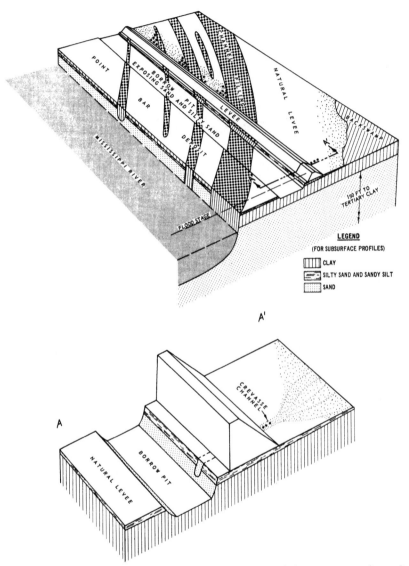

FIGURE 7 *Clay channel fillings and swales crossing beneath levees at an angle. Boils tend to form in point bar deposits within the acute angle between the levee and the clay body. Boils (shown with asterisks) and seepage (shown with a dot pattern) are generally absent in backswamp deposits. A special case is illustrated in the expanded section shown along A-A'. Here a well-developed, semipervious natural levee deposit lies between the backswamp clays and the artificial levee. In such instances seepage may occur in the extreme landward portions of the natural levee and in old natural levee crevasses backfilled with sand.*

107

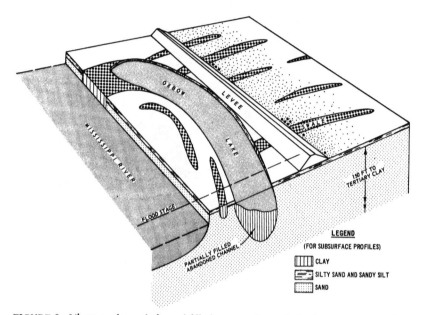

FIGURE 8 *Where swales and channel-fill clays cross beneath the levees at more or less right angles, boils are fairly randomly dispersed and not as frequent or severe as when an acute angle is formed between the levee and the clay bodies. In this case an oxbow lake partially filling an abandoned channel is a serious source for seepage of floodwaters beneath the levee.*

40-mi reach of the river shown in Fig. 3, significant underseepage during the 1973 flood was confined almost entirely to areas where these deposits underlie the levee. What could not be shown in Fig. 3, because of the scale of mapping, are the numerous swales that cross beneath the levees within this reach; although underseepage data in most instances were insufficiently detailed to pinpoint the influence and effect of such minor clay bodies, their effect has been amply demonstrated in previous studies (Mansur et al., 1956).

An important and often critical factor illustrated in Fig. 7 is the effect of borrow pits on the riverside of the levee in initiating or increasing underseepage. Borrow for the levee, particularly during the early years of levee construction, was often taken directly riverside of the levees. Such pits often expose impervious underlying sand and silty sand, and provide ready access for seepage of floodwaters beneath the levee. An important under-seepage preventive measure has been to locate such borrow areas only where they do not expose underlying pervious strata. Where critical underseepage conditions are caused by borrow areas, the areas are often filled with impermeable riverside blanket. No attempt has been made in Fig. 3 to delineate borrow pits that may affect the localization of underseepage.

Natural Levee Deposits

It was stated above that underseepage is generally confined to areas of point bar deposition. An exception to this is where the levee is built on semipervious natural levee deposits.

Natural levees were formed along the migrating Mississippi River channel before the construction of artificial levees. Each year the river topped its banks during floods, the coarsest materials in suspension in the floodwaters were dropped near its banks, and the fines were carried into the low-lying adjacent backswamp areas. With time, well-defined low ridges averaging 10 to 15 ft high were formed, particularly on the outside of bends and along many straight reaches of the river. Continued migration of the river left natural levee segments complexly distributed over the floodplain surface; in many instances, the artificial levees were built on soils readily identifiable as natural levees. Where these deposits overlie point bar deposits, they generally add to the weight of the underlying point bar top stratum and thereby help resist lifting of the top stratum by excessive substratum pressures in the underlying clean sands. In other instances, however, such as in the situation shown in detail along section A-A' in Fig. 7, such semipervious strata form ready paths for seepage. Here natural levee deposits overlie impermeable backswamp clays, and access of water from the borrow pit permits seepage, particularly where the sloping natural levee surface joins the backswamp. The most critical situation occurs where old crevasse channels have been scoured through the natural levee by ancient floods and backfilled with materials even more permeable than the bordering natural levee. Such backfilled crevasse channels provide ready seepage paths and are sometimes the sites for restricted boil formation.

Backswamp Deposits

As briefly mentioned above, backswamp deposits consist chiefly of clay left in suspension in floodwaters as the floods top riverbanks and spread out in the lowlands adjoining the natural levees. Strata ranging from paper-thin to several inches thick gradually accumulate in these low-lying areas, and thicknesses of from 30 to 80 ft of impervious clay are not uncommon. As a rule, backswamp deposits, because of their imperviousness and fairly broad lateral extent, are least troublesome of all the alluvial environments from the standpoint of underseepage. Only in the situation discussed above, where the backswamp forms an impermeable floor for an overlying semipervious natural levee deposit, do moderate underseepage problems develop.

Figure 3 nicely illustrates the effect of backswamp clays on underseepage where such clays underlie the levee. From approximately LS 300 to 1200 on the west bank of the river, the levee is built on backswamp clays. Significant underseepage in this extensive levee reach occurred only from LS 280 to 290, and at Arkansas City between LS 745 and 770. In both instances the levee is so aligned that it extends over small portions of point bar deposits. Note also the extensive borrow pit in the levee setback opposite river mile 570. This illustrates location of riverside borrow pits in deposits, which, because of their thickness and impermeability, have no effect on underseepage.

Channel-Fill Deposits

The thickest and generally the most impervious of the deposits bordering the river are channel-fill deposits, which fill abandoned meander loops of the river. When cutoff of the meander occurs the upper and lower entrances to the loop are often plugged with sandy sediments, and the abandoned channel is left as an oxbow lake in the alluvial plain. As the river migrates away from the point of cutoff, the oxbow lake becomes isolated, often a score or more miles from the active stream, and only the finest of the sediments in overbank flows reach the lake. Eventually, the lakes are completely filled with fine sediment; as a result, significant bodies of clay known as "clay plugs" are found throughout the alluvial plain. These bodies are as deep and as wide as the former cutoff channel, with depths varying from about 100 to 130 ft and with widths averaging about 3,000 ft. Hundreds of clay plugs that preserve the entire abandoned loop and literally thousands of clay plugs that have been partially destroyed by subsequent river meandering have been mapped in the alluvial plain. These significant clay bodies have a marked effect on river meandering, channel stability, and, where they lie beneath the levees, on underseepage.

Figure 9 shows the effect of one such abandoned channel fill on seepage. In this instance we are dealing with a split abandoned channel, one that once contained an island in the cutoff loop, a fairly common occurrence in the Mississippi Valley. Because of the geometry of the clay plug and the angle at which it is crossed by the levee, seepage and sand boils are particularly troublesome in the cul-de-sac represented by that part of the former river island just landward of the levee. Boils are also common where the clay plug or channel fill forms an acute angle with the levee. This is similar to the situation previously described where the smaller clay swales cross beneath the levee at acute angles.

Figure 10 shows a similar situation. In this instance a borrow pit flanks the riverside

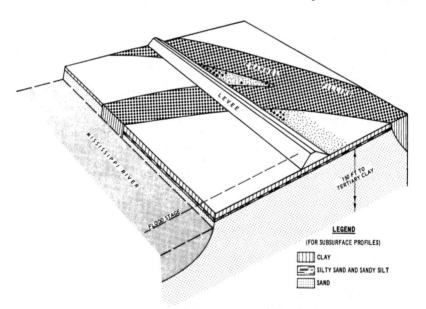

FIGURE 9   *Effect of a split channel filling on localizing seepage and sand-boil formation.*

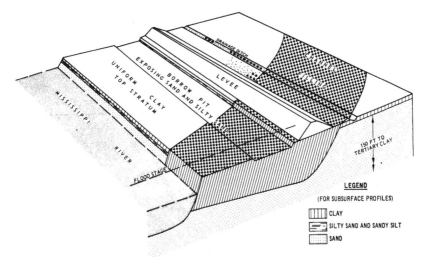

FIGURE 10 *Drainage ditches penetrating fairly permeable materials on the landside of the levees are usually the sites for heavy seepage and boil formation. Borrow pits on the riverside on the levee, which have removed impervious top stratum, greatly accentuate the problem.*

of the levee, and a drainage ditch penetrates fairly permeable material some distance from the landside toe of the levee. Boils and seepage are found in the acute angle made by the channel filling with the levee, but the most pronounced drainage and boil development are in that portion of the drainage ditch which has partially penetrated the clay and silty top stratum.

Because the drainage ditch is at some considerable distance from the levee toe, boils are frequent, but movement of subsurface material to the surface and the danger of piping are negligible.

Figure 8 illustrates a situation somewhat analogous to the seepage problem occasioned by a riverside borrow pit. In this instance the seepage source, however, is a partially filled abandoned channel, an oxbow lake, occurring close to the riverside of the levee. Such partially filled channels permit ready access for seepage beneath the levee, and when point bar deposits flank the landward side of the levee, boils and underseepage are common. Cases in point are seepage reaches 1 through 8 in Fig. 3 where oxbow lakes Beulah and Caulk Point lie just riverward of the levee and furnish a source for underseepage through the levee.

Note that seepage through the clay channel fillings is rare in Fig. 3. However, significant seepage was recorded through the lower arm of the clay plug at seepage reach 11. This is probably due to a thick sand filling in this lower arm of the clay plug at the time of cutoff. Seepage reach 12, between levee stations 3330 and 3340, also occurs in a mapped clay plug. More detailed mapping and a boring or two might clarify what appears to be an anomalous situation.

UNDERSEEPAGE AND LEVEE DESIGN

That the localization of boils and underseepage is due largely to the thickness and distribution of the semipervious and impervious units in the top stratum has been amply demonstrated. The key to the delineation of such units in plan and profile is the geologic environment of deposition. Careful studies involving air-photo interpretation of these environments have proved extremely useful in design for levee underseepage in the Lower Mississippi Valley. Soil borings placed so as to prove out and refine these interpretations are a second important step in levee design. Once the distribution of the top-stratum units has been determined, engineers base their design of underseepage control measures on a variety of parameters. Thicknesses of the substratum sands are determined from available geologic maps or by borings, and permeabilities are determined by field pumping tests or by correlations between the D_{10}, or effective grain size, and permeability. Seepage flow and hydrostatic heads landward of the levee are determined for the project flood. These parameters are based on seepage formulas and/or piezometric data.

Mansur et al., (1956) in summarizing underseepage control measures, list riverside blankets, relief wells, landside seepage berms, drainage blankets or trenches, cutoffs, and sublevees, but state that only the first three methods are considered generally applicable for Mississippi River levees. Sublevees and drainage blankets or trenches are cited as being applicable in certain special situations.

Impervious riverside blankets are soil blankets sealing thin top-stratum areas or seepage into a borrow pit that has uncovered permeable strata. The blanket should be the width of the borrow pit, or from 1,000 to 1,500 ft wide. The thickness of the blanket should be from 3 to 5 ft. The permeability of the blanket should be on the order of 0.01 to 0.1×10^{-4} cm/s. Such blankets reduce both landward substratum pressure and seepage.

Relief-well systems are wells spaced from 75 to 300 ft apart on the landward sides of levees to relieve uplift pressures. Mansur et al. (1956) recommend wells to depths of 60 to 120 ft with screens 40 to 80 ft in length. Such wells reduce substratum pressure and intercept seepage. Disadvantages of relief wells are that they require periodic inspection and maintenance, must be protected from backflooding, and increase the total quantity of seepage about 20 to 40% depending on conditions. These disadvantages can be partially overcome by providing the wells with suitable guards, check valves, and standpipes to prevent flow during low-flood stages.

Landside berms control seepage by increasing the thickness of the landward top stratum so that the weight of the berm and top stratum is sufficient to resist uplift pressures. A berm also lengthens the path of seepage flow, thereby reducing the tendency of failure by piping. The berm should be wide enough so that the head at the berm toe is no longer critical. Thicknesses of these berms at the toe of the levee range from 3 to 10 ft, the width of the berm from 100 to 400 ft. Berms can be used to control seepage efficiently where the landside top stratum is relatively thin and uniform, or where no top stratum is present, but they are not efficient where the top stratum is relatively thick and high uplift pressures develop. Berms may vary in type from impervious to completely free draining. The selection of the type of berm to use should be based on the availability of borrow materials and relative cost of each type. For details on the design of these and other underseepage control measures, see the comprehensive work by Mansur et al. (1956).

CONCLUSIONS

A common problem during floods along the lower Mississippi River is the formation of sand boils on the landward sides of levees. If the hydrostatic pressure in the pervious substratum landward of a levee becomes greater than the submerged weight of the top stratum, the uplift pressure may cause heaving and rupture at weak spots, with a resulting concentration of seepage flow in the form of sand boils. This, in turn, can lead to piping and instability of the levees during critical high-water periods. The disposition of pervious versus impervious floodplain deposits beneath the levee and the angle at which such bodies are crossed by the overlying levees are controlling factors in the localization of sand boils. Thus, recognition of alluvial landforms forming the riverbanks, the types of soils associated with them, and their detailed mapping in plan and profile are important factors in levee design. Corrective design involves (1) detailed delineation of the surface and subsurface geology, (2) careful selection of borrow pits to avoid stripping critically thin top-stratum deposits, and (3) the use of riverside or landside berms or blankets, and/or the installation of relief wells.

REFERENCES

Kolb, C. R., Mabrey, P. R., and Steinriede, W. B. 1958. *Geological Investigation of the Yazoo Basin, Lower Mississippi Valley:* Tech. Rept. 3-480, Waterways Experiment Station, Vicksburg, Miss.

Mansur, C. I., Kaufman, R. I., and Schultz, J. R. 1956. *Investigation of Underseepage and Its Control, Lower Mississippi River Levees,* 2 vols.: Tech. Memo. 3-242, Waterways Experiment Station, Vicksburg, Miss., 421 p., appendixes, and 241 plates.

Mississippi River Commission. 1974. *Flood Control and Navigation Maps of the Mississippi River, Cairo, Ill., to the Gulf of Mexico:* the Commission, Vicksburg, Miss.

USAE District, Vicksburg. 1974. *Report of 1973 Highwater Levee Inspection,* 3 vols.: Corps of Engineers, Vicksburg, Miss.

7

CHANNELIZATION:
ENVIRONMENTAL, GEOMORPHIC,
AND ENGINEERING ASPECTS

Edward A. Keller

INTRODUCTION

Channelization is a controversial engineering practice used most often in an attempt to control flooding or drain wetlands for farming. It has also been used to improve navigation, control stream-bank erosion, or improve the stream alignment relative to a bridge crossing. Terms synonymous with channelization include channel or drainage improvement, channel modification, and channel works; more recently, a new form of channelization, known as channel restoration, is being initiated on stream channels that have been disrupted by previous channelization, catastrophic events, or unwise landuse.

Regardless of what it is called, channelization is today a vigorously debated issue, because in too many cases channelization is associated with or causes environmental disruption that is unacceptable to many people. Until ways are discovered to modify channels that minimize environmental disruption, objections to channelization will continue.

Many concepts concerning the behavior of streams introduced in this chapter are poorly understood by engineers, hydrologists, and geologists, and therefore are controversial. Furthermore, because the manipulation of streams involves philosophical, ethical, and aesthetic questions, and because scientists have the responsibility to express an opinion, the author has rendered several value judgments.

A general discussion of the environmental, geomorphological, and engineering aspects of channelization that might be combined to alleviate some of the adverse impacts associated with man's modification of streams is offered. This involves a review of the objectives, techniques, and adverse impacts of channelization; a discussion of issues of environmental concern with the ways we might design streams to make better utilization of natural stream processes; and a discussion of engineering trends in channelization.

HISTORICAL PERSPECTIVE

Human use and interest in the land has historically included significant drainage modification. Examples from the Old World include canals and ditches to carry water into and sewage away from ancient cities, and diversion of streams for irrigation. In the United States, stream channelization started about 150 years ago, and thousands of miles of river and stream channel had been modified (Arthur D. Little, Inc., 1973).

While American settlers in the nineteenth century claimed the land to feed the people of the rapidly growing eastern cities, towns sprang up along the banks of rivers. This was natural, since a river was a good source of water, a waste-disposal site, and a natural transportation route for the movement of commerce. As development moved westward, one of the first procedures was to clear the land and modify the drainage. Which came first, clearing or drainage modification, depended on whether drainage of wetlands, flood control, or navigational improvement was the objective.

In the midwest areas, such as northern Indiana and east-central Illinois, early settlers found an abundance of vast shallow ponds and mosquito-infested swamps. The land was considered worthless, and one settler early in the nineteenth century reportedly would not trade a horse and saddle for 640 acres, valued at about $1 million in 1974. Early drainage was accomplished with horse-drawn slip scrapers, men with shovels, and in some cases with a gigantic ditching plow that required 68 oxen driven by eight men to pull it. The drainage of the land at this time was a necessary, backbreaking job if the valuable farmland was to be rendered usable (Hay and Stall, 1974).

In western areas such as California, farmers and ranchers in the Coast Ranges straightened and enlarged streams using horse-drawn slip scrapers and Chinese laborers in hopes of preventing floods as early as 1871. This early stream-modification work was seldom well planned, properly engineered, or adequately financed. As a result, much of this early channel work, while perhaps addressing the primary objective of drainage or flood control, was local in extent and did not consider the broader environmental consequences of numerous diverse projects.

The highly fragmented nature of drainage projects of the nineteenth century in the United States became more coordinated in the first few decades of the twentieth century when unified river-basin and watershed planning and development was established (Arthur D. Little, Inc., 1973). Since then most of the responsibility for federally assisted channel modification has been delegated to the Soil Conservation Service and the Corps of Engineers. In addition, some channelization is being carried out by other federal agencies, such as the Bureau of Reclamation, the Tennessee Valley Authority, and numerous state and local organizations.

The development of environmental awareness in the 1950s and 1960s culminated in the 1969 enactment of the National Environmental Policy Act. This act and others, such as the Fish and Wildlife Coordination Act, are the most recent steps in recognizing environmental concerns associated with stream-channel modification projects. The National Environmental Policy Act requires that an environmental impact statement be prepared prior to implementation of federal stream-modification projects; the Fish and Wildlife Coordination Act establishes that planning of stream projects, which control or modify a channel, must consider the effect on wildlife, and planners must consult with the U.S. Fish and Wildlife Service and state agencies that manage wildlife resources.

The historical trend of channelization in this country has been from an early fragmented effort by farmers and ranchers toward better planned and engineered projects involving consideration of the entire river basin and probable environmental impact. Although the trend is positive, too many streams are still being modified for city, county, or state purposes without coordinated engineering design and environmental considera-tion. Furthermore, a better understanding of the behavior of streams along with new approaches and techniques is needed before most adverse effects of channelization can be alleviated. Therefore, until we learn more about streams, pragmatic solutions are (1) to minimize floodplain encroachment by people and structures, thereby reducing the neces-sity for a structural or construction solution for flood control; (2) to minimize the total length of stream channelization necessary to either relocate a channel after a catastrophic event or to stabilize the stream bed and banks where highways or railroads cross streams; and (3) where channelization is shown to be a necessary or practical solution, to design for the least possible amount of channel alteration to obtain the desired objective (don't overdesign). The desired philosophical framework for channelization is, therefore, to minimize the practice and provide the minimum of control to satisfy the desired objective.

Channelization: Operations
in Channel Modification or "Improvement"

In recent years, two types of channelization have emerged: (1) planned projects in which engineering design and environmental impact may or may not be integral parts, and (2) emergency channel work following catastrophic storms, which may often be done without proper planning and engineering. Following the severe storms and floods that struck Virginia in the aftermath of Hurricane Camille in 1969, emergency federal money was used to channel streams in hopes of alleviating future floods. In some cases, local bulldozer operators with little or no instruction or knowledge of streams were contracted to clear and straighten stream channels. The results of this unplanned and unsupervised channelization have been disastrous to many miles of streams in Virginia. On the other hand, catastrophic storms may entirely fill a stream channel with mud, sand, gravel, and debris, while washing out roads, farms, and homes. In these cases, emergency channel work is clearly needed. For example, the Virginia Highway Department and the Soil Conservation Service worked together to restore Hat Creek following Camille's devastation in Nelson County, Virginia. Figure 1 shows Hat Creek soon after the storm turned the small creek into a raging torrent, destroying parts of a highway and buildings. The channel was moved back to its prestorm location, and the road and slopes repaired (Figs.

FIGURE 1 *Hat Creek Valley, Nelson County, Va. Severe bank erosion and flooding from storms generated by Hurricane Camille in 1969, which nearly obliterated the original channel and damaged roads and buildings. (Photo: O. E. Hatcher, Soil Conservation Service.)*

2 and 3). There is little doubt that this type of emergency channel work is not only beneficial but also necessary. However, it should be minimized and confined to stream channels that require immediate attention, as did Hat Creek. Emergency funds should not be considered as a license for wholesale modification of any stream at the request of property owners or others. This is particularly true for streams that, even in flood, threaten no homes, buildings, or livestock.

Operations in channel modification or "improvement" vary with the desired objective. Common methods include (1) widening, deepening, and straightening, (2) clearing and snagging, (3) diking, and (4) bank stabilization. On any one stream project, any variety of these may be used.

Widening, Deepening, and Straightening. is most often associated with channelization. Trees and other stream bank vegetation are removed, and the stream bed, bank, and channel location are changed. Because increasing the size of a channel or increasing its downvalley slope by straightening is likely to disrupt the quasi-equilibrium between the running water, sediment concentration, and physical channel characteristics, this technique causes most of the environmental degradation associated with channelization.

Clearing and Snagging. involves the removal of trash, brush, logs, stumps, and other obstructions to flow. It may or may not include the dredging out of stream-channel deposits, such as sand and gravel, that obstruct flow. No channel straightening or enlargement is involved; and trees and brush along the stream banks need not be, but often are, removed.

Diking. is used primarily for flood control It involves physically increasing the height and/or size of one or both of the stream banks. The objective is to protect critical areas from flooding or induce the stream to carry a higher discharge prior to overbank flow. Compacted earth-fill material, hauled in or dredged from the stream channel, is usually used to construct the dike.

Bank Stabilization. generally involves a combination of (1) sloping stream banks by pulling back the top or building up the toe, (2) implacement of stones to prevent bank erosion (riprap), and (3) planting of vegetation on stream banks to retard erosion. Bank stabilization is commonly used in conjunction with any or all of the other channelization techniques.

ADVERSE IMPACTS OF CHANNELIZATION

The possible adverse environmental impacts due to channelization are well known and documented in numerous case studies. Generally, the adverse effects fall into one of several categories: (1) damage to the physical stream channel and/or floodplain, (2) damage to fish and wildlife, (3) aesthetic degradation, and (4) downstream effects.

Damage to the Channel and Floodplain

Channelization, particularly if it involves widening, deepening, or straightening of a channel, may be associated with considerable bank erosion, which may locally enlarge the width of a channel by two or three times its constructed width (Emerson, 1971; Yearke, 1971). Furthermore, the deepening of a channel effectively lowers the local base level of tributary channels; this may initiate a cycle of erosion in the tributaries as they adjust to new conditions (Daniels, 1960). The erosion may work its way up the tributaries as a series of knickpoints in response to changes in the mainstream. Each knickpoint is

FIGURE 2 *Hat Creek Valley, several years after emergency work restored the stream to its prestorm channel. Photograph taken from approximately the same location as for Figure 1. The stream is to the left and out of the photograph (see Fig. 3). (Photo: R. F. Dugan, Soil Conservation Service.)*

FIGURE 3 *Hat Creek, in the vicinity of Figs. 1 and 2, several years after channel restoration and bank stabilization. (Photo: R. E. Dugan, Soil Conservation Service.)*

associated with an upstream scarp or rise and a lower plunge pool (like a small waterfall). This migrates upstream until it eventually is dampened out into a series of small rapids.

It is emphasized that every stream is different. The soil, topography, and channel material all vary; therefore, the nature and extent of bank erosion caused by channelization will vary with these parameters. For example, channel degradation may occur (Blackwater River in Missouri and Tallahatchie River in Mississippi) even in cohesive alluvium if the clay swells and cracks on drying (J. C. Brice, pers. comm.). Therefore, prediction of the nature and extent of bank erosion caused by channelization requires a detailed analysis, in both the vertical and horizontal direction, of the earth materials that compose the stream banks.

Damage to Fish and Wildlife

Any activity that removes the vegetation, disturbs the bottom sediment, increases the sediment content, or changes the flow characteristics of a stream will affect the fish and wildlife. Although most streams experience some disturbance in various places along the channel every year due to flooding and natural bank failure, channelization often does all the above along an extensive length of channel. In fact, the biological productivity of some streams has been reduced by channel modification to such an extent that some fish and wildlife management experts believe that channelization is completely antithetical to the production of fish and other life forms in streams.

Fish and other stream inhabitants generally require a habitat characterized by the following: (1) a variety of low-flow conditions varying from slow, deep water (pool) to fast, shallow water (riffle) is necessary for feeding, breeding, and cover; (2) a variety of high-flow conditions and shelter areas provides protection from excessive water velocity; (3) as a result of high- and low-water conditions and variable stream velocity, there is a natural sorting of bed-load materials on riffles and point bars; this sorting is helpful in providing a good environment for the bottom-dwelling organisms in streams that fish and other animals depend on for a food supply; and (4) a certain amount of trees, brush, and other vegetation is necessary to provide cover and food for fish. In trout streams, vegetation that hangs out over the banks helps to insulate the water from the sun, thus allowing water temperatures to remain cool.

Channelization may convert a meandering stream with an abundance of long pools separated by short riffles into a straight stream with very few pools and numerous long riffles. Furthermore, the natural sorting of stream-bed material and vegetation on the stream banks may be destroyed (Fig. 4). At high flow the water velocities of the channelized stream may exceed what fish and other organisms without cover can tolerate. Furthermore, because the storm water quickly leaves the river basin rather than infiltrating into the groundwater system, the low-flow discharge of a stream that depends on groundwater discharge into the channel is reduced. Even worse, the water that is available at low flow may be only a few inches deep across the wider channelized stream; because it is shallow and there are no trees to protect the water from the sun, it quickly warms up (Corning, 1975). For the above reasons, streams that have been channelized generally suffer from a reduction in their biological productivity.

Studies of trout streams from the Rocky Mountains to the Appalachian Mountains suggest that the number of trout and their size in a stream is greatly reduced by channelization. Furthermore, the reduction in fish production is directly related to the intensity of channelization. Graham (1975) reported that a study of the Ruby River in

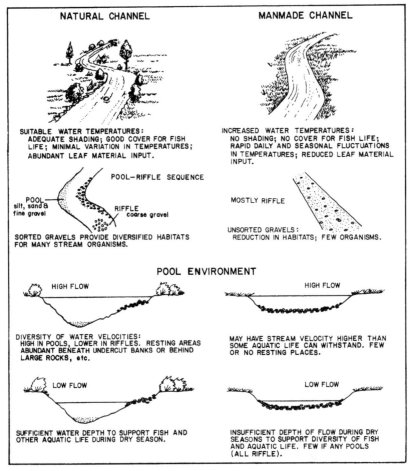

FIGURE 4 *Comparison of a natural channel morphology and hydrology stream versus a channelized stream. (Modified after Corning, 1975.)*

Montana showed that the river in its natural state had a fish (trout) content of 133 lb of fish/1,000 ft of channel. This was reduced to 67 lb/1,000 ft of channel following bank stabilization (riprap); an even lower fish count of 55 lb/1,000 ft of channel was found in a section of the river where the stream bottom was bulldozed and reshaped.

Aesthetic Degradation

Aspects of a landscape that provide scenic and other aesthetic value along streams, in addition to the presence of water in the channel, include diversity of sensual stimulus and physical contrast, i.e., visual and other sensual experiences associated with contrasts such as slow, shaded versus sunlit water; deep, slow water in pools versus fast, shallow bubbling water of riffles, and deep green reeds and rushes near the stream compared to light green or brown leaves of trees and brush along the banks; these provide for varied aesthetic values (Eiserman et al., 1975). Physical or biological characteristics, such as presence or absence of human activity, existence of marshes, natural curvature of the streams (meanders), and the presence of unique topography, such as waterfalls, rugged rock outcrops, or scenic vistas, also affect aesthetic value. Unfortunately, channelization by straightening a channel, removing vegetation and obstructions to flow, and using unnatural bank stabilization structures generally reduces the aesthetic quality of a stream.

Downstream Effects

Anything that happens to change the upstream hydrology and sediment production must affect downstream processes. However, the variety of responses and possible interactions is complex. For example, it is an exaggeration to assume that channelization always causes downstream flooding (Arthur D. Little, Inc., 1973).

Channelization in upstream tributaries of a drainage basin would increase the flood hazard downstream from the channel works if the effect of the channel modification was to move the storm water quickly downstream faster than lower reaches can discharge it, or in such a way that the flood peaks of the tributaries coincide. On the other hand, channelization might actually reduce downstream flooding if the flood peak from a channelized tributary moves out of the basin prior to the arrival of flood peaks from other tributary channels.

Considering what happens on a single stream channel rather than an entire basin, we would expect that upstream channelization would generally increase downstream flooding. This results because channelization tends to remove floodwater quickly and confine more water in the enlarged channel. Therefore, when floodwaters from the modified channel are discharged into the downstream natural channel, with a lesser channel capacity, flooding is likely to occur. Downstream flooding due to upstream channelization has been observed in the Blackwater River in Missouri (Emerson, 1971) among other cases (Arthur D. Little, Inc., 1973). It is emphasized, however, that each stream must be evaluated independently if accurate predictions of downstream flooding are to be achieved. Generalizations remain dangerous.

Sediment pollution is another possible downstream effect of channelization. This is particularly true if the upstream channelization work experiences rapid channel erosion. The deposition of sediment downstream reduces the capacity of the channel, and thus facilitates more frequent flooding. Furthermore, rapid deposition of sediment damages aquatic life (Apmann and Otis, 1965).

Sediment pollution of all types is a serious problem in urban areas, and it is unlikely that any urban channel program will be successful if sediment is not controlled. Therefore, sediment control and flood control are two problems that must be solved simultaneously if either is to be alleviated.

RECOVERY OF CHANNELIZED STREAMS

Morphological and biological recovery of channelized streams is dependent upon complex interrelated factors varying with time; there is not yet a sufficient number of studies following channel modification through recovery to make accurate predictions.

Nevertheless, estimates of the time required for biological recovery on an unmaintained modified stream vary from 50 to 100 yr (Corning, 1975). That recovery may take many years is verified by a study on the Luxapalila River in Mississippi, which concluded that little recovery of fish productivity was observed 43 yr after initial channelization (Arner, 1975).

Morphologic recovery of a modified channel, defined as a return to a more natural distribution of bed forms such as pools and riffles, is likely to be more rapid than biologic recovery, because the diversity of flow provided by pools and riffles is necessary for maximum biologic productivity. Although incipient pools may begin to form soon after channelization (Morisawa, 1974; Keller, 1975), the time necessary for the formation of a morphologically stable series of pools and riffles is not known; in fact, if the stream is severely disturbed and periodically remodified, they may never form.

ENVIRONMENTAL CONCERNS
AND NATURAL FLUVIAL PROCESSES

On the environmental side of channelization, there is good news. More people are aware of environmental problems associated with channelization than ever before, and there is an increase in environmental programs reflecting this awareness. This is fortunate, because the pressures that produce the need for channelization are not going to disappear. Highways will continue to be constructed and maintained, urbanization will continue, and demand for flood control will increase. Furthermore, if, as some scientists predict, the next environmental crisis is a worldwide food crisis, more land will have to be cleared, which will mean that more channel projects will be necessary. The conclusion is that channel alterations will continue; therefore, we need to reconcile development with environmental concerns (Hirsch, 1975).

To develop a strategy to minimize the environmental degradation associated with channelization, we must recognize several concepts. First, the stream and the floodplain are a system. Channelization reduces storage of water on the floodplain and encourages inappropriate landuse (floodplain encroachment) by producing a false sense of security that the flood hazard has been alleviated. Second, a more balanced approach to water-resource projects is needed. All assets and liabilities must be considered, giving equal weight to environmental values. Third, priorities must be set. That is, as a responsibility to future generations, more streams must be inventoried, and decisions must be made to ensure that some streams are protected from future modification by man. Fourth, research must establish more precisely the physical behavior of natural streams so that we shall be better able to recognize the potential impacts of channelization (Hirsch, 1975). Fifth, greater concern must be given to alternatives or stream modifications that minimize environmental degradation and maximize the overall return to man.

Designing with nature is becoming a fundamental principle of landuse planning, and this concept can be extended to designing stream channels. The idea behind this concept is to take advantage of natural systems rather than to oppose them. In the case of streams, we might emulate nature when possible and design more natural channels, rather than attempting to impose arbitrary boundaries, such as a uniform straight ditch. Important aspects of fluvial geomorphology and fluvial hydrology that are now or eventually may be useful in designing stream modification projects are (1) the concept that streams are open systems, (2) the convergent–divergent flow criterion, (3) the identification of geomorphic thresholds in streams, and (4) the identification of the complex relations among erosion, deposition, and sediment concentration in streams.

Streams as Open Systems

Streams are basically open systems in which there is a rough balance, or quasi-equilibrium, between the load imposed and the work done. This means that there are relationships between variables that can be regarded as independent, such as the sediment load and discharge, and dependent variables, such as channel slope and channel morphology, so that any change in the independent variables will be compensated for by a change in the dependent variables. In natural streams, channel form and process evolve in harmony. This is inherent if streams are open systems, and may possibly be taken advantage of in designing streams to minimize some of the adverse effects of channelization.

Adverse changes in channel form and process, such as accelerated bank erosion and sediment pollution caused by changing the fluvial processes and channel form during and after channelization, are well known and documented. However, with the exception of river training, little recent effort has been given to the modification of channel form, particularly the channel cross section, to partially control fluvial processes.

Before we can apply form-process relations to channelization, the hypothesis that needs to be tested is: the modification of channel form can be used to minimize some of the adverse effects of channelization by partially controlling the behavior of the stream. Experiments on Gum Branch Creek, a small stream near Charlotte, North Carolina, were designed to test this hypothesis. The stream was first channeled about 20 yr ago and was being reworked as a part of a county program to clean out streams with a sediment pollution and flood problem. The objective of the experiment was to cause the stream to build a series of point bars along a 500-ft length of channel in desired locations by manipulating the channel cross section. No structures of any type were used. Where a point bar was desired, an asymmetric channel was constructed by cutting a 2-to-1 slope on one bank and a 3-to-1 slope on the opposite bank. The stream bottom was disturbed as little as possible. In between these areas the channel was symmetric, and both banks were cut to a 2-to-1 slope. If the theory was right, the asymmetric channel cross section would converge the flow toward the steeper bank, facilitating the deposition of a point bar on the side of the channel with the 3-to-1 bank slope. The channel was designed so that the point bars would be spaced about every six times the channel width, as in natural streams. Upstream from the experiment the stream was channeled in a more conventional way with a uniform channel cross section.

Following construction in the summer of 1974, point bars formed precisely where planned, and periodic detailed cross sections suggested that they remained morphologically stable for over 1 yr. That is, the bars remained in the same location and changed little in size or shape. During that time, there were four overbank flows, and the bars always emerged following the recession of the floodwaters. This suggested that the process of point bar formation can be controlled by manipulating the channel cross section.

The experimental channel remained morphologically stable until October 1975 when the point bars were suddenly buried by an influx of sediment. The source of the sediment was partially derived from upstream bank erosion along the channeled section, where

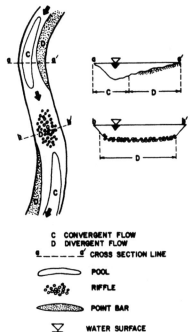

C CONVERGENT FLOW
D DIVERGENT FLOW
‖_____‖′ CROSS SECTION LINE

POOL

RIFFLE

POINT BAR

WATER SURFACE

FIGURE 5 *Idealized diagram showing the convergence–divergence criterion.*

considerable bank failure occurred following a severe storm in May 1975, and intensive upstream construction activities in the drainage basin.

The lesson learned from the experiments on Gum Branch Creek to date is that the concept of recognizing streams as open systems with a mutual adjustment of channel form and process may be applied to induce the stream to erode or deposit in a desired location. However, a program of channel modification with the objective of cleaning out sediment and other obstructions to flow without an adequate sediment-control program and upstream bank stabilization is doomed to failure.

Criterion of Convergent–Divergent Flow

An important concept of fluvial hydrology is that convergent flow is associated with scour and divergent flow with deposition. Furthermore, convergent flow, and thus scour, is characteristic of pools; and divergent flow, and thus deposition, is characteristic of point bars and riffles (Fig. 5; Leliavsky, 1966). This criterion, developed in 1894 by N. de Leliavsky, has been used in river-training projects in Europe and the United States, and is expected to be valuable in channel modification utilizing natural fluvial processes.

Convergence and divergence of flow as used here refers to a natural, three-dimensional constriction or expansion of flow along, across, and in a single channel. It is not generally applicable at tributary junctions nor at the joining of anabranches, where the combining of two channels interact to initiate a more complex set of processes. Furthermore, the association of scour and deposition with convergent and divergent flow, respectively, is not inconsistent with the concept that scour is generally related to bed load transport and specifically to continuity of sediment transport.

The convergent—divergent criterion is significant because many fluvial stream channels with gravel beds are characterized by a regular sequence of pools produced by convergence of flow and scour at high discharge and riffles produced by divergence of flow and deposition at high discharge. Pools and riffles at low flow are, therefore, relic forms, and flow conditions may be just opposite those characteristic of high flow. That is, at low flow, pools are areas with deep, slow-moving water, and riffles are areas with fast, shallow flow (Keller and Melhorn, 1973). As was discussed earlier, these diversities in flow conditions are very significant in producing the type of aquatic habitat necessary for a stream to have high biological productivity. Therefore, designing modified channels to have areas at which flow converges and diverges is consistent with our premise of designing with nature.

Geomorphic Thresholds

Changes in streams are not necessarily continuous and gradual; rather, changes take place after a certain threshold is reached (Schumm, 1973). The changes in response to a threshold are probably a type of discrete feedback inherent in open systems. For example, Schumm discovered that small channels in a flume experienced discrete changes in channel pattern, from straight, to meandering, to braided, in response to increases in channel slope. Interestingly, these changes were not gradual, but occurred quickly after a threshold slope was exceeded. Although it is admittedly dangerous to make direct analogies between laboratory streams and natural streams, it might have important consequences to channelization, if the general concept of thresholds is valid (Keller, 1975). For example, if changes in streams, such as widening of the channel leading to braiding, take place in response to threshold changes in slope, sediment concentration, and other variables, then theoretically a stream might be straightened a certain amount without experiencing widening of its channel. How much straightening could be tolerated by the stream would depend on identifying the correct threshold controlling channel pattern. Therefore, identification of thresholds would be extremely valuable to engineers designing channels, because it would allow them, for the first time, to plan a channel and be more confident that the modified channel would not experience detrimental bank erosion or change of channel pattern. Unfortunately, quantitative identification of the geomorphic threshold necessary to do this has not been, as yet, adequately determined.

Erosion Deposition and Sediment Concentration

Relationships among erosion, deposition, and sediment concentration in natural stream channels with alluvial bed and bank materials are poorly understood. This is unfortunate because these relationships must be known if stable channels are to be designed. At this time there is no agreement on how the straightening of a channel will affect the velocity of the water, and thus the erosion, deposition, and sediment concentration (Arthur D. Little, Inc., 1973). The conventional and generally accepted assumption is that artificial straightening which increases the channel slope of a stream causes an increase in the velocity of the water. The faster water then facilitates erosion of the stream bed and banks, increasing the sediment concentration until a new balance between the load imposed (sediment concentration) and work done to erode, transport, and desposit sediment is established. A contrary hypothesis is that the velocity–slope product in streams is a function of the concentration and median size of the sediment discharge, suggesting that with constant sediment size and concentration, velocity is inversely proportional to slope (Maddock, 1972). These differences of opinion concerning basic relations between velocity and slope reflect our generally poor understanding of the behavior of alluvial stream channels. Obviously, before relationships between erosion, deposition, and sediment concentration can be applied to channelization projects with any hope of success in designing stable channels to minimize adverse effects, more information concerning the precise behavior of alluvial stream channels is necessary.

ENGINEERING TRENDS IN CHANNELIZATION

Environmental Considerations

The importance of minimizing the adverse effects of channelization on biological systems is widely recognized, and much information on techniques to increase the biological productivity of a modified channel is available. Unfortunately, there is little information on techniques to minimize physical and aesthetic degradation associated with channelization.

Most techniques to improve fish productivity and other aquatic life of a modified stream involve the construction of special structures to provide a variety of flow conditions similar to those found in pools and riffles of natural streams (Soil Conservation Service, 1971). Figure 6 shows an idealized diagram of how this might be accomplished. Although such a system may have merit, it appears that pools formed by an artificial sill in the channel would have to be constantly maintained or they would fill with sediment. This might be alleviated if the pools were designed to converge the flow of water at high discharge and scour the bottom of the pool. Such self-cleaning action of pools is inherent in natural streams and should be designed into the plans for artificial pools.

Other possible ways to minimize the adverse effects on biologic systems are to channelize only one bank of a stream, or to construct a special flood-flow channel (Fig. 7). Both methods may be used in a particular project. Of the two, the diversion of floodwaters away from critical areas by using a flood-flow or bypass channel will cause less environmental degradation and, in addition, will maintain the scenic quality of the natural stream. However, one-side construction, which uses the existing vegetation of the unaltered channel bank for stability and cover, is certainly preferable to modifying both banks.

The main philosophical thread holding all the environmental considerations together is that the best channelization project is the one which is absolutely necessary and involves the minimum modification of the natural channel. It is emphasized that this does not suggest that all channelization is bad. For some urban hydrology problems and in the necessary drainage of some wetlands, it may be the only practical solution. Furthermore, many streams that are modified are often nothing more than ditches that have long been adversely affected by human use and abuse in urban areas, and little is lost in their further modification. However, this attitude must not be extended to natural channels with high scenic value and biologic productivity.

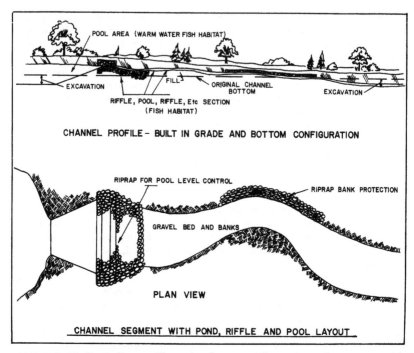

FIGURE 6 *Idealized diagram illustrating how specially engineered structures might create a diversity of flow conditions similar to natural pools and riffles. (From* Planning and Design of Open Channels, *Chapter 7, Technical Release 25, Soil Conservation Service, 1971.)*

River Training

River training is a complex engineering procedure, and a complete discussion of it is beyond the scope of this chapter. Nevertheless, because river training merges imperceptibly with channelization, some principles of river training may be applicable to channelization of small streams; for completeness, selected aspects are appropriate for discussion.

River training, in part, is the art of inducing a river to scour or fill in desired locations. It generally involves using artificial structures, such as permeable, pile dikes, to converge the flow of a river and cause the channel to scour. The basic idea is to apply de Leliavsky's convergent–divergent criterion.

River training is not a new practice. It has long been used to control rivers in Europe and has been extensively used on the Missouri River in the United States. The technique has not been used much in this country on small rivers and streams. This is unfortunate, because river training involves the use of the river itself to maintain a channel and thus is closer to a true design with nature.

The training of hundreds of miles of the Missouri River in conjunction with upstream flood-control reservoirs must rank as one of the most ambitious channel-work projects in history. The channel work began in the mid-1930s, and maintenance and new work are still going on today. The primary objective of the project is to use the energy of the river to develop and maintain a navigation channel. Prior to training, the river was often characterized by a wide channel with numerous shifting islands and rapidly changing banks that eroded valuable farmland (Fig. 8).

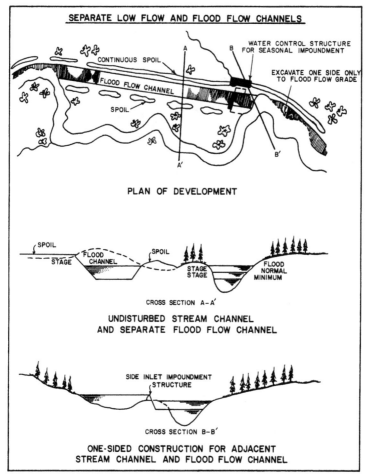

FIGURE 7 *Idealized diagram illustrating one-bank channelization and construction of a special flood-flow channel. (From* Planning and Design of Open Channels, *Chapter 7,* Technical Release 25, *Soil Conservation Service, 1972.)*

FIGURE 8 *Upstream view of the Missouri River approximately 25 mi upstream from Sioux City, Iowa. This section of the river has not been trained. (Courtesy of the Omaha District Office, U.S. Army Corps of Engineers.)*

The channel of the Missouri River was trained from its original semibraided state into a single channel that meanders in a series of gentle bends by construction of numerous permeable pile dikes, which are designed to reduce the velocity of water and induce deposition of sediment downstream from the dikes on the inside of bends. As the sediment builds up, it eventually emerges, and willows and other vegetation become established. This process of burying the dikes in sediment that becomes vegetated generally takes 5 to 8 yr following construction. The dikes effectively converge the flow of the river, and beyond the end of the dikes, on the outside of bends, the velocity of the water increases, which effectively scours out a pool while eroding the channel deeper until a state of quasi-equilibrium is reached. In addition, bank-protection structures, similar to riprap, are often necessary to stabilize the outside of bends from excessive erosion. Figure 9 shows how the dikes, which are aligned normal to the stream bank, are angled downstream, or are L shaped, are used on the Missouri River to induce the power of running water to maintain the desired channel.

The method of constructing the dikes is shown in Fig. 10. It is basically a triple row of pile clumps connected by a double line of pile stringers. The structure is securely anchored to the bank and protected from washing out by erosion. The clumps are 15 to 18 ft apart and consist of three piles, each arranged in a tripod and driven into the river bottom to a depth of at least 20 ft. Wire cable is used to lash the clumps and stringers together, and the flexibility allows the piles to move a little under stress. This is an advantage, because the structures are exposed to moving ice and floating debris. The last step in constructing the permeable pile dike is the placement of stone blocks around and beneath the clumps for additional strength.

FIGURE 9 *Downstream view showing a section of the Missouri River that has been trained using permeable pile dikes and bank stabilization structures. (Courtesy of the Omaha District Office, U.S. Army Corps of Engineers.)*

The training of the Missouri River has generally functioned as designed. It has produced a single navigable channel, which has assured that a significant tonnage of cargo may be hauled annually on the Missouri River. It has also stabilized the banks of the river, thus protecting farmland and in some cases actually creating new land for cultivation. Although no major environmental degradation or hazard has been produced or recognized

FIGURE 10 *Photograph (1934) showing how permeable pile dikes were constructed to train the Missouri River. (Courtesy of the Omaha District Office, U.S. Army Corps of Engineers.)*

by the training of the Missouri River, there are at least two areas of concern; both are associated with the constriction of the river caused by the dikes and other structures. First, there is some evidence that the loss of surface area and bank-full discharge capacity may be facilitating a flood hazard (Belt, 1975). However, river training is probably not the only cause, and some of the flood problem is probably related to historic change in landuse as natural wetlands were drained for farming. Wetlands are natural sponges that retard surface runoff, and their drainage greatly reduces the time necessary for runoff while increasing the total runoff and, therefore, the flood peak.

The second concern associated with the training of the Missouri River is that the biological productivity of the river has been reduced. Narrowing and deepening of the river have reduced both the areal extent and variety of aquatic habitat. Furthermore, the trained river generally lacks the numerous weedy sloughs and quiet side channels (chutes) that characterize the inside of bends on unmodified portions of the river (Morris et al., 1968). Recently, the design of the dikes has been changed to include gaps or notches (Fig. 11). It is hoped that these notches will allow an increase flow of water through the dike, preserving the backwater areas and providing more habitat diversities (P. D. Barber, pers. comm.).

Even though there are differences in behavior between large rivers and small streams, in part because the range in the properties of bed and bank materials is much less than the range in magnitude of flows (J. C. Brice, pers. comm.). principles of river training could be applied to channelization projects. For example, structures to converge or diverge flow might be constructed to provide the diversity of flow conditions necessary for high biologic productivity, while increasing the scenic value of modified streams.

Construction of More Naturally Appearing Channels

Design critera for channel-modification projects are changing. The old criterion of straight ditches with a uniform gradient and uniform trapezoidal channel cross section is giving way to new channel designs that produce more natural appearing streams. However, it is emphasized that the new criteria, as with the old criteria, remain based on an incomplete knowledge of stream behavior; therefore, success may be sporadic. Two examples will be discussed: (1) Whiskey Creek near Grand Rapids, Michigan, where aesthetic aspects were given prime consideration, and (2) drainage projects on the Coastal Plain of North Carolina, which attempt to restore meandering channels previously degraded by logging.

Whiskey Creek. Grand Rapids, Michigan, in recent years has experienced rapid residential and commercial development; as hundreds of acres of farmland were converted to urban uses, a storm-water runoff problem quickly developed. The Whiskey Creek

DIKE CONSTRUCTED WITH NOTCH

FIGURE 11 *Idealized diagram showing a new design for dikes that will allow an increased flow of water and hopefully provide more diversified wildlife habitat. (Courtesy of the Omaha District Office, U.S. Army Corps of Engineers.)*

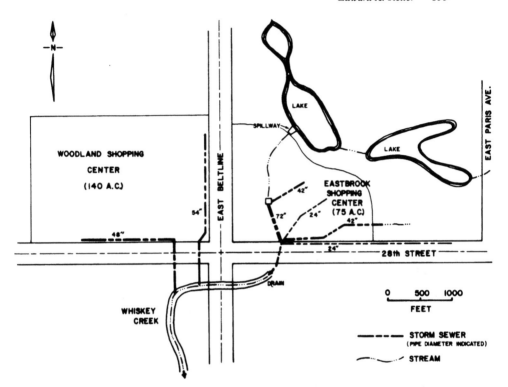

FIGURE 12 *Map showing the upstream part of the Whiskey Creek drainage basin. (Courtesy of Williams & Works, Grand Rapids, Mich.)*

drainage project is designed to convey the storm-water runoff from a 3 mi² area, which includes two large shopping centers and substantial residential and commercial properties. Figure 12 shows the upstream portion of the drainage system, which includes two ponds, numerous storm-water sewers, and Whiskey Creek.

The engineering design of the Whiskey Creek drain is interesting and significant because it is one of the few cases where aesthetics was given prime consideration. The objective was to construct a channel that would have water flowing in it the entire year and appear as a meandering country stream (Fig. 13) (anonymous, 1972).

The construction was completed in the fall of 1972. Sheet piling and rustic rock (riprap) were used to construct structures designed to maintain a low-flow discharge (Fig. 14); although these structures functioned as designed, there was partial failure of the sheet piling by the summer of 1975 (Fig. 15) caused by severe bank erosion. Sheet piling and stone were also used to control bank erosion on bends (Figs. 16, 17, and 18). This procedure, after 3 yr of operation, appears to adequately control erosion. The sheet piling and rock on the inside of bends is in places being covered by sediment as point bars form, and is probably unnecessary for the stability of the channel. In one instance, sheet piling, in conjunction with a concrete layer over wooden forms, was used for control of bank

FIGURE 13 *Construction of meanders in Whiskey Creek. (Courtesy of Williams & Works, Grand Rapids, Mich.)*

erosion on a bend. Although the concrete layer has survived 3 yr, it is now experiencing erosion. Stream water has worked beneath and behind the structure, and it is cracking along its entire length, as indicated by a line of vegetation along the main zone of fractures (Fig. 19). This suggests that riprap is a better protection against bank erosion on the outside of bends.

FIGURE 14 *Whiskey Creek in 1972. The structure in the center of the photograph is designed to help maintain the low-flow discharge. (Courtesy of Williams & Works, Grand Rapids, Mich.)*

FIGURE 15 *Whiskey Creek in 1975. Photograph taken from the same location as Fig. 14. Although part of the sheet piling on the left bank has failed and is in need of repair, the structure to maintain the low-flow discharge is functioning as designed. (Photo: E. M. Burt, courtesy of Williams & Works, Grand Rapids, Mich.)*

The Whiskey Creek drain is probably as attractive as an open urban drain can be. Certainly it is an improvement over an open straight ditch. Hopefully, in the future, additional drains will be designed that are attractive as well as functional.

Coastal Plain Drainage Work. As part of a mosquito-abatement program, the state of North Carolina is involved in a certain amount of channel work designed to drain upland swamps on the Coastal Plain. In some cases the channel work is perhaps best described as "channel restoration," because the objective is to reestablish streams that have been literally plugged from debris left following logging operations that ended many years ago.

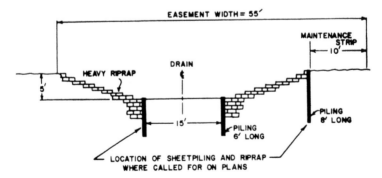

FIGURE 16 *Generalized cross section showing the engineering design for Whiskey Creek. (Courtesy of Williams & Works, Grand Rapids, Mich.)*

FIGURE 17 *Whiskey Creek during construction. Sheet piling and piled rock, special riprap, were used to stabilize the channel banks. (Courtesy of Williams & Works, Grand Rapids, Mich.)*

FIGURE 18 *Stabilized bend (downstream view) on Whiskey Creek 3 yr after construction. The sheet piling and piled rock on the inside of the bend, while aesthetically pleasing, are not necessary for stream-bank stabilization. (Photo: E. M. Burt, courtesy of Williams & Works, Grand Rapids, Mich.)*

FIGURE 19 *Downstream view of Whiskey Creek showing the use of a concrete pad in conjunction with sheet piling to stabilize the stream bank on the outside of a bend. Photograph taken 3 yr after construction. The line of vegetation delineates a continuous fracture, which threatens the stability of the structure. (Photo: E. M. Burt, courtesy of Williams & Works, Grand Rapids, Mich.)*

Some of the wetland areas may in fact be man-made swamps. There is some evidence to suggest that, prior to logging, free-flowing streams drained areas that today are characterized by sluggish streams so filled with debris that drainage is nearly impossible.

The evidence that these streams were once incised and free flowing, and that logging on the Coastal Plain may have caused some of the drainage problem, at least in Onslow County, North Carolina, is based on field observations: (1) excavation of the channels uncovers numerous hand-cut cypress logs, and (2) numerous handmade cypress shingles can be found along the newly excavated channels.

The primary objective of the channel work in Onslow County is drainage improvement. Since flood control is not a primary concern, a large straight channel is neither needed nor desired. Straight channels through sandy coastal-plain soils can be disastrous because they tend to erode their banks very rapidly. Engineers have observed that, if they excavate in black organic-rich soils, bank erosion is much reduced. These organic-rich soils probably formed along the original drainage prior to the logging. To follow the original drainage and the organic soil, a meandering channel is constructed (Figs. 20 and 21). This satisfies the objective of drainage while providing a more natural channel with diversified aquatic habitat.

The channel-restoration work on the Coastal Plain is a step in the right direction, but it is not sufficiently documented to be useful in developing design criteria in other areas. However, this channel work indicates the importance of understanding the properties of the materials in which the channel is excavated. In Onslow County, this information has been utilized; this is the essence of designing with nature.

FIGURE 20 *Channel restoration during construction in Onslow County, N.C.*

SUMMARY AND CONCLUSIONS

Channelization is a controversial issue, and, although more environmental aspects are being considered today than ever before, the pressure to modify channels will increase in the future. Therefore, it is important to discover ways to modify channels that will minimize environmental degradation and maximize the overall return to man.

Two types of channel work are currently being done: (1) emergency work on streams, following catastrophic events, that too often is poorly planned and executed, and (2) channelization that is planned but may or may not be properly engineered and include consideration of environmental impact.

FIGURE 21 *Objective of channel restoration in Onslow County, N.C., is to produce a meandering channel with stable banks and provide a diversity of aquatic habitat.*

Objectives of channelization include flood control, drainage of wetlands, control of bank erosion, navigation improvement, and the establishment of a more favorable flow alignment at bridge crossings. Techniques of modification include widening, deepening and straightening, clearing and snagging, diking, and bank stabilization.

Environmental degradation due to channelization may include damage to the channel and floodplain, damage to the aquatic habitat, aesthetic degradation, and downstream flooding and sediment pollution. Furthermore, once modified, morphologic and biologic recovery of a stream may be very slow.

Several aspects of fluvial geomorphology and hydrology that are now or may eventually be helpful in designing more natural channels are (1) the concept that streams are open systems in which channel form and process evolve in harmony, (2) utilization of the convergent–divergent criterion, (3) identification of geomorphic thresholds for stream channels, and (4) recognition of the complex relationships between deposition, erosion, and sediment concentration.

Engineering trends in channelization include constructing structures to provide diversity of aquatic habitat, the continuing use of river-training procedures, the construction of more natural appearing channels, and channel restoration of streams degraded by adverse landuse.

The state of the art of channelization still suffers from lack of understanding concerning the behavior of alluvial streams; therefore, it is recommended that future necessary channelization be confined to the shortest possible length of channel and provide the least amount of artificial control necessary to meet the desired objective of the project.

Examples of new design ideas from Michigan and the Coastal Plain of North Carolina suggest that new engineering trends are producing urban and rural stream channels that are functional and aesthetically pleasing, while causing minimal environmental disruption. This is the essence of designing with nature.

ACKNOWLEDGMENTS

Critical review and suggestions for improvement of the manuscript by James Brice, Thomas Maddock, Jr., and Nelson R. Nunnally are gratefully acknowledged. Channelization research is sponsored by the University of North Carolina Water Resources Research Institute, Project B-089-NC.

REFERENCES

Anonymous. 1972. Whiskey Creek improved in looks and flow: *Michigan Contractor and Builder,* Oct. 14, 1972.

Apmann, R. P., and Otis, M. B. 1965. Sedimentation and stream improvement: *N.Y. Fish Game Jour.,* v. 12, no. 2, p. 117–126.

Arner, D. H. 1975. Report on effects of channelization modification on the Luxapalila River: Symposium on Stream Channel Modification, Aug. 15–17, 1975, Harrisonburg, Va.

Belt, C. B. 1975. The Effects of Constricting the Missouri River Since 1879: *Geol. Soc. America Abstr. with Programs,* v. 7, no. 7, p. 994.

Corning, R. V. 1975. Channelization: shortcut to nowhere: *Virginia Wildlife,* Feb. 1975, p. 6–8.

Daniels, R. B. 1960. Entrenchment of the Willow Drainage Ditch, Harrison County Iowa: *Amer. Journ. Sci.,* v. 258, p. 161–176.

Eiserman, F., Dern, G., and Doyle, J. 1975. *Cold Water Stream Handbook for Wyoming:* Soil Conservation Service and Wyoming Game and Fish Department, 38 p.

Emerson, J. W. 1971. Channelization: a case study: *Science,* v. 173, p. 325–326.

Graham, R. 1975. Physical and biological effects of alterations on Montana's trout streams: Symposium on Stream Channel Modification, Aug. 15–17, 1975, Harrisonburg, Va.

Hay, R. C., and Stall, J. B. 1974. History of drainage channel improvement in the Vermillion River watershed, Wabash Basin: *Illinois State Water Survey, WRC Report 90,* 42 p.

Hirsch, A. 1975. Keynote address: Symposium on Stream Channel Modification, Aug. 15–17, 1975, Harrisonburg, Va.

Keller, E. A. 1975. Channelization: a search for a better way, *Geology,* v. 3, no. 5, p. 246–248.

____, and Melhorn, W. N. 1973. Bedforms and fluvial processes in alluvial stream channels: selected observations: in Marie Morisawa, ed., *Fluvial Geomorphology,* State University of New York, Binghamton, N.Y., p. 253–284.

Leliavsky, S. 1966. *An Introduction to Fluvial Hydraulics:* Dover, New York, p. 102–106, 162–166.

Little, Arthur D., Inc. 1973. *Report on Channel Modification:* report submitted to the Council on Environmental Quality, v. 1, 394 p.

Maddock, T., Jr. 1972. Hydrologic behavior of stream channels: *Transactions of the Thirty-Seventh North American Wildlife and Natural Resources Conference,* Wildlife Management Institute, Washington, D.C., p. 366–374.

Morisawa, M. 1974. Readjustment of an "improved" stream channel: *Geol. Soc. America Abstr. with Programs,* v. 6, p. 877.

Morris, L. A., Langemeier, R. N., Russell, T. R., and Witt, A., Jr. 1968. Effects of main stem impoundments and channelization upon the limnology of the Missouri River, Nebraska: *Amer. Fisheries Trans.,* v. 97, no. 4, p. 380–388.

Schumm, S. A. 1973. Geomorphic thresholds and complex response of drainage systems: in Marie Morisawa, ed., *Fluvial Geomorphology,* State University of New York, Binghamton, N.Y., p. 299–310.

Soil Conservation Service. 1971. *Planning and Design of Open Channels,* Chapter 7: *Tech. Release 25.*

Yearke, L. W. 1971. River erosion due to channel relocation: *Civil Eng.,* v. 41, p. 39–40.

8

DRAINAGE BASIN CHARACTERISTICS
APPLIED TO HYDRAULIC DESIGN
AND WATER-RESOURCES MANAGEMENT

John F. Orsborn

INTRODUCTION

Floods, average annual flows, and low flows of streams are used for various hydraulic design and water-resources planning purposes. For example, the frequency of floods and their magnitude are the bases upon which risk is considered in the design of the capacity of a spillway. At the other flow extreme, the frequency of annual minimum flows is of importance in the evaluation of water-right applications and the protection of instream flow values, such as fisheries, recreation, and aesthetics.

There are many U.S. Geological Survey gaging stations with adequate records available for use, but there is considerable uncertainty involved in predicting flows at sites some distance from those gaging stations. One has to move upstream of the uppermost gaging station in a drainage basin only past the first major tributary before reaching the point where all streams can be considered as being ungaged. Stream-gaging systems have paralleled the development pattern of our water resources, and now development pressures are outstripping the gaging system. There is a paucity of stream-gage information available in remote and uninhabited areas where current and potential environmental impacts exist.

More often than not, the need to know hydrologic information occurs at sites where available gaging station data are not directly applicable. Therefore, hydrologists usually face two major problems in hydrologic analyses of drainage basin systems:

1. The lack of precipitation (input) and streamflow (output) data, and data on everything that happens (throughput) to the input before part of it becomes output.

2. Our inability to be able to effectively and confidently simulate and transfer correlations from one hydrologic province to another, and from "small" drainage basins to "large" drainage basins because of what we call "scale effects."

In more applied terms, these two problems mean the following:

1. Rarely are there enough stream-gaging (or precipitation) stations of sufficient length of record and propinquity to the problem area to provide direct determinations of the desired streamflow (flood, average, or low flow).

2. Input—output models with which we attempt to reproduce the myriad of interrelated hydrophysical processes require both simplifying assumptions and real data for verification both during their development and after their application.

A new look must be taken at the drainage basin system if we are to improve hydraulic engineering design, thus reducing construction and operation costs, and to provide more complete information to water-resource planners while operating our hydrologic-data-acquisition system economically. Considering the drainage basin as the integrator of all hydrologic processes, then an output–output model that relates a flow of a certain type and frequency (e.g., 2-yr flood) to another flow of the same type (50-yr flood), in terms of basin and flow stability characteristics for a series of gaging stations in a province, should provide both a reproducible and transferable prediction model for ungaged streams in that province. The model provides a natural relationship with which man-made influences (such as storage effects on floods) can be evaluated. Topics to which the output–output, hydrogeologic model concepts have been applied and tested for gaged and ungaged streams include low flows; floods; average annual flows; duration curves; low, average, and flood interrelations; and sediment transport. Other possible areas of application include natural water quality, the effects of urbanization, variability in average annual flows, and the determination of average annual precipitation on a drainage basin from its geomorphic parameters.

Better information on ungaged streamflows is needed to solve such problems as the following:

1. Design of hydraulic structures and the solution of floodplain and shoreline management problems.

2. Allocation of waters among instream and offstream uses.

3. Determination of natural low flows in order to evaluate the existing effects of offstream uses.

4. Evaluation of methods for determining relationships of low flows and fisheries habitat to reduce field measurements and maintain credibility.

5. Determination of natural low flows at points of diversion within national forests and at the forest boundaries, so that reservation rights can be established for lands in the public domain.

6. Establishment of flows for the protection of instream uses (fisheries, wildlife, recreation, aesthetics) under state laws (Orsborn et al., 1973).

7. Availability of supplies for new municipal, industrial, irrigation, and energy conversion.

As more development pressures are brought to bear, the need to know the natural amounts of and variability in ungaged streamflows, without having to resort to extensive and expensive stream gaging programs, is becoming increasingly important. Some limited samples of gaging are required for verification and possible adjustments of the hydrogeologic models, but at much less cost. Also, planning and/or design could proceed without having to wait several years for miscellaneous measurements to be made and correlated against a nearby gage, assuming the existence of a nearby, long-term gage and that the correlation proved to be reliable. The hydrogeologic modeling method, coupled with a limited gaging program for verification, can provide the rapid and accurate methods needed for determining the regime of ungaged streamflows. This is not meant to imply that long-term reference gaging stations are not needed, but these new methods provide an opportunity to shift stream-gaging emphasis to hydrologically complicated areas without increasing the number of gages.

AVAILABLE METHODS
OF HYDROLOGIC ANALYSIS

The dichotomy that exists between methods of analyzing the prediction of runoff generated from precipitation was discussed by Amorocho and Hart (1964). One school of

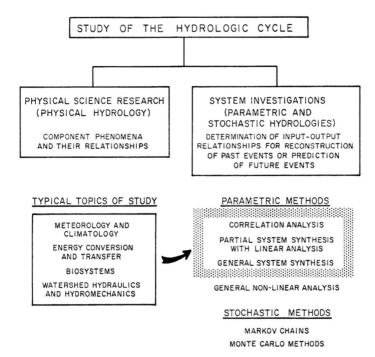

FIGURE 1 *Summary of hydrologic analysis methods. (After Amorocho and Hart, 1964.)*

thought has pursued the scientific, component-by-component approach; the other has utilized parametric relationships to solve applied problems. A summary of these two approaches is diagrammed in Fig. 1.

A goal of many hydrologic engineers and scientists is to better understand, and to be able to predict, various components of precipitation—runoff (or basin system input—output) relationships. Hydrologic scientists account for various intrabasin operations such as sublimation, evaporation, infiltration, transpiration, percolation, and storage. Some models synthesize these components and daily hydrographs, but, as pointed out by Benn (1972), ". . . their use requires considerable 'juggling' of input parameters to obtain the desired fit. Further, the input parameters are usually seldom explicit measurable environmental factors. This seriously limits the use of the model(s) in ungaged watersheds."

Important advances have been made in scientific and engineering hydrology, but the solution to the problem of being able to confidently and economically estimate ungaged streamflows with available means still evades us. One could say that scientific hydrologists select watersheds and then explain the variability in the data. Hydrologic engineers have the watersheds selected through the design of public works, and then explain the rationale behind their "rational" runoff equations. The science and art of hydrology have developed along lines of approach that include the following:

1. Precipitation and streamflow recording.

2. Detailed observations of small watersheds.

3. Analysis of geomorphic parameters as map coverage improved and aerial photographs became available.

4. The creation of synthetic hydrographs.

5. Various regression analyses of extreme and average flows as functions of precipitation and a host of basin and climatic factors.

6. The generation of long-term, anticipated flow experiences using stochastic processes.

Methods for predicting ungaged streamflows that have evolved out of past hydrologic studies are summarized in Fig. 2, using ungaged low flows as an example. The general

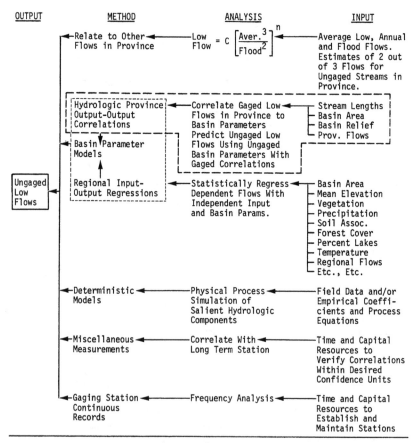

FIGURE 2 *General schematic diagram of alternative methods for determining ungaged low flows.*

processes in Fig. 2 are applicable to the determination of any ungaged flow. It will be many, many years (if ever) before hydrologists will be able to admit that they can predict streamflows accurately enough to only need stream gaging records for checking the prediction. Therefore, gaging station records, although they require the largest investment in time and capital, are given the basic position at the bottom of the diagram. As discussed by Riggs (1968), certain lengths of records are required before certain types of analyses can be completed with desired accuracy limits.

One of the quickest and least expensive approaches for determining low-flow and flood characteristics is to correlate miscellaneous records against concurrent values at a long-term station in the vicinity. This method is more readily applicable to low-flow investigations as discussed by Riggs (1972). Crest-stage measurements for floods require a higher degree of calibration onsite, installation of a stage-recording device, generally more analysis time, and thus the expenditure of more resources than do low-flow investigations. Also, frequencies of events often do not correlate well between the miscellaneous and long-term gaging stations, and care must be taken in assigning any recurrence intervals or probabilities to the miscellaneous site flow measurements (Riggs, 1961; Benson, 1968).

An area of hydrologic analysis that is not shown specifically in Fig. 2 deals with the "rational" equations for predicting floods. These equations are essentially forms of deterministic models that require ground-truth verification with stream-gaging records. The benefit of these models is that they may be used to evaluate landuse changes, such as urbanization, and to incorporate subroutines for the evaluation of other problems, such as water quality, flood routing, and drainage systems.

The more sophisticated the model, the more detailed the data requirements, and the more sensitive will be the interactive and interfacing aspects of the total analysis. Amorocho (1969) noted that "we can state that so far it has never been possible to formulate any real hydrologic situation in strictly deterministic terms." Even 5 yr later, our inability to know that a deterministic model is suitable for a real-world watershed and all its heterogeneities was addressed by Dasman (1973), who said,

> Today natural diversity still baffles us. Even the simplest natural communities escape our comprehension. We abstract and simplify them intellectually with energy flow charts or systems diagrams. When we understand the pictures and formulae, we delude ourselves into believing we understand reality.

Regression models for predicting ungaged floods on a regional basis have proved successful on many occasions. This type of mathematical-statistical approach can be either an input–output or an output–output model. The basic relationships are

$$Q = f(P, A, L, G, \ldots) \tag{1}$$

or

$$Q = f(R, A, L, G, \ldots) \tag{2}$$

where Q is the mean annual flood, P the average annual precipitation, R the average annual runoff (usually for a base period of record), A the drainage area, L the percentage of drainage area in lake surface, and G a geographical factor.

Equation (1) relates the mean annual flood to precipitation, and Eq. (2) relates it to mean annual runoff. Numerous other variables (examples listed in Fig. 2) have been used by many investigators; but, in essence, this method of analysis tests the independent variables through regression, and selects the combination of factors with the strongest correlation to yield an equation of the form

$$Q = 0.638A^{0.889}R^{1.135}L^{-0.037}G \tag{3}$$

This sample equation was derived by Bodhaine and Thomas (1964) for Pacific Slope basins in Washington. Charts are used to determine R and G; A and L are measured from maps or aerial photographs, and the terms are entered into a nomograph to determine the mean flood. Larger floods are estimated based on ratios to the mean in geographic areas.

Reich (1971) conducted an extensive study of floods for solving highway drainage problems in over 100 watersheds smaller than 200 mi^2 (518 km^2) in the East and Southwest. He evaluated numerous methods for flood prediction and found that regional flood frequency analyses, based on stratification of hydrologically homogeneous samples, worked best. Homogeneity was determined on the basis of lithology, topography, climate, and local hydrology. Probably the most detailed regression study of flood equations yet conducted was performed for the Highway Research Board by Bock et al. (1972). Basin and climatic variables were tested and stratified for hundreds of small watersheds across the country. The use of regression models, and their benefits and shortcomings, have been thoroughly evaluated by Crippen (1974).

Although some success has been achieved in predicting ungaged floods with regional regression equations, their application to the prediction of ungaged low flows has been disappointing. As stated by Thomas and Benson (1970) in their report on four diverse regions of the United States, "Low-flow relations are unreliable in all study regions; they can provide only rough estimates of low-flow characteristics at ungaged sites."

In his discussion of methods for analyzing low-flow data and estimating ungaged low flows, Riggs (1972) states,

> Attempts have been made to regionalize low-flow characteristics by multiple regression on several basin characteristics, including geologic indexes. Some of these regressions showed the geologic parameters to be statistically significant, but the standard errors of these regressions were too large to justify application of the relations to ungaged sites. One of the better regressions, which was derived for Connecticut, related 7-day, 10-day low flow to drainage area, mean basin elevation, and percentage of basin covered by stratified drift; it has a standard error of 68 percent (Thomas and Cervione, 1970). The principal roadblock to regionalization of low-flow characteristics is our inability to *describe quantitatively the effects of various geological formations on low flows*—even where detailed geologic maps are available.

Mathematical expressions which relate average annual flow (QAA) to factors such as average annual precipitation (P) and drainage area (A) are widely used (U.S. Geological Survey, 1972). Within relatively homogeneous hydrologic provinces, Riggs (1964) has shown that average annual flow correlates strongly with drainage area alone. In Finland,

Mustonen (1967) applied regression analyses and found that annual runoff correlated most strongly with seasonal precipitation and mean annual temperature, but not with soil type or vegetation indexes. Lull and Sopper (1966) determined that in the northeastern part of the United States, average annual and seasonal runoff were most closely related to *isohyetal annual precipitation,* percentage of forest cover, elevation, latitude, July mean maximum temperature, and percentage of swamp.

The existence of consistent relationships between average annual flow and floods of various recurrence intervals has been demonstrated (Cönturk, 1967). This relationship was shown in Eq. (3), where average annual runoff (R) was combined with other parameters to improve predictability. The existence of this relationship between average and flood flows will be explored further in a subsequent section on correlations between low, average, and flood flows at a site. This special output–output–output model is shown at the top of Fig. 2 (Orsborn et al., 1975a).

In the next section of this paper the development of the new hydrogeologic method (denoted by long, dashed lines in Fig. 2) for predicting ungaged streamflows will be described, and several sample applications of the methodologies will be discussed.

DEVELOPMENT OF HYDROGEOLOGIC METHODS

The powerful simplicity of John Keill's *1698* concepts of the hydrologic cycle forms an early basis for the development of the hydrogeologic, output–output methods of predicting ungaged streamflows:

> And as for Rivers, I believe it is evident, that they are furnished by a superior circulation of Vapours drawn from the Sea by the heat of the Sun, which by Calculation are abundantly sufficient for such a supply. For it is certain that *nature never provides two distinct ways to produce the same effect, when one will serve.* But the increase and decrease of Rivers, according to wet and dry Seasons of the year, do sufficiently show their Origination from a Superior circulation of Rains and Vapours (quoted from White, 1968).

These three sentences state three basic areas of inquiry that have been and are currently receiving widespread attention from students of physical watershed processes: (1) open-system concepts, (2) principles associated with the minimum expenditure of energy by river flows, and (3) input–output relationships of deterministic, physical-process, mathematical models.

The investigation of the landform features of drainage basins has developed from the early works of philosophers and geologists. The more specialized fields of physical geography and hydrology were combined into the concepts of a morphometric system during the past 30 yr (Chorley, 1969). The drainage basin open-system concept and mathematical analogies have advanced the determination of precipitation–runoff relationships in gaged and ungaged streams. Improved map coverage and aerial surveys have assisted in these phases of hydrologic inquiry. Considering that fluvial activity and the form of the land must be related, Horton (1945) provided a very strong tool for landform analysis.

Another step in the development of the hydrogeologic method of streamflow analysis was provided by Strahler (1958) through his application of dimensional analysis to both landform and streamflow characteristics. It was demonstrated that various geomorphic characteristics of the basin could be combined with a flow term to develop dimensionless parameters. These dimensionless ratios of forces can be used to derive physical process equations. For example, the hydrogeologic Froude number is expressed as

$$F = \frac{Q^2}{Hg} \qquad (4)$$

where Q is a flow rate per unit area, H the basin relief, and g the gravitational acceleration term. From the continuity equation, Q is a measure of velocity and thus the inertia of the flow. The basin relief, H (elevation difference between the headwaters and the outlet), is a measure of the potential energy or driving force due to gravity acting on all flows in a basin. Thus, the Froude number Q^2/Hg represents a ratio of inertia to gravity forces in any basin, and it is used extensively in the relations developed between streamflows and basin characteristics in the hydrogeologic method of ungaged streamflow analysis.

The ideas for considering drainage basins as integration devices for all input–output, or inflow–outflow, relationships came from a study of the seasonal levels in Lake Michigan–Huron, as reported by Pierce and Vogt (1953). By determining that winter maximum and summer minimum lake levels follow natural probability distributions, prediction relationships were developed between the two extremes for 50 and 95% confidence limits. In this case, the various hydrologic processes (throughputs) in the Lake Michigan–Huron watershed were simply evaluated as an integrated change in lake-level response.

The same method of relating one extreme occurrence to another was applied to the analysis of piezometric levels in water table and artesian aquifers by Orsborn (1966). This study and a later one (Orsborn, 1970) confirmed that hydrologic maxima and minima can be related to each other, including maxima and minima of different frequencies, both directly and indirectly as a function of basin characteristics.

The final verification of the feasibility of relating basin outflows to each other through the use of various combinations of basin and flow characteristics was achieved as a result of a study conducted in connection with implementation of minimum-flow legislation in the state of Washington (Orsborn et al., 1973). Although the original application of the hydrogeologic, output–output methodology was applied to low flows in Washington, its basic principles were later applied and found to be valid for analyzing floods and average annual flows in hydrologic provinces throughout Washington, Idaho, and Oregon (Orsborn and Sood, 1973; Orsborn et al., 1975a, 1975c).

Besides the work of Strahler (1958) on basin and flow characteristics, numerous investigators have evaluated basin characteristics with respect to various flows and relations among the basin parameters. For example, Julian et al. (1967) found that the 25-yr mean annual specific yield (flow per unit area) in regions of Wyoming and Utah correlated most strongly with the elevation above which 50% of the basin area lies, longitude, latitude, and mean basin gradient. Onesti and Miller (1974) conducted a principal components analysis of 16 channel and basin variables within stream orders, bank-full discharge being the only flow parameter. The results of their study showed that, as basin size (and stream order) increased, there was a downstream trend of increased interdependence among the variables. Carlston (1963) and Orsborn (1970) have demonstrated that drainage density (DD) can be related both to floods and low flows. The

TABLE 1 *Sample of Linear Geomorphic Characteristics Used in Hydrogeologic Analysis*[a]

Property, symbol	Dimensions	Relates to:
Stream length, *LS*	*L*	Channel networks, percentage of input becoming surface runoff (output), soil type, geology, basin storage, contribution to low flow
Basin length, *LB*	*L*	Aspect ratio *LB/WB*; concentration time
Basin perimeter, *PB*	*L*	Basin shape, concentration time[b]
Basin relief, *H*	*L*	Potential energy, form of precipitation, ground cover, etc.
Basin width, *WB*	*L*	Rectangular equivalent derived from *A/LB*
Basin area, *A*	L^2	Catchment size, volume of input, distribution of precipitation input
Drainage Density, *LS/A*	L^{-1}	Soil types and runoff conditions, method of determination not standardized
Channel slope, *SC*	—	Average rate of expenditure of energy as flow moves through basin[b]

[a]See Strahler (1958) for detailed list of basin properties and references.
[b]Not used in these methods, but listed for comparison with other methods.

relation between stream length (*LS*) and drainage area (*A*), as developed by Hack (1957), was

$$LS = 1.4A^{0.6} \tag{5}$$

for basins in the Shenandoah Valley. A similar relationship is utilized in a later section of this chapter to demonstrate how average annual flow, and thus average annual precipitation, can be determined from the stream length in a watershed using a correlation graph developed from isohyetal and watershed maps.

Only the linear basin characteristics listed in Table 1 are needed to develop correlations with gaged streamflow records to apply the hydrogeologic method for predicting ungaged flows within a province. Emphasis is placed on primary terms that can be quantitatively evaluated from maps or aerial photos, and which provide the best correlations with streamflows.

Low flows are determined using combinations of basin characteristics, such as total stream length (*LT*), lengths of first-order streams (*L1*), basin relief (*H*), drainage density (*LS/A* or *LT/A*), and basin area (*A*). Flood flows have been found to be related to drainage area (*A*), relief (*H*), and basin length (*LB*) to width (*WB*) ratio, rather than using circularity ratio. There are problems associated with not having maps of consistent scale, but the following observations have been found to hold regarding the determination of basin characteristics from Geological Survey topographic maps:

1. Solid blue lines are used to determine stream lengths, and thus drainage density, from 1:24,000 and 1:62,500 maps.

2. Map scales may be mixed usually without causing too much of a shift in provincial correlations between flows and basin characteristics. The influences of map scale on basin stream characteristics have been evaluated by Yang and Stall (1971).

3. Drainage basin area, length, width, and relief may be determined from 1:250,000 scale maps for basins that are not so small as to have these values distorted.

4. Basin length (LB) is scaled along the major axis, and basin width (WB) is derived from basin area for an equivalent rectangular basin. This eliminates the necessity for making numerous measurements and then averaging them.

5. Basin relief (H) is determined by subtracting the basin outlet (or gage) elevation from the elevation of the highest continuous contour at the head of the basin. The physical significance of relief (or potential energy) is discussed in detail by Yang (1971).

The definition of drainage density is a special topic that needs to be addressed and to have guidelines established for eventual definition and standardization. Drainage density values are relative to the methods and map scales used to determine them; as long as the method is applied consistently, the values have been found to be adequate for describing the developed relationships.

The ungaged flows for which methods have been developed using the principles previously described are listed below in the order in which the methods will be briefly discussed. This is the order in which the methodology has been developed, and the first three types of flow (low, average annual, and flood) can be used to define the duration curve at the ungaged site.

$Q7L2$	2-yr, 7-day low flow	$QF2P$	2-yr peak flood
$Q7L20$	20-yr, 7-day low flow	$QF50P$	50-yr peak flood
QAA	average annual flow	QS	sediment flow

One flow is considered to be functionally related to another flow of the same type but of different recurrence interval (RI) by the expression

$$QX1 = f(QX2) \qquad (6)$$

For low flows, this general relationship is

$$QL1 = f(QL2)^{-n} \qquad (7)$$

with $-n$ denoting that the slope of the relationship is negative.

The concepts for determining ungaged low flows provided the basis for similar ungaged flood procedures, and originated as part of a minimum-flow study in the state of Washington (Orsborn et al., 1973). The primary objectives were to develop methods that could be (1) transferred from one hydrologic province to another, and (2) operated by an engineer with normal hydrologic training. Referring to the recurrence interval (RI) graph in Fig. 3, the modified (log-log) 7-day average low-flow RI curve was used as the basis for the analytical development. The reference basin outflow is $Q7L2$, the 7-day average low flow with a 2-yr RI. The 7-day average low flow with a 20-yr RI ($Q7L20$), is the second basin outflow used to define the gaged and ungaged RI graphs. This notation ($Q7L2$, for example) was designed for two reasons: (1) it can be used to consistently

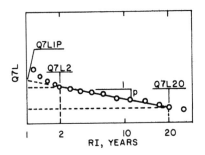

FIGURE 3 *Nomenclature sketch for low-flow analysis.*

abbreviate flow terms with respect to Q (flow), 7 (averaging time period), L (type of flow), and 2 (recurrence interval); and (2) this notation fits computer format and is much easier to type than subscripts.

The U.S. Geological Survey has developed measures of low-flow characteristics that use some of the flows (Fig. 3) as a basis for comparing streams (Nassar, 1973). A measure of the low-flow yield is the minimum 7-day average discharge at the 2-yr RI (Q7L2) per unit of drainage area, and is called the *low flow index* (LFI). The LFI provides a quick appraisal of the low-flow yield of streams having different drainage areas and geologic conditions.

A measure of the year-to-year variability of low flows is the *slope index,* the ratio of the minimum 7-day average discharges at 2- and 20-yr recurrence intervals (Q7L2/Q7L20). Variations in slope indexes of frequency curves at different sites are due to differences in basin characteristics. A high value of the slope index indicates a steep RI curve and large variations in low flow. The slope index is used in a modified form as an indicator of low-flow stability in comparing basins in the new hydrogeologic method.

The 7-day, low-flow frequency (or RI) curve was converted from log-probability to log-log scales in Fig. 3 because (1) the slope index can be defined by the dimensionless slope p of the graph between 2 and 20 yr, and (2) the equation for the relationship between the 2- and 20-yr flows can be used for both analysis and simulation throughout the analysis. Between the 2- and 20-yr low-flow RI's, the plotted points form a straight line on log-log paper, whereas usually it is curved on probability paper. In Fig. 3, the projected 7-day average low flow at a 1-yr RI (Q7L1P) is used to compare the synthetic "maximum" low flows between basins, thus providing a basis for comparison of streams with respect to groundwater reservoir size. The slope of the graph p is related to the slope index mentioned earlier, and is an excellent indicator of basin storage stability, and thus geology. The relationship between the U.S. Geological Survey slope index (Q7L2/Q7L20) and the slope p in Fig. 3 is shown in Fig. 4 for a sample of stations in southwestern Washington. The equation of the relationship indicates that it would apply to any stream, being merely a mathematical conversion from a ratio to a gradient.

Geomorphic parameters and combinations were tested and selected, based on how well they correlated with Q7L2 and Q7L20. Some of the best combinations were $(3/2 \cdot L1 \cdot H)$, $(LT)(H)$, $LT(H)^{0.5}$, and $(\sqrt{DD} \cdot L1)$. L1 is the length of first-order perennial streams, H is relief (potential energy) of the basin, LT is total length of streams, and DD is drainage density. A detailed study of parametric consistency within hydrologic prov-

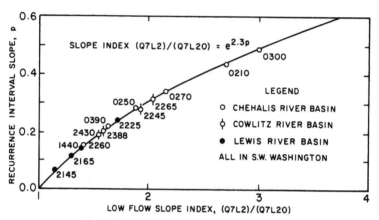

FIGURE 4   *Low-flow slope index related to slope (p) of log-log, 7-day average low-flow recurrence interval graph.*

inces has not been conducted, but other studies of certain provinces have shown that, within certain altitude levels (and thus geology and precipitation ranges), some combinations of basin parameters are more consistent than others.

The basic equation developed for relating gaged and ungaged flows to the log-log RI graph in Fig. 3 was in the form of Eq. (7) as

$$Q7L(\text{RI}) = C(\text{RI})^{-p} \tag{8}$$

In this equation, the left side is the low flow at a particular RI, and the coefficient $C = Q7L1P$ at RI = 1.0. Therefore, for the reference 7-day 2-yr low flow, Eq. (8) becomes

$$Q7L2 = \frac{Q7L1P}{(2)^p} \tag{9}$$

For known $Q7L20$ values, p becomes the difference in the logs of $Q7L2$ for $Q7L20$ (over one log cycle from 2 to 20 yr).

Using this definition of the RI log-log graph, and recalling that the slope p is an indication of natural low-flow stability, a new relationship was developed for the average of the statewide data sample such that

$$Q7L2 = 23\left[\frac{(AH)^{0.5}\,(Q7L1P)}{300p}\right]^{0.5} \tag{10}$$

or

$$Q7L2 = 23\chi^{0.5} \tag{11}$$

The 300 in the bottom of Eq. (10) comes from $AH/3 \div 100$ for convenience. After development of these relationships for gaging stations in a province, the ungaged $Q7L2$ is estimated using the best basin parameters (such as $3/2 \cdot L1 \cdot H$), and $Q7L1P$ is estimated from the ratio $(Q7L1P)/(Q7L2)$ for the gages in the province. Then Eq. (10) is solved for the slope of the RI graph (p). Next $Q7L20$ is estimated by the ratio of $(Q7L20)/(Q7L2)$ for the province, and checked by correlation with the best basin parameters for the province. The $Q7L2$ and $Q7L20$ values are plotted on log-log paper, and p is measured graphically or solved for mathematically. If it agrees with the slope p determined in Eq. (10), the solution is complete. The p values for the ungaged stream can be compared with those at gaging stations in the province as a further verification.

An additional method for checking $Q7L2$ was developed using the estimated values of $Q7L1P$ and p from Eq. (10) in the relationship

$$Q7L2 = 10 \left[\frac{(LT)H^{0.5} \ (Q7L1P)}{1000p} \right]^{0.6} \qquad (12)$$

where the coefficient 10 and the power of 0.6 vary from province to province. If the value of $Q7L2$ determined from the basin characteristics does not agree within reason with the value calculated by Eq. (12), the first estimates should be checked, the geology of the area should be investigated for anomalies, and field measurements should be made during the low-flow period if no miscellaneous records are available. Procedures for estimating low-flow RI curves for ungaged streams are presented in detail by Orsborn and Sood (1973).

Methods for determining average annual flows are relatively simple compared with determining extreme flows, such as floods and minimums. The most commonly used approach has been to set QX equal to some function of the drainage A:

$$QX = C(A)^n \qquad (13)$$

where C and n are the average coefficient and power determined from a plot of streamflow data and drainage areas. Simple modifications of this equation provide good estimates of average annual flow (QAA) in many different geographic areas.

Equation (13) can take the form

$$QAA = C'(P) \cdot A^{n'} \qquad (14)$$

where P is the average annual precipitation, and C' and n' are new coefficients and powers derived from regression analyses of the data. Average annual flows and their variability can be readily correlated for gaged streams and then used to estimate QAA in ungaged streams in the same hydrologic province. An example of such a correlation is shown in Fig. 5, for which $C' = 0.052$ and $n' = 1.0$. The Deschutes River basin data will be used later in an example to demonstrate how average annual flow and precipitation can be determined from stream length and drainage area. The Rainier gage in Fig. 5 has the longest period of record.

The introduction of precipitation from a specific storm tends to increase the potential variability in the precipitation–runoff relationship. Long-term average annual precipitation tends to be stable. Improvements in statewide isohyetal maps will provide for increased accuracy in estimating average annual flow. A simple input–output model is expressed in Eq. (14).

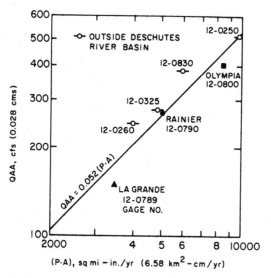

FIGURE 5 *Average annual flow (QAA) related to average annual precipitation (P) and drainage area (A) for Deschutes River stream gages and other stream gages in the region of western Washington, South Puget Sound.*

The hydrogeologic approach for floods has yielded final equations similar to those developed for low flows [e.g., Eq. (10)]. Beginning with the basic output—output concept as in Eq. (6),

$$QF2 = f(QF50) \qquad (15)$$

Referring to Fig. 6, Eq. (15) can be rewritten as

$$QF2 = \frac{f^{-1}(QF50)}{p} \qquad (16)$$

Floods with 2- and 50-yr RI's are used because they are the most readily available from the U.S. Geological Survey records. Very few stations have more than 50 yr of records, and the recurrence interval graph can be adequately defined if these two values are determined for ungaged streams. The slope p is measured directly from the graph.

In evaluating f', geomorphic parameters of area (A), basin relief $(H,$ basin length (LB), and basin width (WB) were introduced such that

$$QF2 = f'' \left[\frac{(AH)^{0.5} (LB/WB)^{0.5} (QF50)}{10,000p} \right]^{m} \qquad (17)$$

which is comparable to Eq. (10) for low flows.

$$Q7L2 = 23 \left[\frac{(AH)^{0.5} (Q7L1P)}{300p} \right]^{0.5} \qquad (10)$$

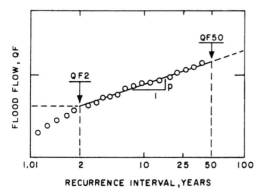

FIGURE 6 *Nomenclature for flood recurrence interval graph.*

The basin area and relief terms have been discussed previously, and the basin length-to-width ratio accounts for variations in the time of concentration in basins of different shapes. A graphical example of Eq. (17) is shown in Fig. 7 for basins in southwestern Washington. For this province, f'' and m in Eq. (17) are equal to 430 and 0.6, respectively, and these have been found to be applicable to areas in other Pacific Northwest states as well. Primary geomorphic relationships for QF2 and QF50 with basin characteristics (e.g., $L1 \cdot H$ and $LT \cdot H$) have not worked as they did for making independent first estimates of low flows. Therefore, an alternative approach was developed using data from the Lewis River study area in southwestern Washington (Orsborn and Sood, 1973).

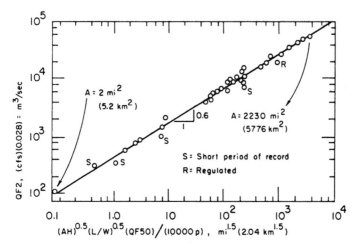

FIGURE 7 *Two-year floods related to basin and flood parameters for southwestern Washington streams.*

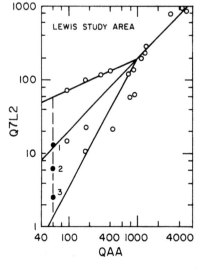

FIGURE 8 *Low flow related to average flow.*

It was observed that three streams with the same average annual flow (and about the same area and precipitation) had widely varying low flows, as noted by points 1, 2, and 3 in Fig. 8. Considering that the several basins with the same average annual runoff (but with different low flows) must have differences in geology and therefore surface runoff, Figs. 9a and 9b were developed showing a relationship of floods, low flows, and average flows. Multiplying the 2-yr low flow at each gage by its 2- and 50-yr recorded peak flood flows squared, the very strong correlations in Fig. 9 appeared. The scattered points 1, 2, and 3 in Fig. 8 are clustered at the lower end of the graphs in Figs. 9a and 9b.

In equation form, Fig. 9a is

$$QF2P = 1,000\left[\frac{QAA}{50.0(Q7L2)^{0.32}}\right]^{1.56} \tag{18}$$

and Fig. 9b is

$$QF50P = 1,000\left[\frac{QAA}{33.0(Q7L2)^{0.32}}\right]^{1.56} \tag{19}$$

The methods for determining the average annual and low flows in ungaged streams and in streams above gages have been outlined. Once these flows have been estimated, the 2- and 50-yr floods for the ungaged portion of a stream can be determined from Eqs. (18) and (19). From these two flood flows, a flood RI graph for the ungaged stream can be prepared and the slope checked against gaged RI curves in the province. The estimated value of the 100-yr flood flow (or higher) can be determined by extrapolation.

If the graph in Fig. 9a is redrawn in the form of Eq. (18), it results in the relationship shown in Fig. 10, using 2-yr peak floods as an example. Once the estimates of QF2P and QF50P have been made for an ungaged stream, they can be checked using the hydrogeologic output–output model in Eq. (17). Details of this procedure have been reported by Orsborn et al. (1975a, c).

Using the predicted 2-yr flood, average annual flow, and 2-yr low flow, and the shape characteristics of duration curves for stream gages in a hydrologic province, an average duration curve can be generated for the ungaged stream. This is accomplished by knowing the mean percentage of time that the average annual flow is equaled or exceeded, and using the three flows to guide the fitting of a typical duration curve for the province. The mechanics can be done either graphically or by computer.

The Deschutes River southeast of Puget Sound in Washington, which was used in the average annual flow example, has been studied in some detail to determine its suspended sediment characteristics (Nelson, 1973). The geomorphic characteristics of the three basins are listed in Table 2. The average suspended sediment concentration graphs of the Deschutes River near LaGrande, Rainier, and Olympia were analyzed in terms of river basin geomorphic parameters and averages of precipitation and streamflow (Orsborn et al., 1975b). The three gaging station average suspended sediment concentrations (QS) are related to the concurrent instantaneous river discharge (QI) at those stations. Equations

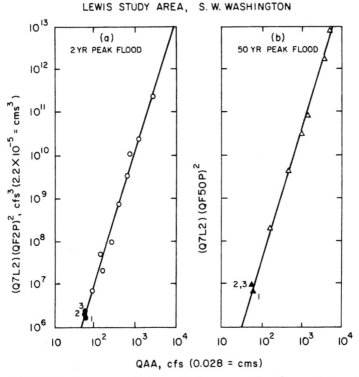

FIGURE 9 *Average annual flows related to average low flows and 2- and 50-yr peak floods.*

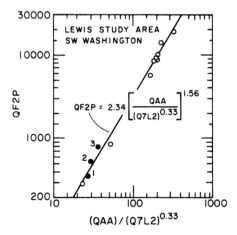

FIGURE 10 *Peak flood flow related to a ratio of average flow and low flow in southwest Washington (all flows in cfs).*

for the three average suspended sediment rating curves at river discharges of 1,000 through 7,000 ft^3/s are presented in Table 3.

The suspended sediment concentrations at the river discharge values between 1,000 and 7,000 ft^3/s were then plotted for each station against a "river parameter" as shown in Fig. 11. The estimated bed load transported by the Deschutes River amounts to only about 10% of the suspended load. The river parameter combines first-order stream length ($L1$), total stream length (LT), basin relief (H), and drainage area (A) at each station.

TABLE 2   *River Basin Parameters of Deschutes River, Washington*[a]

Gage station (no.)	$L1$ (mi)	LT (mi)	Upper elev. (ft)	Gage elev. (ft)	Relief, H (mi)	Basin area, A (mi²)
LaGrande (12078902)	38.6	61.5	2,550	549	0.38	56.2
	$\Sigma = 38.6$	$\Sigma = 61.5$				
Rainier (12079000)	11.2	24.1	2,550	350	0.42	89.8
	$\Sigma = 49.8$	$\Sigma = 85.6$				
Olympia (12080000)	8.9	29.1	2,550	95	0.47	160.0
	$\Sigma = 58.7$	$\Sigma = 114.7$				

[a]*Nomenclature:* $L1$, length of first-order (unbranched perennial streams); LT, total length of perennial streams; upper elevation, highest average contour around headwaters; H, relief: difference in elevation between headwaters and gage (or outlet, for ungaged basin); and A, drainage area defined by topographic divide above gaging station or basin outlet. Conversions: mi(1.61) = km; ft(0.305) = m; mi² (2.59) = km².

TABLE 3 *Suspended Sediment Concentration and Discharges at Three Stations in the Deschutes River Basin (Mean Values)*

Station	QS (mg/liter)	1,000	2,000	3,000	4,000	5,000	6,000	7,000
		QI (ft³/s) · 0.028 = m³/s						
LaGrande	$QS = 0.00034(QI)^{1·83}$	105.4	374	785	1,374	1,999	2,788	3,719
Rainier	$QS = 0.02(QI)^{1·55}$	89.7	274	492	788	1,080	1,436	1,828
Olympia	$QS = 0.000082(QI)^{1·93}$	49.4	187	411	742	1,102	1,564	2,112

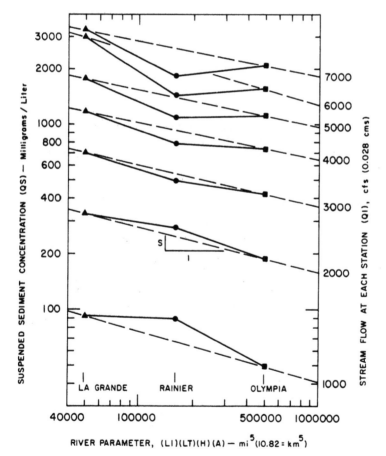

FIGURE 11 *Suspended-sediment concentration discharge related to river parameter (first-order stream length, total stream length, relief, and basin area) for the three stations in the Deschutes River basin, Wash.*

Considering the slopes (s) of the dashed lines in Fig. 11, each representing an instantaneous river discharge, each line can be written as

$$QS = C[(L1)(LT)(H)(A)]^{-s} \begin{bmatrix} 7,000 \\ 1,000 \end{bmatrix} \tag{20}$$

Or, using the abbreviation RP for the river parameters ($L1, LT, H, A$), Eq. (20) can be written for each river discharge as

$$QS = \frac{C}{(RP)^s} \tag{21}$$

The coefficient C in Eq. (21) varies as a function of the instantaneous discharge, as do the slopes s of each of the dashed discharge lines in Fig. 11. The relationships between the coefficients C, the slopes of the graphs s, and river discharge QI are

$$C = 0.35 \times 10^{-6}(QI)^{2.33} \tag{22}$$

and

$$s = \frac{1.10}{(QI)^{0.19}} \tag{23}$$

In Eq. (23), s decreases with discharge, which indicates that, as river discharges get larger, there is less variation in sediment concentration between the upper and lower portions of the Deschutes River basin. When the daily flows of the average duration curve are substituted for QI in Eq. (20), the average annual sediment load can be calculated. The flow-duration curve is converted to a sediment-duration curve.

SAMPLE APPLICATIONS OF THE HYDROGEOLOGIC OUTPUT–OUTPUT MODELS

A brief description of some relationships that have been developed recently will serve as examples of how the hydrogeologic methods can be applied to determine various flows. Several of the examples are from the Coeur d'Alene, St. Joe, and St. Maries River basins, which provide the major inflow to Lake Coeur d'Alene in northern Idaho (Orsborn et al., 1975c). The five long-term gaging stations in the three basins used to develop the necessary correlations for predicting low, average, and flood flows are listed in Table 4. The gaging station numbers are used to identify plotted data points in subsequent graphs.

TABLE 4 Long-Term Gaging Stations on Major Streams Entering Lake Coeur d'Alene

Station	River	Location	Symbol
12-4110	Coeur d'Alene	Near Prichard	□
12-4130	Coeur d'Alene	At Enaville	△
12-4135	Coeur d'Alene	Near Cataldo	○
12-4145	St. Joe	At Calder	▽
12-4150	St. Maries	At Lotus	⊙

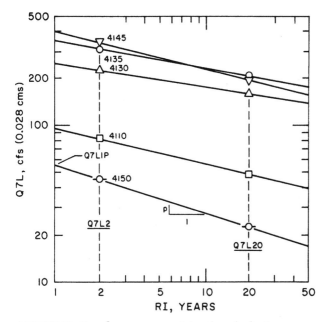

FIGURE 12   *Low flow–recurrence interval graphs for Coeur d'Alene, St. Joe, and St. Maries rivers in north Idaho.*

For low-flow analysis, the values of the 2-yr (Q7L2) and 20-yr (Q7L20), 7-day average low flows were determined from Geological Survey records, and then plotted in Fig. 12. The values of the 1-yr projected low flows (Q7L1P) and the slopes p of the low flow–recurrence interval graphs were determined at the gaging sites. These basin characteristics above each station were tested to determine the "best basin parameter" for making the first estimate of the two flows in ungaged streams.

As shown in Fig. 13, the combination ($LT \cdot H^{0.5}$) was the most consistent basin parameter combination for estimating Q7L2 and Q7L20. The single data point for the St. Maries River would appear to provide a dubious prediction equation, even though it was assumed parallel to the relationships for the Coeur d'Alene and St. Joe rivers. The St. Maries watershed lies at a generally lower elevation than the other two basins, and has a different geologic structure, lower low flows, and higher flood flows. But, as will be shown next, this best basin parameter correlation is used only to make first estimates, which are later verified against other parameters, flows and flow variability (p, slope of the RI graph).

The next step in the analysis is to combine basin characteristics A, H, and LT with Q7L1P and p into χ and χ' relationships similar to Eqs. (10) and (12), as shown in Fig. 14 for the gaging stations. The prediction models for low flows are complete with Figs. 13 and 14. To test the model, the St. Joe River was divided into three sequential ungaged subbasins above the gage at Calder, and flows were estimated for the three subbasins.

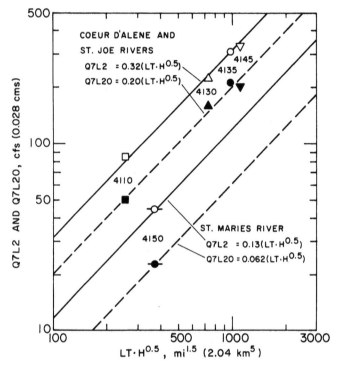

FIGURE 13 *Low flows related to best basin parameter for Coeur d'Alene, St. Joe, and St. Maries rivers in north Idaho.*

Headwater elevations in the side basins are higher than in the headwaters of the total basin, and the two lower subbasins consist of short, steep feeder streams to the main river channel. The predicted 7-day, 2-yr low flow (Q7L2) determined from the correlations developed in Figs. 13 and 14 was 305 ft^3/s (8.5 m^3/s), and the recorded flow at gage 4145 was 339 ft^3/s (9.5 m^3/s) for the period of record (1920–1973). Direct estimates for the total basin above the gage were equal to 320 ft^3/s (9.0 m^3/s), or an error of about 6%.

Low flows predicted for the South Fork of the Coeur d'Alene River were found to be only about 60% of the values recorded during an 8-yr period (1966–1973). The average low flows during this period at other stations with longer records were found to be close to Q7L2. It would appear that mine and mill pond drainage along the South Fork are augmenting the natural low flow of the river, and this is being investigated.

Correlations between average annual flows, average annual precipitation (as determined from isohyetal maps published by the Weather Service), and drainage area yielded the following relationships. For the Coeur d'Alene and St. Joe river gaging stations,

$$QAA = 0.052(P \cdot A) \tag{24}$$

and for the St. Maries River,

$$QAA = 0.030(P \cdot A) \qquad (25)$$

Several smaller watersheds in the higher elevations of the Coeur d'Alene basin yielded the expression

$$QAA = 0.060(P \cdot A) \qquad (26)$$

for the period of record 1966–1973.

As mentioned in the discussion of average annual flows in the Deschutes River basin in Washington, correlations can be developed between average annual precipitation P and the basin parameters of total stream length LT and drainage area A, as shown in Fig. 15. By incorporating the provincial model for average annual flow QAA in the form of Eq. (14) in Fig. 5, the average annual flow can be estimated on the basis of stream length. Note the similarity between the equation of the lower line in Fig. 15 and Hack's (1957) relationship in Eq. (5).

Flood-flow correlations for the three north Idaho rivers were developed as Eqs. (17) and (18), and are shown in Figs. 16 and 17, respectively. There were no unused gaging station data available to check the predictability of the relationships in Figs. 16 and 17, but the "low-average-flood" (LAF) equations have been tested in other provinces, as shown in Table 5 (Orsborn et al., 1975a). Two stations in Washington, one in southwest

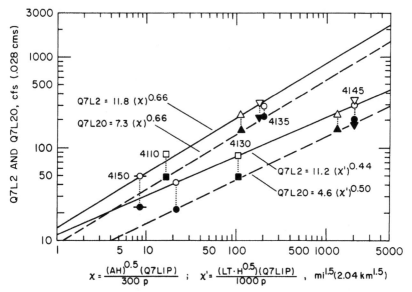

FIGURE 14 *Low flows related to combined basin and flow parameters for Coeur d'Alene, St. Joe, and St. Maries rivers in north Idaho.*

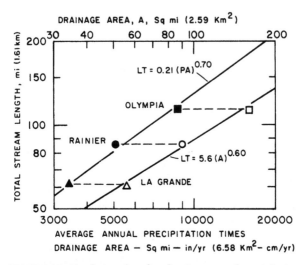

FIGURE 15 *Total river length related to annual precipitation and drainage area for Deschutes River basin study area.*

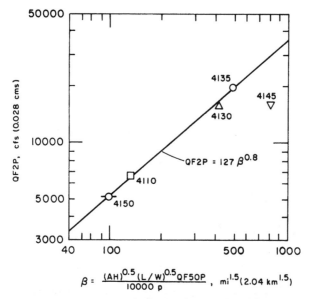

$$\beta = \frac{(AH)^{0.5}(L/W)^{0.5}QF50P}{10000\,p}, \quad mi^{1.5}(2.04\,km^{1.5})$$

FIGURE 16 *Two-year peak floods related to basin characteristics, 50-yr peak flood and flood stability for the Coeur d'Alene and St. Maries rivers in north Idaho.*

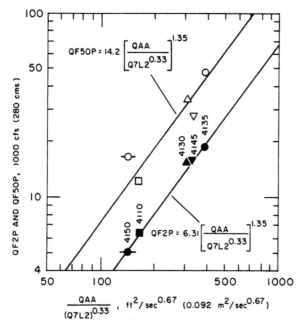

FIGURE 17 *Peak flood flows related to average annual and low flows for Coeur d'Alene, St. Joe, and St. Maries rivers in north Idaho.*

province 3 and one in southeast province 8, were selected to test the prediction equations. The low flows ($Q7L2$) were estimated using the hydrogeologic basin parameter methods described previously. The average annual flows (QAA) were estimated using regional correlations of gaged flows with average annual precipitation and drainage area [$QAA = C(P \cdot A)$]. Neither station was used in the provincial regression analyses.

Inspection of Table 5 shows that the predicted peak flood flows compared very favorably with the recorded values, especially for $QF2$ values and for the recorded $QF50$ value at the Asotin Creek gage, which was based on 44 yr of record. The province 8 equations were based on the records at only four gaging stations. The "recorded" $QF50$ value at the Salmon Creek gage was based on a statistical extrapolation of only 28 yr of record. Note also that the natural, predicted $Q7L2$ values at both stations are larger than the recorded values, and that there are diversions above both stations.

As part of the LAF study (Orsborn et al., 1975a), the Oregon flood data were tested for both the maximum average annual day flood series (QFD) and the annual instantaneous peak flood series (QFP). The two values are defined in Fig. 18, and the relationship of

$$QFD = 1.20(QFP)^{0.95} \tag{27}$$

is shown graphically in Fig. 19. Both 2- and 50-yr floods fit the relationship in Eq. (27)

TABLE 5 *Prediction of 2- and 50-Year Peak Floods in Washington Provinces 3 and 8 (Orsborn et al., 1975a)*

Factor	Units	Salmon Creek, Prov. 3, SW Wash. (sta. 14-2120)	Asotin Creek, Prov. 8, SE Wash. (sta. 13-3345)
Drainage area	mi^2	13.8	156.0
Average precipitation	in./yr	58.0	28.0
Basin relief	mi	0.23	0.76
Recorded QAA	ft^3/s	61.0	68.4
Predicted QAA	ft^3/s	62.6	61.0
Recorded Q7L2[a]	ft^3/s	3.0	27.5
Predicted Q7L2[a]	ft^3/s	3.3	30.0
Predicted QF2P[b]	ft^3/s	810	331
Recorded QF2P	ft^3/s	847	328
Predicted QF50P[b]	ft^3/s	2,100	1,010
Recorded QF50P	ft^3/s	1,720[c]	1,120[d]

[a]Diversions above both gaging stations.
[b]Using computer regression analyses for provinces with 95% C.I. data.
[c]Extrapolated based on 28 yr of record.
[d]Based on 44 yr of record.

and Fig. 19. The only station that did not fit was in an arid part of Oregon where the stream received most of its flow from springs. The relationship for Oregon provinces 3 and 4 (the Willamette Valley and East Cascades Slope) is the same as in Eq. (27). Their data were plotted in a separate graph to avoid congestion and are not included here.

SUMMARY AND CONCLUSIONS

In the past few years new methods of hydrologic analysis have been developed in the state of Washington. The methods have been applied and tested in hundreds of watersheds in the Pacific Northwest, ranging in size from 2 to 2,200 mi^2 (5.2 to 5,698 km^2),

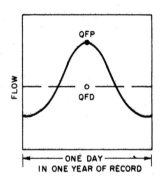

FIGURE 18 *Nomenclature sketch for definition of peak and maximum daily floods.*

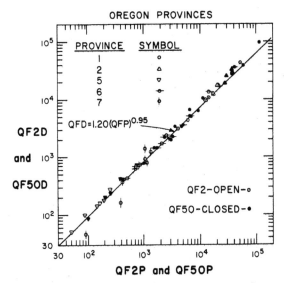

FIGURE 19 *Maximum daily floods related to peak floods in Oregon provinces 1, 2, 5, 6, and 7.*

with a range in average annual precipitation of 10 to 120 in. (25.4 to 304.8 cm) per year. Low flows analyzed have been as small as 0.05 ft^3/s (0.014 m^3/s, and the methods have been verified over four orders of magnitude for floods as large as 100,000 ft^3/s (2,800 m^3/s).

The hydrogeologic methods of predicting ungaged streamflow were generated from two concepts:

1. That drainage basin systems integrate all hydrologic inputs and yield outputs which are the net result of all the hydrologic and geologic processes acting on those inputs.

2. That one basin-integrated output can be correlated against another basin-integrated output in terms of drainage basin geomorphic characteristics.

It may have been more appropriate to pose this question in the Introduction: "Why should we predict runoff for ungaged drainage areas?" The same question was asked by Snyder and Knisel (1974); in answering it, they developed a very thorough reexamination of the philosophy regarding the synthetic generation of streamflow data. Perhaps it is sufficient to say that ungaged streamflows are needed for planning, design, and management of water resources because of the following:

1. Water resources are not developed in a planned, centrally controlled process, but by many agencies and organizations in a somewhat predictable but random fashion.

2. Water-resources development usually involves the utilization of remote and/or unadjudicated sources.

3. As a result of the first two conditions, the planning and installation of data-acquisition systems lag development of water resources.

To help alleviate this situation, various alternatives are available to the hydrologist, ranging from the establishment of a gaging station (assuming some data are better than none) to the generation of a mathematical model. A less arduous task would involve the use of rational equations or transfers of estimated flows from a "nearby" gage in another basin using such relationships as "flow per unit area." But now we are treading on even shakier ground, because we know from experience that the scaling up of flow relationships from smaller to larger watersheds, and the transfer of flow relationships outside the regions for which they were developed, have not been valid methods of hydrologic analysis.

A large part of our "lack of predictability" has come from the problems inherent in our precipitation–runoff approach to the development of the simulation models. How can one expect a mathematical input–output relationship for one basin to be transferable to another that has different geology, soils, vegetation and climate, even when we attempt to assign indexes to these variables?

Instead, as suggested in this chapter, why not let the basins tell us how they integrate all inputs and throughputs to yield a series of outputs such as minima, maxima, and averages? Then, through the application of basic physical principles such as those of potential energy and inertia, relationships can be determined between the flows of interest and basin characteristics that are indicative of the fluvial interaction between precipitation, landform, and runoff. For example, one can reason that low flow should correlate with length of perennial streams in a basin, because this is usually a measure of the interface between the aquifer and the stream. Also, the stream network intensity varies directly with the relative percentage of precipitation that appears in streams as surface runoff, and it is thus indicative of larger floods and smaller low flows.

Examples of hydrogeologic, output–output models developed utilizing these concepts have been described and applied to diverse hydrologic provinces in the Pacific Northwest. Topics of interest for which the methods have been verified for ungaged streams include low-flow and flood-flow recurrence-interval graphs, average annual flows, duration curves, and sediment loads. Some emerging areas that have not been discussed include correlations for predicting natural water quality, determining the effects of urbanization on floods and low flows, and the use of satellite imagery and direct digitization to do all the output–output model generation, analysis, and verification (Rango et al., 1975). With increased interaction between geomorphologists and hydrologists, and the casting out of a few of the more traditional approaches to hydrologic analyses, within a few years we may be able to slide a satellite photograph (or digitized tape) into a black box and see a printout on a scope (in a matter of seconds) which says,

> Ungaged stream (X) in hydrologic province (Y) at site (Z) has a mean annual flow of . . . and a maximum reasonable flood of . . . and a minimum flow (adjusted for existing and allowable diversionary rights) of . . . and a variability in these flows of. . . . Records show that the nearest stream gage is 30 km to the north with no strong correlation to the ungaged site. Confidence limits are ±5% for flows at site.

Impossible, you say? I think not; in fact I would say it is very probable, if we don't let disciplinary traditions stand in our way.

REFERENCES

Amorocho, J. 1969. Deterministic, non-linear hydrologic models, *Proceedings 1st International Seminar for Hydrology Professors,* Vol. I: University of Illinois, Urbana, Ill., p. 420–472.

Amorocho, J., and Hart, W. E. 1964. A critique of current methods in hydrologic systems investigation: *Trans. Amer. Geophys. Union,* v. 45, no. 2, p. 307–321.

Benn, B. O. 1972. Regional planning potential of deterministic hydrologic simulation models: Proceedings, Seminar on Hydrologic Aspects of Project Planning, HEC, Corps of Engineers, Davis, Calif., March 7–9, p. 13–26.

Benson, M. A. 1968. Uniform flood-frequency estimating methods for federal agencies: *Water Resources Res.,* v. 4, no. 5, p. 891–908.

Bock, P., et al. 1972. Estimating peak runoff rates from ungaged small rural watersheds: *Natl. Coop. Hgwy. Res. Program Rept. 136,* Highway Research Board, Washington, D.C., 85 p.

Bodhaine, G. L., and Thomas, D. M. 1964. Magnitude and frequency of floods in the United States: part 12 in Pacific Slope Basins in Washington and Upper Columbia River Basin, *U.S. Geol. Survey Water Supply Paper 1687,* 337 p.

Carlston, C. W. 1963. Drainage density and streamflow: *U.S. Geol. Survey Prof. Paper 442-C,* 8 p.

Chorley, R. J., ed. 1969. *Introduction to Fluvial Processes:* Methuen, London, 218 p.

Cönturk, H. 1967. Mean discharge as an index to mean maximum discharge: *Proceedings, Leningrad Symposium on Floods and Their Computation,* IASH-UNESCO-WHO, v. 2, p. 826–833.

Crippen, J. R. 1974. Basin-characteristic indexes as flow estimators: National Water Resources Engineering Conference of the American Society of Civil Engineers, Los Angeles, Calif., Jan. 21–25, 23 p.

Dasman, R. C. 1973. A rationale for preserving natural areas: *Jour. Soil Water Conserv.,* v. 28, no. 3, p. 114–117.

Hack, J. T. 1957. Studies of longitudinal stream profiles in Virginia and Maryland: *U.S. Geol. Survey Prof. Paper 294-B,* 97 p.

Horton, R. E. 1945. Erosional development of streams and their drainage basins; hydrophysical approach to quantitative morphology: *Geol. Soc. America Bull.,* v. 56, p. 275–370.

Julian, R. W., Yevjevich, V., and Morel-Seytoux, H. J. 1967. Prediction of water yield in high mountain watersheds based on physiography: *Hydrology Papers,* no. 22, Colorado State University, Ft. Collins, Colo., 20 p.

Lull, H. W., and Sopper, W. E. 1966. Factors that influence streamflow in the northeast: *Amer. Geophys. Union Water Resources Res.,* v. 2, no. 3, p. 371–379.

Mustonen, S. E. 1967. Effects of climatologic and basin characteristics on annual runoff: *Amer. Geophys. Union Water Resources Res.,* v. 3, no. 1, p. 123–130.

Nassar, E. G. 1973. Low flow characteristics of streams in the Pacific slope basins and lower Columbia river basin, Washington: *U.S. Geol. Survey, Open-File Rept.,* 68 p.

Nelson, L. M. 1973. Sediment transport in streams in the Deschutes and Nisqually River Basins, Washington, Nov., 1971–June, 1973, USGS: Prepared in cooperation with the state of Washington, Department of Ecology, Draft Open File Rept., Tacoma, Wash., 46 p.

Onesti, L. J., and Miller, T. K. 1974. Patterns of variation in a fluvial system: *Amer. Geophys. Union Water Resources Res.*, v. 10, no. 6, p. 1178–1186.

Orsborn, J. F. 1966. The prediction of piezometric levels in observation wells based on prior occurrences: *Amer. Geophys. Union Water Resources Res.*, v. 2, no. 1, p. 139–144.

_____. 1970. Drainage density in drift-covered basins: *Amer. Soc. Civil Engr. Hydr. Div. Jour.*, v. 96, no. HY1, proc. paper 7033, p. 183–192.

_____. 1975. A geomorphic method for estimating low flows: American Society of Civil Engineers Annual Meeting, Denver, Colo., Nov. 3–7, 38 p.

_____, and Sood, M. N. 1973. Technical supplement to the hydrographic atlas, Lewis River basin study area: Washington Department of Ecology, State Water Program, 64 p.

_____, et al. 1973. A summary of quantity, quality and economic methodology for establishing minimum flows: *Washington Water Res. Center Rept. 13*, v. 1, p. 21–43.

_____, et al. 1975a. Relationships between low, average and flood flows in the Pacific Northwest: OWRT Project A-074-WASH, Department of Civil and Environmental Engineering, Washington State University, Pullman, Wash., 61 p.

_____, et al. 1975b. Hydraulic and water quality research studies of Capitol Lake sediment and restoration problems: College of Engineering, Washington State University, Pullman, Wash., 315 p.

_____, et al. 1975c. Surface water resources of the Coeur d'Alene, St. Joe and St. Maries rivers in northern Idaho, App. D: in Preliminary investigation of the water resources of the northern part of the Coeur d'Alene Indian Reservation, Department of Civil and Environmental Engineering, Washington State University, Pullman, Wash., p. 160–197.

Pierce, D. M., and Vogt, J. E. 1953. Method for predicting Michigan–Huron lake level fluctuations: *Amer. Water Works Assoc. Jour.*, v. 45, p. 502–520.

Rango, A., Foster, J., and Salomonson, V. V. 1975. Extraction and utilization of space acquired physiographic data for water resources development: Preprint X-913-75-3, Goddard Space Flight Center, Greenbelt, Md., 28 p.

Reich, B. M. 1971. Runoff estimates for small rural watersheds: *Fed. Hgwy. Comm. Contract FH-11-7429 Final Rept.*, Pennsylvania State University, University Park, Pa., 132 p. without appendixes.

Riggs, H. C. 1961. Frequency of natural events: *Amer. Soc. Civil Engr. Hyd. Div. Jour.*, no. HY1, proc. paper 2706, p. 15–26.

_____. 1964. The relation of discharge to drainage area in the Rappahannock River Basin, Virginia: *U.S. Geol. Survey Prof. Paper 501B*, p. B165–B168.

_____. 1968. Frequency curves: in *Techniques of Water-Resources Investigations of the U.S. Geological Survey*, chap. A2, book 4, 15 p.

_____. 1972. Low-flow investigations: in *Techniques of Water-Resources Investigations of the U.S. Geological Survey*, chap. B1, book 4, 18 p.

Snyder, W. M., and Knisel, W. G., Jr. 1974. Why should we predict runoff for ungaged drainage areas?: *Jour. Soil Water Conserv.*, Sept.–Oct., p. 229–232.

Strahler, A. N. 1958. Dimensional analysis applied to fluvially eroded land forms: *Geol. Soc. America Bull.*, v. 60, p. 279–299.

Thomas, D. M., and Benson, M. A. 1970. Generalization of streamflow characteristics from drainage-basin characteristics: *U.S. Geol. Survey Water Supply Paper 1975,* 55 p.

U.S. Geological Survey. Circa 1972. Determination of streamflow for shoreline management; streams of Washington under the requirements of shoreline management act of 1971: 51 p.

White, G. W. 1968. John Keill's view of the hydrologic cycle, 1698: *Amer. Geophys. Union Water Resources Res.,* v. 4, no. 6, p. 1371–1374.

Yang, C. T. 1971. Potential energy and stream morphology: *Amer. Geophys. Union Water Resources Res.,* v. 7, no. 2, p. 311–322.

____, and Stall, J. B. 1971. Note on the map scale effect in the study of stream morphology: *Amer. Geophys. Union Water Resources Res.,* v. 7, no. 3, p. 709–712.

III

RESOURCE ENGINEERING

9

KINZUA DAM
AND THE GLACIAL FORELAND

Shailer S. Philbrick[1]

INTRODUCTION

Kinzua Dam and Allegheny Reservoir provide the opportunity to see the effect of the geomorphology of the upper Allegheny Basin on the construction of a major reservoir project. My connection with the project commenced in 1936 and continued, except for two interruptions totaling only 12 months, until final settlement of the construction contract claims in 1974. Allegheny Reservoir is a flood-control reservoir located in the Allegheny River valley in northwestern Pennsylvania and southwestern New York. It extends upstream about 31 mi from about 9 mi east of Warren, Pennsylvania, to the downstream end of Salamanca, New York (Fig. 1). The dam site is at the downstream side of a preglacial col between a westerly flowing stream, a tributary of the ancient Conewango, and a northerly flowing tributary of the ancient Kinzua Creek; both eventually joined the preglacial northerly flowing upper Allegheny River that joined the St. Lawrence River (Carll, 1880). Thus, the reservoir is in a reversed valley in the unglaciated foreland of the Illinoian and Wisconsin glaciers. The foreland is the area south of the glacial margin but was affected by the glacier. The foreland had a well-developed dendritic drainage pattern, with a relief of about 1,000 ft, incised in very gently south dipping Devonian to basal Pennsylvanian shales and sandstones of the Allegheny Plateau.

The reversal of the original direction of flow of these streams was first recognized by Carll (1880) of the Second Pennsylvania Geological Survey, whose studies of the oil and gas wells in the valleys showed that the surface of bedrock sloped north, opposite to the slope of some of the stream beds. He depicted the preglacial drainage in his Plate 2 (1880); later workers with much more information have found it generally correct, although the route of the preglacial Allegheny as shown in Fig. 1 is different because it is based on records of wells drilled in the century since he began his studies. Papers by Leverett (1902), Butts (1910), Leggette (1936), and Muller (1963, 1975) have been especially informative and helpful. The writer's studies in connection with the Allegheny Reservoir confirmed that the preglacial Allegheny River flowed north from Steamburg, New York, via the present Conewango and Cattaraugus valleys into the ancient St. Lawrence valley, now occupied in part by Lake Erie.

The dam structure consists of a concrete gravity-type dam with a gated overflow spillway and an earth wing. The dam is 1,897 ft long and stands 179 ft above stream bed. The reservoir covers 21,180 acres at a maximum flood-storage elevation of 1,365 ft above sea level. Relocations included 83 mi of highways and 37 mi of railroad of which the Erie–Lackawanna Railroad relocation in the upstream part of the reservoir was the most interesting. The purpose of the project was to provide flood control and low-flow

---

[1] Geologist, Retired, U.S. Army Corps of Engineers.

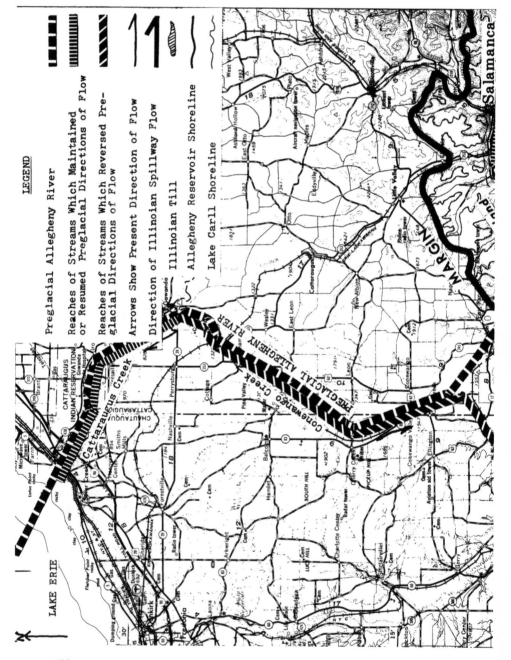

LEGEND

Preglacial Allegheny River

Reaches of Streams Which Maintained or Resumed Preglacial Directions of Flow

Reaches of Streams Which Reversed Preglacial Directions of Flow

Arrows Show Present Direction of Flow

Direction of Illinoian Spillway Flow

Illinoian Till

Allegheny Reservoir Shoreline

Lake Carll Shoreline

176

FIGURE 1 *Present and preglacial drainage important to Kinzua Dam. (Glacial margin in Pennsylvania from Shepps et al., 1959, and in New York from Muller, 1975.)*

regulation for the Allegheny valley and recreation on the 12,000-acre summer pool. A later addition to the project was the development of pumped storage hydroelectric power of about 400,000 kilowatts by public utilities unrelated to the government. The investment by the United States in this project is about $109 million of which the dam accounts for about $22.6 million. These were mainly 1958 to 1966 dollars. The reservoir benefits to September 1975, in less than 10 years of operation, were $256 million or 2.35 times the cost of the project. If the $166 million of benefits attributable to the prevention of damages by Hurricane Agnes of 1972 are excluded, the benefits in less than 10 yr of operation were about 84% of the total cost of the project. The value of the project has been amply demonstrated.

This project is described briefly in "Geology: Science and Profession," a brochure of the American Geological Institute, 1965, as an example of the value of applying geology to engineering. The geologic section in that description swings away from the thalweg to cross the buried shelf and incorrectly suggests that there is a cross-valley ridge in the bedrock surface of the valley bottom at the dam site.

Our steps in the planning, design, and construction were seriously affected by the necessity to accommodate our efforts to the conditions in the valley resulting from its complex history. The engineering problems arising from the geomorphic processes in the glacial foreland included the following:

1. Selection of damsite.
2. Selection of type of dam.
3. Permeability of the foundation of the embankment section of the dam.
4. Sources of fill for the earth embankment.
5. Sources of concrete aggregate.
6. Foundation and slope problems of highway and railroad relocations.

GEOMORPHIC PROCESSES
IN THE GLACIAL FORELAND

Reversal of Drainage

Prior to the first glaciation in the Pleistocene epoch, the drainage of northwestern Pennsylvania and southwestern New York flowed north to the St. Lawrence valley, as illustrated in Fig. 1. The streams in the Allegheny Valley area had cut deep, well-developed dendritic valleys, the direction of flow of which was demonstrated by the pattern of the valleys. "The best-laid schemes o' mice and men, Gang aft agley" (Burns, 1785). And in this case the intervener was the Illinoian or an earlier ice sheet that moved south and east, covering the downstream sections of these northwesterly flowing valleys until the water that entered the heads of the drainage systems from the southeast had no outlet to the northwest. The valleys were filled with water, creating a series of lakes, which rose in elevation until they overtopped the interstream divides that paralleled the courses of these streams. As suggested by A. L. Bloom (pers. comm., 1975), the blocking of the streams was not simultaneous. Because the streams flowed northwest and the ice moved south, those streams to the northeast were blocked first. Thus, a valley to the east filled with water before a valley to the west, so that the water which overtopped a western divide flowed down the west side of the divide either into a north-flowing ice-free stream or into a valley in which ponded water stood at a lower elevation than in the valley in which the water originated. The overtopped divide was eroded on its western slope by the down-rushing water. This process was repeated in several places as the waters escaped to the west to create a continuous stream subparallel to the ice front, which is now the Allegheny River. Because the surface of the Allegheny Plateau sloped to the north, the lowest points on the interstream divides south of the ice front were close to the ice front. The new Allegheny River valley was created by the erosion of the lowest saddles south of the ice front, and thus was close to the ice front. Eventually the Illinoian ice sheet, as mapped by Muller (1975, in press), crossed the new Allegheny Valley about 4 mi southeast of Steamburg, New York, to dam the by-then-reversed river flow, until a chain of spillways was developed through the low hills on the east side of the valley on the line and location of the spillway flow arrow on Fig. 1.

All this is pertinent, because the damsite for the Allegheny Reservoir is located very close to an ancient interstream divide, called by Muller (1963) the Kinzua Col, between tributaries of the preglacial Kinzua Creek on the northeast and the preglacial Conewango Creek on the west, both of which creeks were then north-flowing streams.

Lake Carll

I believe that it is entirely appropriate to designate as Lake Carll, the proglacial lake that filled the ancient Allegheny River valley to about elevation 2,020 and eventually drained westerly as its overflowing waters eroded the Kinzua Col, in respect to the distinguished geologist, John Franklin Carll, 1828–1904, who first recognized its existence about 100 yr ago. Its outside dimensions were northerly about 36 mi and easterly about 40 mi, but less than half this suggested area would have been water surface because of the numerous headlands projecting into the lake. That part of its shoreline on Fig. 1 shows that it was a many-fingered and embayed lake with numerous islands lying off the promontories. It would have been an ideal shoreline to which to run for protection in a sudden storm.

Erosion of the Allegheny Valley

The first stage in erosion of the valley after blockage of its outlet was the reduction of the cols by erosion of their downstream or western slopes. The crests of the cols retreated upstream, generally eastward into the glacial lakes, which were lowered as the cols were lowered. This process continued until the outwash gravels, being deposited upon the early lacustrine clays, advanced to cover the cols. In the foreland, Illinoian outwash gravels occur south of the Kinzua Col, and also south of those cols farther downstream. The Kinzua Col must have been eroded at least as low as elevation 1,400, the upper surface of the Illinoian outwash upstream of the col, and actually much lower as discussed later. In the warmer Sangamon epoch, the Allegheny River removed the outwash in the valley bottom except for scattered, infrequent remnants, and scoured the channel to a depth of more than 100 ft below the elevation of the present stream bed. The cols retreated some more during this epoch and became only knickpoints in the thalweg. This has been called the deep stage erosion.

The Wisconsin ice sheets were the sources of great quantities of outwash, which filled the main deep stage valley progressively from north to south, but did not spread laterally up the south or east side valleys. Those ice-free side valleys which were adjusted during Sangamon time to the gradient of the main valley were dammed by the outwash and became temporary lakes in which lacustrine sediments were deposited. The upper surface of the Wisconsin outwash is now covered with fine alluvial sediments except where the stream is flowing in or close to the top of the outwash.

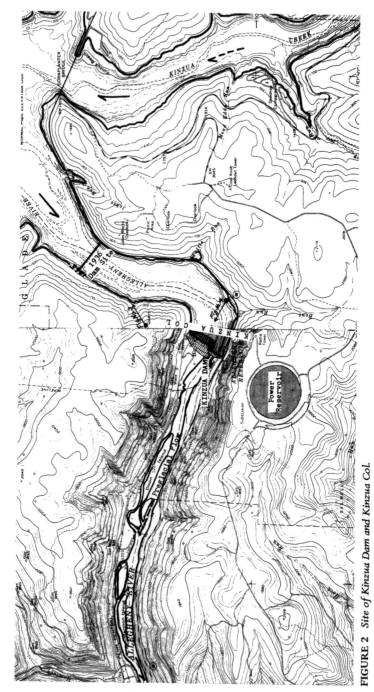

FIGURE 2   Site of Kinzua Dam and Kinzua Col.

Site of Kinzua Col

The Kinzua Col was first described by Carll (1880), who located it at Big Bend, then known as Great Bend, on the Allegheny River about 3 mi south of the old village of Kinzua. The col was about 1,000 feet upstream of Kinzua Dam, as shown in Figs. 1 and 2. Carll related its position to the massive sandstones that cap the highlands trending across the Allegheny in the vicinity of Big Bend. This high plateau-like ridge still forms a divide between tributaries to the present-day Kinzua Creek to the east and tributaries that join the Allegheny in the vicinity of Warren, Pennsylvania, some 9 mi to the west.

Carll offered three valid arguments for the existence of the divide at Big Bend:

(1) The narrowness of Great Bend cut as compared with the valleys both above and below it. If the ancient current which excavated the deep and broad valleys above and below passed through the bend, why this contraction of the valley at this point where there is no conspicuous change in structure to cause it?

(2) Northeast of the bend the lateral streams come in from a southerly direction corresponding with a northerly flow of the Kinzua (now a part of the Allegheny), and the contours of the hills at the intersection with the main valley, point in the same direction.

(3) West of the bend, features of a similar character indicate a westerly flow for the drainage in harmony with the present current.

The Kinzua tributary whose divide was overtopped and breached was the western branch of Bent Run, now a steep gully entering the Allegheny at the bend. The upper branches of Bent Run on the plateau are similar in orientation to those of its eastern neighbor, Dewdrop Run, another tributary to ancient Kinzua Creek. Bent Run probably reached the old valley floor of Kinzua Creek at the upstream end of the present gorge near the north end of Lenards Island, now submerged beneath the reservoir. The west-flowing stream on the west of the divide was a tributary of, and parallel to, Browns Run [called Hooks Run on Carll's (1880) Plate 2, not the same Hooks Run as on Fig. 2], which it joined along with Dutchman Creek near the eastern end of Warren. Figure 3 is a northerly view upstream of the gorge on the upstream side of Kinzua Col from the pleateau point on the inside of Big Bend. Figure 4 is an aerial oblique looking downstream through the damsite from above the site of the Kinzua Col.

Erosion of Kinzua Col

The Kinzua Col probably had a summit elevation comparable to the low divides on the plateau in the vicinity of Big Bend (Fig. 2). There are several narrow saddles between elevations 2,000 and 2,020 with widths of about 500 ft between opposing 2,000-ft contours. Assuming that there has been only slight lowering of these saddles since the Illinoian glaciation, the Kinzua Col may have been near elevation 2,020 and suffered a total of nearly 940 ft of erosion to reach the Sangamon thalweg of elevation 1,087, now buried beneath the Wisconsin outwash at the damsite. Not all this erosion occurred in a continuous process, because it was interrupted by the deposition of the Illinoian outwash gravels. Evidence is not clear as to the depth of erosion prior to these gravels, none of which are identifiable at the damsite. However, the bedrock shelf on the left bank of the river at the site (Fig. 5B), lying between elevations 1,190 and 1,170, is on the outside of the Big Bend and has no lithologic controls defining its limiting elevations. It may

FIGURE 3   *View north, upstream, from Kinzua Col into preglacial Kinzua Valley July 1963.*

represent a fluvial erosion surface developed during a period when down cutting was less
active than lateral planation. This would be before the Illinoian outwash, which entered
the valley mainly north of the state line, had reached this far downstream, and therefore
obviously before the deep stage erosion of the Sangamon. Thus, the shelf is probably
middle to late Illinoian in age. Its upper, landward edge is close to the 1,193 elevation of
the shallower buried channel in bedrock at the western side of Kinzua valley beneath the
Cornplanter Bridge (Fig. 6). I do not believe that the similarity of these two elevations is
coincidental but, rather, they represent a gradient of ancient Illinoian Kinzua Creek,
reasonably adjusted to the bedrock channel of the ancient Illinoian Allegheny River,
which join only about 1 mile downstream from the Cornplanter Bridge.

The Sangamon erosional interval resulted in reduction in elevation of the thalweg at
the damsite from between elevations 1,190 and 1,170 down to elevations 1,087 or
slightly lower (Fig. 5B). As this erosion progressed, the knickpoint, or remnant of the col,
regressed at a nonuniform gradient. There is a sharp rise of more than 40 ft in the thalweg
to a point opposite the crest of the nose inside of Big Bend. Could this represent the
upstream side of an old plunge pool almost beneath the position of the Kinzua Col? The
location of the remnant divide in the thalweg of the bedrock surface cannot be fixed with
as much certainty as can that of the Kinzua Col. In the vicinity of the 1936 damsite (Fig.
2), the bedrock surface was well established by many drive sample and core borings into

FIGURE 4 *Aerial view west, downstream, through Kinzua damsite from above Kinzua Col (initial stage of embankment nearly completed), Sept. 12, 1961.*

rock at about elevation 1,133 (Fig. 5A). This is only a rise of 13 ft in 1 mi and suggests that headward erosion was not as rapid as downstream near the nose. Thus, the remnant divide was somewhere near the 1936 damsite, and retreat to the Kinzua valley was nearly completed in Sangamon time. Still, the bedrock floor was a strongly sloping thalweg down which the Wisconsin meltwater must have coursed at first in a tumbling, turbulent torrent, but probably in not nearly as spectacular display as did the ponded waters of the Illinoian time, which roared down the west slope of the col to carve the gorge and reduce the bedrock divide to transportable debris.

History of Kinzua Valley

Figure 6 provides a basis for speculation on the history of the Kinzua Valley. The double channels in bedrock and the materials overlying the bedrock surface, even if classified according to the engineering soils classification of the U.S. Bureau of Public Roads, furnish some clues as to episodes in the history of the valley.

The present channel of Kinzua Creek lies at the foot of the west valley wall, where it has been scouring the bedrock during the Holocene. Further to the east, the valley bottom is a broad floodplain about 40 ft beneath which is a 300-ft-wide channel in bedrock at about elevation 1,193. Aside from the upper few feet of topsoil and silty soils, the old channel is filled with gravelly and sandy soils carrying less than 35% passing the No. 200 sieve. There are a few feet of fine soil interbedded with the coarser soil, but the fill fits the general category of materials to be expected in outwash soils. Thus, this channel erosion seems to be Illinoian in age, now filled with later Illinoian outwash. It is believed that this channel is correlative with the previously mentioned bedrock shelf at

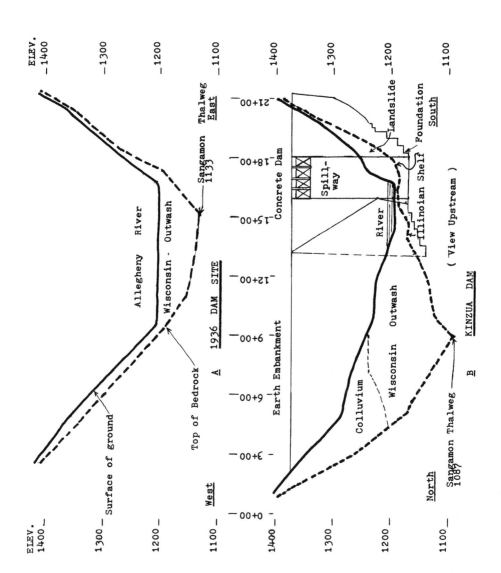

ELEV.
—1400
—1300
—1200

Sangamon
Thalweg
East
—1100
1133

Allegheny   River

Wisconsin - Outwash

Top of Bedrock

A   1936 DAM SITE

Surface of ground

West

ELEV.
1400 —

1300 — Surface of ground

1200 —

1100 —

Earth Embankment

Concrete Dam

Spill-
way

River

Landslide
—1200

Foundation
South

Illinoian Shelf

—1100

Colluvium

Wisconsin   Outwash

North

Sangamon Thalweg
1087

(View Upstream)

B   KINZUA DAM

—1400

—1300

—1200

—1100

184

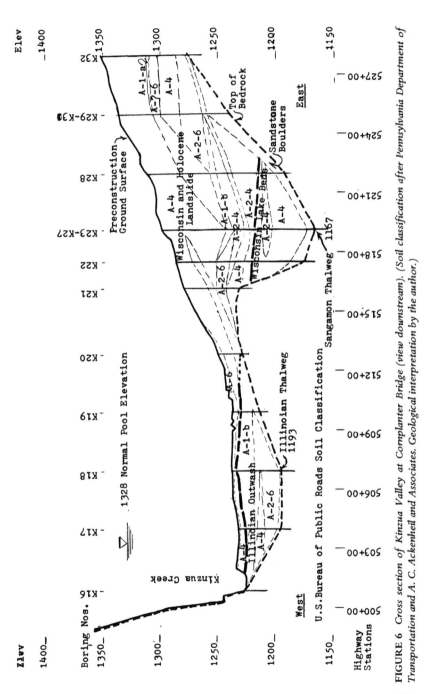

FIGURE 6 Cross section of Kinzua Valley at Cornplanter Bridge (view downstream). (Soil classification after Pennsylvania Department of Transportation and A. C. Ackenheil and Associates. Geological interpretation by the author.)

the damsite. The greater part of the Illinoian outwash, which may have accumulated to elevation 1,420 in this vicinity, as suggested by the terrace at the Old State Road picnic area on the east side of the reservoir shown in the northeast corner of Fig. 2, was eroded during Sangamon time. Kinzua Creek, behaving as a superposed stream on the outwash, cut its Sangamon channel on the east side of its valley down to about elevation 1,167. This is about 34 ft above the elevation of the Sangamon channel at the 1936 damsite, whereas the present Kinzua Creek here is about 17 ft above the river bottom at the 1936 damsite. Even if the Sangamon channel of Kinzua Creek was not fully adjusted to the gradient of the Sangamon channel of the Allegheny River, there is no evidence for a major waterfall at the junction of the two streams.

The erosion of the Sangamon channel continued downward to about elevation 1,175 until the Wisconsin outwash, mainly entering the valley north of the state line, advancing down the Allegheny Valley, reached an elevation such that Kinzua Creek was blocked from free flow and could no longer transport clayey or silty sediments. Deposition of this fine material continued with minor interruption until the accumulation reached about elevation 1,220. This is about the elevation of the top of the outwash in the Allegheny Valley in the vicinity of the mouth of Kinzua Creek. The top of the Wisconsin outwash, where I have encountered it in the Allegheny Valley in this area and upstream in the vicinity of Salamanca, New York, occurs as a clearly defined separation between coarse sandy and silty gravels and the overlying fine alluvium. Such a distinction occurs around the mouth of the Kinzua Creek at about elevation 1,220, so that the top elevation of Wisconsin outwash at 1,220 is reasonably reliable. The great thickness of variably dipping soil layers, shown in the exaggeratedly scaled Fig. 6, is believed to be a mass of colluvium deposited during the Wisconsin and the Holocene and is analogous to the less-well-layered landslide that displaced the Allegheny River toward the north bank at the site of Kinzua Dam (Fig. 5B).

Glacial and Interglacial Deposition

The glacial foreland at the time of the advance of the Illinoian or earlier ice and the blockage of the old north-flowing valley became an area of sedimentation important to the future development of the Allegheny River. We might first consider the earliest glacially related sediments, those in the proglacial lake in the ancient Allegheny Valley. This lake received alluvial sediments, which, downstream from the deltas where the coarse materials were dropped, were clays and silts. These fines were deposited as varved lacustrine sediments over the floor of the lake, the ancient valley floor. Their presence is not known except in a few places, because the old valley floor lies well below the present valley floor in the northern end of the reservoir. They were uncovered at the foot of the north slope of the valley across from Red House (Fig. 1) with a top elevation of about 1,380 and a thickness of about 35 ft. About 10 mi farther southwest near the state line and on the east side of the valley at the Riverview–Corydon Cemetery, a water well, drilled from the surface of the Illinoian outwash terrace at about elevation 1,400, penetrated about 100 ft of gray clay between approximate elevations 1,200 and 1,300 before reaching water-bearing sand and gravel at a depth of over 200 ft. The thickness of the gray clay, although not exact, indicates that the proglacial lake received sediments for a long period of time before being overrun by Illinoian outwash. There is no indication that this clay is till. Muller (1975, plate 2) shows no till closer than 2 mi to the north of this well.

The glacially dammed sections of the preglacial valleys were choked with tills and other glacial deposits. When the ice margin retreated northward at the end of the Illinoian glaciation, the streams could not resume the interrupted northward flow, because their valleys had been filled to an elevation above that of the valley floor eroded during the Illinoian by the newly developed Allegheny–Ohio River. During the Sangamon interglacial epoch, erosion continued in the newly formed valley to a depth of several hundred feet below the surface of the outwash gravel of the Illinoian. Most of the outwash gravels were eroded, and only patches of them remain on the hillside and on protected rock-based terraces in the tributary streams. In the steep-sloped valley walls of the river systems, these terraces have been the sites for farming and manufacturing, and villages, towns, and cities.

During deepening of the valleys, soil blankets of lean sandy clays were formed on the hillsides where, because of numerous perched water tables, local zones of instability developed.

The maximum relief occurred during the time of Sangamon entrenchment, and the bedrock plateau and valley walls reacted to the resulting increased stress. In the plateau rocks, vertical joints opened, and blocks of sandstone glided slowly toward the valleys where they cantilevered over far enough to drift down the slopes. Along the sides of the ravines the blocks moved toward a common line so then streams of rock were formed in the ravines. The stress relief in the valley bottom by erosion of the valley allowed the rock in the valley bottom to spring upward with the development of small thrust faults. On the valley walls the side rock lost its restraint through the action of the valley-bottom rock and moved laterally, creating zones of open joints along the valley walls. Small, thin zones of gouge developed at the bases of the in situ blocks as they moved toward the river. Ferguson (1967) has described this process excellently.

With the advent of another glacier, the Wisconsin, with its boundary close on the boundary of the Illinoian ice sheet (Shepps et al., 1959; White, 1969), the action in the valleys was again one in which ice stood to the northwest and water entering from the southeast moved southwest parallel to the ice front. By this time the valley system was already established and had been deepened during the Sangamon interval. Thus, the Wisconsin outwash gravels formed local barriers to the movement of alluvial sediments down one valley, and quite probably similar events occurred at the mouth of Willow Creek near the state line and Sugar Run, farther south toward Kinzua Creek. Eventually, the outwash gravels filled the Allegheny Valley to a depth of from 75 to 100 ft above the Sangamon channel bottom. The gravel, having been deposited in an aggrading stream, is crossbedded, channeled, and quite variable in texture. Borings within the valley bottom have shown a tremendous variation in the types of materials. Much is well sorted (poorly graded), being essentially gravel; the interstices may be filled with fine to medium sand. This openwork or gap-graded gravel is extremely pervious. Because of the poor distribution of grain sizes, the sands are poorly packed and free to move. Thus, piping (selective transport of certain sizes) can easily occur, leaving the gravels unsupported and subject to reduction in gross volume and consolidation with resulting settlement.

Periglacial effects occurred on the valley walls. Colluvial materials must have been formed during the Wisconsin interval as well as during the Holocene, because the soil blanket is so great that construction of such thickness in the short period since the end of the Wisconsin seems improbable. The lean sandy clays, subjected to the effect of the perched water tables, became unstable and moved downslope to create slump slides of considerable volume. Backward rotation of segments and clearly marked scarps are characteristic of these landslides. Block slides of soil appeared as long sheets dipping parallel to the hillslope and extending upward several hundred feet from the valley floor. Rock glaciers were developed in the ravines, probably during Wisconsinan time, but are masked by the thick vegetation. The toes of some of these slides rest on outwash gravels and some on the buried Illinoian bedrock terrace near the elevation of the present streambed, and thus are Holocene.

During the Holocene, downslope activity and deposition continued. Alluvial sands and silts were spread across the top of the outwash. We see this alluvium as islands in the braided stream, which drift from year to year in a downstream direction as the head is eroded and the sediment is deposited at the foot of the island.

Summary of the Geomorphic Processes

As a result of these geomorphic processes, the valley in which the Allegheny Reservoir was to be constructed had certain characteristics, some of which we knew before we commenced the project, all of which we know about now. The damsite was located at a divide that had been eroded as a consequence of reversal of stream drainage. The bedrock floor of the valley was 113 ft below the stream bed at the damsite, nearly 40 ft lower than at the eroded location of the Kinzua Col. Outwash gravels filled the valley to stream bed, above which were finer alluvial sediments in the valley bottom and colluvium on the hillslopes. The erosion of the valley had occurred in three stages, two of which were glacial times, which were separated by a warmer deep erosion stage. The valley walls showed no terraces at the damsite, but one bedrock terrace was buried near stream-bed elevation. A few remnant gravel terraces were in the reservoir, covering the first lacustrine sediments deposited prior to the reduction of the old divide early in the glacial history of the area. Some of these lacustrine sediments had been distorted and broken by landslide activity, probably during the maximum relief of Sangamon time. The lower side walls of the valleys were blanketed with unstable colluvium, chiefly lean sandy stony clays in which well-developed slickensided planes of failure underlay the landslide masses of Wisconsinan to Holocene age.

ENGINEERING PROBLEMS ARISING FROM GEOMORPHIC PROCESSES

Selection of Site and Type of Dam Within the Kinzua Col

The gorge downstream of Kinzua, Pennsylvania, was a textbook illustration of a site for a dam, with the valley widening upstream to provide a reservoir of much greater width than at the damsite. The dam would be the cork in the bottle, as the simile goes. The problem was simple: find the narrowest part of the gorge and put the proposed 150-ft-high concrete dam there. Plane table surveys and visual examination formed the basis for the judgment that the middle of three lines of proposed 3 in. or NX core borings to 50 ft into bedrock across the narrowest section of the valley would be the approximate location of the proposed dam in 1936, as shown on Fig. 2. The transverse profiles based on the borings of 1936–1937 revealed that bedrock was about 75 ft deep below stream bed, and the valley floor prior to deposition of the Wisconsin outwash gravels was relatively flat in cross section, as shown in Fig. 5A. The dam had to be increased in planned height from an apparent 150 ft above stream bed to about 225 ft above bedrock. With the triangular up- and downstream section of a concrete dam and an optimistic base width of 0.75 times height, the rise in height from 150 to 225 ft increased the quantity of concrete by more than 115%, thus doubling the cost of the proposed dam.

At this time (1936–1937), the Commonwealth of Pennsylvania and the State of New York would have been required by the Flood Control Act of 1936 to bear the cost of all lands, damages, and rights of way including relocations. A single track line of the Pennsylvania Railroad (Penn Central) ran the length of the reservoir. The main line of the Erie–Lackawanna Railroad crossed the upper end of the reservoir for several miles in New York State. There were highways, several villages, some excellent valley-bottom farmland, and the Cornplanter Indian Reservation in Pennsylvania and a part of the Seneca Indian Reservation in New York State within the reservoir area. All these factors plus the greatly increased cost of the dam made construction of the Allegheny Reservoir a very unattractive project at that time. It was postponed.

Funds that became available in the summer of 1938 were used to continue surveys and to drill a series of borings at quarter-mile intervals downstream from the original 1936 damsite for the purpose of identifying the elevation of bedrock beneath the stream bed. It was hoped that the actual preglacial divide might exist at a higher rock elevation somewhere farther downstream in the narrow valley. Where a boring encountered rock at a higher elevation than that at the 1936 site, laterally offset borings were drilled, looking for deeper rock, which was found in all such cases. So it was established in 1938 that the position of the 1936 damsite was at about the location of the highest rock surface in the narrow reach of the valley. It seems probable now that the 1936 location represents the position of the recessive remnant waterfall and later rapids originally located at the topographic divide between the Kinzua and Conewango drainages.

Further fieldwork was suspended until 1955, when the right of the United States to Seneca Nation lands in New York State for reservoir purposes had been established by legislative action of the U.S. Congress and successfully defended in the courts up to and including the U.S. Supreme Court.

Between 1938 and 1955, we had designed and been associated with the construction of several cuts 200 to 300 ft high in rocks weaker than those at the site. Restudy of the site suggested the possibility of constructing an earth embankment dam with a spillway cut across the nose within the bend. This would be the classic side channel spillway scheme with which the Pittsburgh District of the U.S. Army Corps of Engineers was quite familiar. The chief difficulty at this site would be the extreme height of the landside of the spillway cut, about 600 ft above the spillway floor, more than twice the height of any such cut in the Pittsburgh District. After preliminary study and estimate, this scheme was dropped as being too expensive and not a desirable undertaking.

In 1938, borings had showed also that a little farther downstream the bedrock was shallow on the south (left) side of the river, so there was hope for a less expensive type of dam there than at the 1936 site.

To establish the depth and extent of a rock shelf in the vicinity of Big Bend, seismic investigations were performed in 1955 and supplemented by drive sample and core borings in 1956. These investigations found a bedrock shelf about 450 ft wide on the south (left) side of the valley, but nearly half the shelf lay beneath the south valley wall. The shelf sloped riverward from about stream-bed elevation 1,190 to elevation 1,170

(Fig. 5B). The lowest point on the shelf appeared to be about 37 ft higher than the surface of bedrock at the 1936 site. The main channel of the stream during Sangamon time was located beneath the broad piedmont slope on the north (right) side of the river, almost midway between the valley walls.

The narrowness of the stream at Big Bend was the result of a large ancient landslide on the south bank, which had slumped over the shelf and buried almost half of it (Figs. 4 and 5B). The borings penetrating through the landslide, which was still creeping as indicated by the irregularities of the surface and grade of State Route 59 passing through the area, defined the surface of bedrock and the subparallel surface of sound rock that would govern the maximum elevation of the foundation of a concrete gravity dam. The instability of the landslide made it mandatory to assume that any dam at that site would require the removal of the creeping slide prior to or during excavation for the dam, because the slide would move if any of its toe were removed to permit the required work in the river bed.

The assumption of removal of the landslide to its base on the bedrock surface promptly opened the possibility of shifting the river southward to the foot of the newly to be exposed bedrock hillside. This permitted the alignment of the south side of the spillway of a concrete gravity dam with the foot of the bedrock hillside. And this in turn pulled the north end of the concrete abutment toward the river and upward and away from the deep, buried Sangamon channel. This raised the foundation and greatly reduced the quantity of concrete, which varies exponentially with the height of the triangular section of a concrete gravity dam. The closure to the north valley wall would be provided by an earth embankment, giving the basis for the term "combination" to this dam, it being a combination of earth and concrete sections.

Comparative estimates of a combination dam at this site with a straight concrete dam at the 1936 site, where the valley is too narrow to construct a combination dam, showed anticipated savings of more than $4 million (1957 dollars) for the combination dam at this site. Actual cost of construction substantiated the estimated savings.

If one visits the site, one will note that the south side of the valley next to the highway is almost all bare rock, which was stripped of soil during excavation of the landslide. If one looks downstream, the shifted river swings away from the south valley wall to join with the preconstruction channel downstream from the site, showing the depth to which the spillway has been embedded into the formerly landslide covered south slope. Not visible, however, is the sloping rock terrace, high above the old Sangamon channel, into which the stepped foundations of the concrete section of the dam have been founded. It was a combination of this terrace, the adequate width of the valley, and the availability of nearby natural construction materials (earth and rock) that made this site desirable.

Our interpretation of the subsurface conditions for the concrete gravity section were correct except for minor adjustments resulting from the opening of some vertical joints subparallel to the hillside by release of stress, and similarly caused by horizontal fractures in the valley bottom, which necessitated the lowering of the foundation elevation of one monolith by a few feet. An early concern in the subsurface investigations for design was the presence of steep bedding in some very well cemented siltstone found occasionally in the otherwise horizontally bedded shales of the foundation. Eventually, similar bedding was found in outcrop and identified as convolute bedding, known locally as flow rolls. These were found in profusion during construction near the elevation of the spillway floor and beneath the riverward end of the rockfill upstream of the concrete section, as big as 10 ft long with a cross section of about 20 sq ft.

Cutoff Wall

The earth embankment extending from the spillway northward to the north valley wall was to be founded over the deepest part of the outwash gravel. The total depth of alluvial, colluvial, and outwash soils amounted to as much as 175 ft, because the thalweg of the Sangamon valley was located beneath the piedmont slope of the north side of the valley. Subsurface investigations by borings had indicated that the outwash materials were variably stratified, with layers of silt-choked sandy gravel interspersed with layers of much more permeable sands and gravels. The exact nature of these materials and their final treatment was to be determined when exposed by excavation during construction of the northern part of the concrete part of the dam. When exposed, it was apparent that some layers were gap graded, and were openwork gravels that might permit piping and removal of the fines after development of the reservoir head. The blanket of impermeable soil that was to be spread and compacted over the north side of the valley was not long enough to reduce the seepage velocity through the highly permeable layers to a speed less than a piping velocity. It was apparent that a cutoff structure extending from the riverward side of the upstream cutoff wall landward to above full pool, or to impervious materials, would be required to prevent piping through the foundation of the dam. Consequently, studies were undertaken to establish the most desirable type of cutoff structure. These led to the selection of a concrete cutoff wall.

The 1,066-ft-long, 2.5-ft-thick concrete cutoff wall built by the ICOS method extended from the top of the average ground surface to a minimum embedment of 2 ft into bedrock, for a total depth ranging from 53 to 175 ft. It extended from the upstream cutoff wall landward to the north abutment some 60 ft landward of the landward end of the pervious zone. It was topped with a cover of impervious material, which was continuous with the impervious blanket extending beneath the upstream section of the embankment to tie into the impervious section of the dam.

Earth Fill

The economical construction of an earth embankment requires that the earth-fill material be obtainable within a reasonable distance of the site of the embankment. The cost of the 3 million cu yd of earth in this embankment had a major effect on project cost. Although there was a large quantity of earth suitable for fill on the lower slopes of the nose within the bend of the river and adjacent to the embankment site, experience had demonstrated that removal of such earth blanket from the underlying bedrock would expose open fractures to reservoir water, and provide channels for leakage and potential piping of the soil foundation of the embankment, if not the embankment itself. Thus, these adjacent materials were not available for fill. The next nearest source of embankment material became the colluvial soils on the opposite side of the stream upstream from the dam site. This borrow pit would lie down the landslide-prone slope from the relocation of State Route 59, thus requiring that the upper limit of the borrow pit be located outside of the stable slope line from the toe or riverward side of the relocation of the highway. The soils were lean sandy clay with some rock fragments, and proved to be entirely satisfactory for fill for the impervious section and a large part of the random section of the embankment. However, the quantity was insufficient to serve the full needs of the dam and an additional colluvial source was developed about 2 mi upstream on the west or right bank of the river, well beyond the limit of concern for percolation of reservoir water through the bedrock of the nose.

Fill for Upstream Cutoff

To keep water from the river channel from passing beneath the upstream blanket and through the embankment foundation, an impervious cutoff along the river edge of the blanket and extending to bedrock was to be constructed in a trench along the north side of the river. Unwatering of this trench would increase the cost of the cutoff immensely; therefore, a search was made for fill that could be placed under water and would become impervious in place. For this, two sources of fine alluvial sand were investigated: Dixon Island downstream of the dam and other islands farther away but upstream of the dam. The upstream materials were placed under water and have proved entirely satisfactory.

Sources of Concrete Aggregate

The dams contain about 450,000 cu yd of concrete, which clearly shows that concrete aggregate was of major importance in the cost of the dam. Over the decade before construction, the sources of concrete aggregate in the Pittsburgh District had been thoroughly investigated for other projects, with the result that most of the river gravels had been tested and relegated by the Corps of Engineers for use as aggregate in concrete that was to be buried and not exposed to weathering. The undesirable materials were the weathered softer sandstone and shale particles and, to a lesser degree, a very small percentage of potentially reactive chert. In the years immediately before construction, the producers had modified their aggregate processing plants to reduce the quantity of softer particles, and thus upgraded the product to the point where it was approved for general use in structures such as Kinzua Dam. Accordingly, the local gravels, Wisconsin outwash, were reinvestigated, and samples of these gravels from commercial sources as well as at the damsite were tested and approved up to 2.5-in top size. With a depth of 90 ft of outwash both at and immediately downstream from the site from which acceptable concrete aggregate up to 2.5 in in size could be produced, it was believed that the low bidder for the dam, who had shown great interest in this late testing and approval, would open a pit, build a processing plant adjacent to the damsite, and import only the 3- to 6-in size aggregate. Then there occurred one of those happenings which make engineering geology a unique endeavor. The contractor made a deal with the Pennsylvania Railroad to convey all the coarse aggregate at a very low freight rate 165 mi to the damsite from a limestone quarry at Pleasant Gap, Pennsylvania. Thus, *no* gravel was obtained from the nearby outwash deposits for use as concrete aggregate.

Foundation and Slope Problems
of Railroad and Highway Relocations

Although about 37 mi of railroad relocations were required by the project, most were outside the reservoir area, being modifications to existing single-track lines of the Pennsylvania Railroad (Penn Central later) to fit them for increased traffic resulting from abandonment of the line through the reservoir. The majority of the work was relocation of the Erie–Lackawanna Railroad for about 10 mi along the north shore of the reservoir between Steamburg and Salamanca, New York (Fig. 1). The highway relocations ranged in class from township roads to primary state highway, like N.Y. 17 on the south side of and across the reservoir from the Erie–Lackawanna relocation, likewise between Steamburg and Salamanca. The foundation conditions in the valley were generally excellent because of the thin recent alluvium, only a few feet thick, and the strength and stability of the underlying outwash sands and gravels for embankment foundations. Major bridges were founded on piles. The major geologic problems were related to conditions along the valley walls.

The relocation of the Erie–Lackawanna Railroad encountered two major problems. On the hillside or valley wall between Meetinghouse Run and Sunfish Run, across the river from the old hamlet of Red House, the new railroad line was projected in a 25-ft-deep cut through the Illinoian outwash gravels. Above the railroad, a township road was to be rebuilt in a side hill cut of about 55-ft depth in the gravel. Both cuts were being made at the same time, with the upper cut much deeper and further along toward completion than the railroad cut below. The shallow lower cut had exposed only a few feet of clay beneath the Illinoian gravel when the riverward portion of the high road for a distance of several hundred feet dropped a few inches, with a clearly defined crack marking the landside of the slumped mass. The clays were somewhat varved, with water content slightly above the plastic limit, plastic index of 27% and liquid limit of 51%. The preconstruction borings showed the clay to occur between elevations 1,345 and 1,380, although no boring penetrated the full 35 ft of indicated thickness. They rested on brown stony silt or brown silty sandy gravel. Geologically, the section seems to indicate an alluvium, overlain by lacustrine clay and in turn by outwash gravels.

In excavations for drainage structures across the railroad there, the clay displayed highly contorted bedding, in which the strike of the bedding and the fold axes therein were subparallel to the trend of the hill. The distortion of the bedding was far more intense than the few inches of motion could account for. I believe that a fossil slide had been reactivated. The original slide had probably occurred during the Sangamon time when the river had removed the outwash gravels and lacustrine clays lying riverward of the line of the relocation.

Prior to the design of the relocation, 1939 air photos were examined. They showed on the steep slope of Jimmerson Hill between Sunfish Run and Sawmill Run, about 1.5 mi east of the above slide and across from Red House, several landslides that had removed the trees earlier and exposed the soil in the forest. Jimmerson Hill is composed of nearly horizontally bedded Devonian shale, which has a slight dip toward the river, overlain by a thin layer of colluvium through which numerous springs issue in such volume and continuity that swampy vegetation grows well locally on the steep slope. The colluvium had been moving for a great period of time, and the railroad had endeavored to stabilize this distressing and dangerous condition by the insertion of horizontal drainage pipes into the hill and the construction of drainage ditches on the hillside. A fence with electric signals had been built at the foot of the slope. It was quite clear that it would be unwise to make any cuts into Jimmerson Hill.

Accordingly, a temporary bypass or shoofly was constructed on a fill in the river. Eastbound traffic was then diverted to the shoofly and westbound traffic rerouted to the old eastbound track. Thereafter, a fill was begun on the vacated westbound or hillside subgrade and continued up the hill until the new subgrade elevation, above reservoir-full elevation, was reached. Track was laid on this fill. The westbound traffic was diverted to the new track and the fill continued over the old eastbound subgrade. Thus, when construction was finished, both tracks were on a higher level on fill that was based on the old roadbed and the new fill on the solid Wisconsin outwash gravels. The toe of the hill had not been removed and the stability of the hill had not been decreased.

Farther to the south, relocation of highways involved both cut and fill to provide perimeter roads, much of which were on the valley slopes around the edge of the reservoir. This construction required the raising of roadways to about elevation 1,370, well above full-pool elevation. Because of the steepness of the slopes, in some cases fills were constructed from river level up to proposed grades, but in other cases cuts were constructed through the hillsides. Landslides occurred not from failure of the foundation of the fills resting on the gravels in the valley bottom, but from failure of fill probably resting on lake clays in the valley bottom or on colluvium on the lower hillslopes. Many slope failures occurred, however, because the toes of then quiescent slides above the highway were undercut and the unsupported slide masses moved downslope.

Our first encounter with such problems came with the relocation of Pennsylvania State Route 59 as it passed the damsite. This highway was in the valley through the damsite and was raised to permit uninterrupted traffic between Warren and Bradford, Pennsylvania, prior to any construction on the left bank for the dam. The relocation at the dam was mainly in a deep rock cut where tectonically generated joints were parallel to the hillside, and at right angles to the hillside around Big Bend toward the north of the dam. The cuts were accomplished without much difficulty, except for the falling of blocks of road bounded by joints parallel to the roadway.

Cuts in soil that had accumulated at the toe or lower slopes of the hillsides caused considerable difficulty. Hillsides up to the point where the slope of the hill paralleled the slope of the bedrock surface were quite troublesome, because their flattened slope was the result of the accumulation of soil by soil slides. Above that point, the soil thickness is not great enough and drainage is strong enough so that the thin soils remained in place. In the soil, each slump slide, as well as the block slides, bottomed on gray, impervious, soft clay layers, which are effective groundwater barriers of low shear strength. The removal of even part of the supporting toes of the ancient landslides was sufficient to start them moving. In many cases movement continued long after the road had been paved, resulting in boulders, soil, and trees falling into the cuts and interrupting traffic.

It was clear very quickly that the materials in the ancient slides with their underlying weak shear planes were unsuitable for the foundations of fills by which the road would have to cross the ravines crossing the line of relocation. Accordingly, the slides were excavated to the bedrock surface, which was benched, and the fills built from this stable foundation up to grade. In some localities, stability in the fills was accomplished by building buttress fills at the toe of slope.

SUMMARY

The Kinzua Dam was sited at the location of a preglacial col in a well-developed, dendritic drainage system in the Allegheny Plateau of Northwestern Pennsylvania. The Illinoian ice sheet blocked the northwestern course of the ancient Allegheny River, ponding its waters in front of the ice in the glacial foreland until they overtopped the col and eroded a new valley to the ice-free southwest to form the present Allegheny Valley. Lacustrine, outwash, colluvial, and alluvial soils were deposited in the valley during Illinoian and Wisconsinan times. However, during the intervening Sangamon interglacial period, the valley was cut to more than 100 ft below the elevation of the present thalweg.

This complex geomorphic history made selection of site and type of structure for Kinzua Dam and the highway and railroad relocations for Allegheny Reservoir difficult and quite different from those usually encountered in the unglaciated Allegheny Plateau.

The Kinzua Dam and the Allegheny Reservoir Project were unique in western Pennsylvania and southern New York because of the complex geomorphic history of the area. Project planning, design, and construction spread over a long period from 1936 to 1966, during which time the art of dam and reservoir building progressed tremendously. We took advantage of this progress, as well as contributing to it. The serious application of geology to the investigation of the damsite resulted in the saving of more than $4 million in the cost of dam construction alone. It is probable that with our greater present knowledge of geomorphology we could have added more savings in the reservoir area, if the project were being undertaken now.

ACKNOWLEDGMENTS

The people of the Pittsburgh District, U.S. Army Corps of Engineers, have supported the efforts of their geologists ever since the beginning of the geologic studies for this project in 1936, and to them I am deeply indebted. The continuous help of Harry F. Ferguson during the studies and in reviewing this paper, and more recently of Frank Pehr, who found the old maps and data, is much appreciated. The kindness of Dr. Ernest H. Muller and Dr. James F. Davis in making available the map of the glacial geology of the Niagara Sheet of the Geological Map of New York State facilitated the preparation of Fig. 1, which has been reviewed by Dr. Muller. Dr. Arthur L. Bloom, by his review, continued my training in geomorphology.

REFERENCES

American Geological Institute. 1965. *Geology; Science and Profession:* The Institute, Washington, D.C.

Burns, R. 1785. To a Mouse: in *Complete Poetical Works of Robert Burns,* Cambridge Ed., 1897, Houghton Mifflin, Boston, p. 31–32.

Butts, C. 1910. Warren folio: *U.S. Geol. Survey Folio 172.*

Carll, J. F. 1880. The Geology of the Oil Regions of Warren, Venango, Clarion and Butler Counties: *Second Geological Survey of Pennsylvania: 1875–1879.* III, Report of Progress, p. 352–353.

Ferguson, H. F. 1967. Valley stress relief in the Allegheny Plateau: *Assoc. Eng. Geologists Bull.,* v. 4, p. 63–71.

Leggette, R. M. 1936. Ground water in northwestern Pennsylvania: *Pennsylvania Geol. Survey 4 Ser. Bull. W3,* 215 p.

Leverett, F. 1902. Glacial formations and drainage features of the Erie and Ohio Basins: *U.S. Geol. Survey Mono. XLI,* p. 127–132.

Muller, E. H. 1963. Geology of Chautauqua County, New York, Part II, Pleistocene geology: *N.Y. State Museum Sci. Serv. Bull. 392,* 60 p.

———. 1975. Pleistocene geology of Cattaraugus County, N.Y.: in Irving H. Tesmer, *Geology of Cattaraugus County, N.Y.,* v. 27, Buffalo Society of Natural Sciences, Buffalo, N.Y.

———. 1975 (in press). Surficial geology of the Niagara sheet: *N.Y. State Museum Sci. Serv.*

Shepps, V. C., White, C. W., Droeste, J. B., and Sitler, R. F. 1959. Glacial geology of northwestern Pennsylvania: *Pennsylvania Geol. Survey 4 Ser. Bull. G-32,* 59+ p.

White, G. W. 1969. Pleistocene deposits of the northwestern Allegheny Plateau, U.S.A.: *Quart. Jour. Geol. Soc. London,* v. 124 (for 1968), p. 131–157.

APPENDIX 1: SOURCES USED
IN COMPILATION OF BASES FOR MAPS
AND FIGURES OR UNCITED IN TEXT

Pennsylvania Department of Transportation (formerly Department of Highways), undated, Traffic Route 59, Legislative Route 209, Sec. 5

> *Sources of Figure 6:*
> Index map, Sheet 2 of 22
> Warren County Bridge over Kinzua Creek core borings S-5314, Sheet 18 of 18
> Untitled Soil Profile and Borings Sta. 499+00 to 542+00, Sheets 2–6 incl. of 12

U.S. Army Corps of Engineers, Pittsburgh District

> *Sources of Figure 5:*

1936	Allegheny River Dam Site, Kinzua, Pa., Sheet No. 2
1937	Upper Allegheny Dam Site, Preliminary Geologic Sections G-Line, K-Line, O-Line Profiles
1938	Logs of Core Drilling Nos. 7, 7A, 8, 9, and 10
1957	Allegheny River Reservoir, Design Memorandum 2, Selection of Site and Type of Dam
1960	Allegheny River Reservoir, Design Memorandum 7, Geology Text
1960	Allegheny River Reservoir Dam, Geologic Section A–A (Axis) 038-R15-10/6
1960	Columnar Geologic Section, Plate 1
1961	Allegheny Reservoir Dam Borrow Area Core Borings Plan 038-U1-10/101.1
	Core Borings Plan 038-U1-10/102
	Core Borings Plan 038-U1-10/103
1962	Allegheny Reservoir Relocation of Erie–Lackawanna R.R.
	Plate 1, Location and Vicinity Map 038-R13-66/16
	Plate 17, Geologic Map and Core Boring Plan 038-R13-10/1
	Plate 18, Record of Exploration 038-R13-10/2
	Plate 19, Record of Exploration 038-R13-10/3
	Plate 21, Record of Exploration 038-R13-10/8
	Plate 28, Record of Exploration 038-R13-10/18
	Plate 29, Record of Exploration 038-R13-10/19
1965	Allegheny Reservoir Dam, Final Report of Upstream Concrete Cutoff Wall
1968	Allegheny Reservoir, Current Reservoir Data–F/C6, 3 sheets

U.S. Geological Survey

    Sources of base for Figure 1:

      U.S. Series of Topographic Maps, Scale 1:250,000

      1962   Buffalo

      1967/1969   Warren

    Sources of base for Figure 2:

      7½ Minute Series (Topographic), Scale 1:24,000

      1964/1971   Clarendon Quadrangle, Pennsylvania

      1966/1973   Cornplanter Bridge Quadrangle, Pennsylvania

10

TIMBER HARVESTING, MASS EROSION, AND STEEPLAND FOREST GEOMORPHOLOGY IN THE PACIFIC NORTHWEST

Douglas N. Swanston and Frederick J. Swanson

INTRODUCTION

Forest operations in mountainous regions of the Pacific Northwest have a major impact on soil-erosion processes. The mountains of the region are youthful, the area has undergone recent tectonic activity, and west of the crest of the Cascades annual precipitation may exceed 375 cm. Consequently, natural erosion rates are high. Heavy forest vegetation and the high infiltration capacity of many of the forest soils protect slopes from surface erosion. The combination of these factors results in mass erosion processes generally being the dominant mechanisms of sediment transport from hillslopes to stream channels. The principal mass erosion processes are slow, downslope movement involving subtle deformation of the soil mantle (creep) and discrete failures, including slow-moving, deep-seated slump–earthflows; rapid, shallow soil and organic debris movement from hillslopes (debris avalanches); and rapid debris movement along downstream channels (debris torrents) (Fig. 1A, B). In many areas, forest vegetation plays an important role in stabilizing slopes and reducing the movement rate and occurrence of these mass erosion processes. When timber is removed from marginally stable slopes, whether by natural processes such as wildfire and wind or by the activities of man, a temporary acceleration of erosion activity is likely.

Accelerated erosion due to forest land management activities may result in reduced productivity of forest soils over sizable portions of affected watersheds, damage to roads, bridges, and other structures, and adverse impacts on the stream environment downstream.

FACTORS CONTROLING MASS EROSION PROCESSES

Geologic, hydrologic, and vegetative factors control the occurrence and relative importance of mass erosion processes. In the Pacific Northwest, areas of clay-rich bedrock and deep, cohesive soils are characterized by dominance of the slow mass movement processes of creep and slump–earthflow; notably in the extensive areas of soft sedimentary rocks of the Klamath Mountains, Oregon, and northern California Coast Range and the volcaniclastic rocks in the Cascade Range. Debris avalanches dominate mass erosion processes in terrain typified by steep slopes, cohesionless soils, and relatively competent bedrock, such as large areas of the coast ranges of Oregon, Washington, British Columbia, and Alaska.

A

FIGURE 1 *Examples of principal mass erosion processes operating on steep forested slopes in the Pacific Northwest. (A) creep and slump−earthflow terrain on Franciscan sediments in the upper headwaters of the North Fork, Eel River, northern California. The entire slope is undergoing creep deformation but note the discrete failure (slump−earthflow) marked by the steep headwall scarp at top center and the many small slumps and debris avalanches triggered by surface springs and road construction. (B) debris avalanches developed in shallow soils overlying compact till in southeast Alaska. Debris torrent developed below debris avalanche at center of photo due to channeling of debris, undercutting of sideslopes, and addition of material by secondary avalanching into channel.*

Periods of high-intensity precipitation, storm events, and condensation snowmelt commonly trigger or accelerate mass wasting events on steep forest slopes (Bishop and Stevens, 1964; Fredriksen, 1965; Dyrness, 1967; Swanston, 1969). These factors directly influence the moisture content of the soil and determine the presence or absence of active piezometric levels in the subsurface. Moisture content and piezometric level affect the weight of the soil mass and control the development of positive pore pressures. These factors act to reduce the resistance of the soil mass to sliding by either mobilization of clay structures, primarily through adsorption of water into the clay mineral structure, or by reducing the frictional resistance of the soil mass along the failure surface.

Vegetation cover in general helps control the amount of water reaching the soil and the amount held as stored water, largely through a combination of interception and evapotranspiration. In high-rainfall areas of the Pacific Northwest, interception is negligible during large storm events important to mass soil movement (Rothacher, 1963). Evapotranspiration (ET) has its principal effect on soil strength by reducing the length of time that soils remain saturated. ET reduces soil moisture during dry months, reduces the degree of saturation that can result from the first storms of the fall recharge period, and accelerates the rate of soil moisture removal at the end of the wet season. Once the soil is recharged as the wet season begins, the effect of ET loss becomes negligible. Also of importance is the depth of withdrawals by ET. Deep withdrawals may require substantial recharge to satisfy the soil-water deficit, delaying the attainment of saturated soil conditions several months or through a number of major slide-producing events. Shallow soils, on the other hand, will recharge rapidly, possibly attaining saturated conditions and maximum instability during the first major storm. ET has also been linked with at least a temporary increase of shear strength and stability during dry summer months through soil moisture depletion and buildup of soil moisture stress (Gray and Brenner, 1970; Manbeian, 1973; Gray, 1973).

Forests may also moderate the rate at which moisture enters the soil. This is particularly important in the case of warm-rain-on-snow events in which forest vegetation may influence the amount of snow collected on the land surface and the rate of melting, where advection and condensation melting are important (Anderson, 1969).

A crucial factor in the stability of active slopes is the role of plant roots in maintaining the shear strength of soil mantles. Roots add strength to the soil by vertical anchoring through the soil mass into fractures in the bedrock, and by laterally tying the slope together across zones of weakness or instability. In shallow soils, both effects may be important. In deep soils the vertical rooting factor will become negligible, but lateral anchoring may remain important. In some steep areas in the western United States, rooting strength may be the dominant factor in maintaining the slope equilibrium of an otherwise unstable area (Bishop and Stevens, 1964; Croft and Adams, 1950). Zaruba and Mencl (1969) have reported the stabilizing effect of tree roots in landslide areas in Czechoslovakia, and Nakano (1971) reports similar effects from unstable areas in Japan.

IMPACTS OF FOREST-MANAGEMENT ACTIVITIES ON MASS EROSION PROCESSES

Timber-harvesting activities, including clearcutting and road construction, modify factors influencing mass erosion in a variety of ways. Some of the most important impacts of these activities are summarized in Table 1.

TABLE 1 *Impacts of Engineering Activities on Factors That Influence Slope Stability in Steep Forest Lands of the Pacific Northwest*

Factors	Engineering activities[a]		References
	Deforestation	Roading	
I. Hydrologic influences			
A. Water movement by vegetation	Reduce evapotranspiration (–)	Eliminate evapotranspiration (–)	Gray (1970) Brown and Sheu (1975)
B. Surface and sub-surface water movement	Alter snowmelt hydrology (– or +)	Alter snowmelt hydrology (– or +) Alter surface drainage network (–) Intercept subsurface water at roadcuts (–)	Anderson (1969) Harr et al. (1975) Megahan (1972)
	Alter concentrations of unstable debris in channels (–) Reduce infiltration by ground surface disturbance (–)	Alter concentrations of unstable debris in channels (–) Reduce infiltration by roadbed (–)	Rothacher (1959) Froehlich (1973)
II. Physical influences			
A. Vegetation			
1. Roots	Reduce rooting strength (–)	Eliminate rooting strength (–)	Swanston (1970) Nakano (1971) Swanston (1969)
2. Bole and crown	Reduce medium for transfer of wind stress to soil mantle (+)	Eliminate medium for transfer of wind stress to soil mantle (+)	
B. Slope			
1. Slope angle		Increase slope angle at cut and fill slopes (–)	Parizek (1971), O'Loughlin (1972)
2. Mass on slope	Reduce mass of vegetation on slope (+)	Eliminate mass of vegetation on slope (+)	Bishop and Stevens (1964)
		Cut and fill construction redistributes mass of soil and rock on slope (– or +)	O'Loughlin (1972)
C. Soil properties		Reduce compaction and apparent cohesion of soil used as road fill (–)	

[a]Influence that usually increases slope stability denoted by (+); influence that usually decreases stability denoted by (–).

The principal impacts of forest removal by clearcutting are to reduce rooting strength and to alter the hydrologic regime at the site. In Japan, Kitamura and Namba (1966, 1968) have described a period of greatly reduced soil strength attributable to rooting beginning about 3 yr following cutting, when there has been significant decay of root systems, and attaining a minimum strength 15 yr after cutting. Similar loss of roting strength on unstable sites has been reported from coastal Alaska[1] and British Columbia (O'Loughlin, 1974), with minimum rooting strength attained 3 to 5 yr after cutting.

The hydrologic impacts of clearcutting include modification of annual soil-water status and changes in peaks of soil water held in detention storage during periods of storm runoff. Increased peak flows can generate active pore-water pressures, triggering shallow debris avalanches and debris torrents. Reduced ET due to clearcutting results in the soil-water status remaining at higher levels for several months longer than it would under forested conditions (Gray, 1970; Rothacher, 1971). This modification in the soil-water regime may result in prolonged periods of active creep and slump—earthflow movement during a single season or reactivation of dormant terrain. Water-yield studies in experimental watersheds in Oregon (Rothacher, 1971; Harr, 1975) suggest that this effect may continue for more than a decade after cutting.

Timber removal may also increase peaks of soil water by accelerated snowmelt during warm-rain-on-snow conditions (Anderson, 1969). This phenomenon may also increase total surface runoff if rain and snowmelt are synchronized (Rothacher and Glazebrook, 1968).

The principal impacts of road construction are to interrupt the natural balance between the resistance of the soil to failure and the downslope stress of gravity by disturbance of marginally stable slopes and alteration of subsurface and surface-water movement. Disturbance results from careless or improper cutting of marginally stable slopes, poor construction and placement of fills on steep slopes, and improper drainage design. Roads alter the routing of water by interception of surface water at cut slopes and surface drainage from roads and by carrying this excess water through ditches (Megahan, 1972; Harr et al., 1975). Mass erosion commonly occurs where natural and artificial drainage systems are inadequate to handle this excess water.

CREEP

Characteristics

Creep is the slow, downslope movement of the soil mantle in response to gravitational stress. The mechanics of creep have been investigated experimentally and theoretically by a number of workers (Terzaghi, 1953; Goldstein and Ter-Stepanian, 1957; Saito and Uezawa, 1961; Culling, 1963; Haefeli, 1965; Bjerrum, 1967; Carson and Kirkby, 1972; and others). Movement is by quasi-viscous flow, occurring under shear stresses sufficient to produce permanent deformation, but too small to result in discrete failure. Mobilization of the soil mass is primarily by deformation at grain boundaries and within clay mineral structures. Both interstitial and absorbed water appear to contribute to creep movement by opening the structure within and between mineral grains, thereby reducing

[1] Douglas N. Swanston and W. J. Walkotten, tree rooting and soil stability in coastal forests of southeastern Alaska. Study No. FS-NOR-1604:26 on file at PNW Forestry Sciences Laboratory, Juneau, Alaska.

friction within the soil mass. This permits a "remolding" of the clay fraction, transforming it into a slurry, which then lubricates the remaining soil mass. In local areas where shear stresses are great enough, discrete failure may occur, resulting in development of slump—earthflow due to progressive failure of the mantle materials.

Movement Rate and Occurrence

Creep movement generally occurs at rates of a few millimeters to a few centimeters per year. Therefore, long periods of observation and moderately sophisticated instruments are necessary to characterize creep. For these reasons, creep research has been focused on applied engineering problems in major construction activities (e.g., Wilson, 1970), and there have been few studies in the forest environment (Kojan, 1968; Barr and Swanston, 1970; Swanston, unpublished data). Recent measurements of creep rate using an inclinometer in flexible plastic pipe are summarized in Table 2. Also shown are examples of creep velocity profiles through soil masses.

The data shown in Table 2 are preliminary values based on 2 to 3 yr of monitoring and may not adequately represent long-term creep rates for these sites. Significant creep has been observed, however, and some interesting indications of the character of natural creep activity are apparent. Natural creep rates monitored in different geological materials in the western Cascade Range and coast ranges of Oregon and northern California indicate rates of movement between 7.1 and 15.2 mm/yr. The zone of most rapid movement usually occurs at or near the surface, although a zone of maximum displacement is usually present at depths associated either with incipient failure planes or zones of groundwater movement. The depth over which creep is active is quite variable and is largely dependent on parent material origin, degree and depth of weathering, subsurface structure, and soil-water content.

Movement rates are variable, as would be expected under natural conditions, but as a general rule lie within the range of 0 to 15 mm/yr. The maximum measured creep rates in Table 2 average 10.1 mm/yr. At many of the sites, movement takes place primarily during the rainy season when maximum soil-water levels occur (Fig. 2A), although creep may remain constant throughout the year in areas where the water table does not undergo significant seasonal fluctuation (Fig. 2B). This is consistent with Ter-Stepanian's (1963) theoretical analysis, which showed that the downslope creep rate of an inclined soil layer was exponentially related to the piezometric level in the slope.

Creep is generally the most persistent of all mass erosion processes. It operates at varying rates in clayey soils at slope angles of even just a few degrees. Therefore, in small watersheds developed in cohesive materials, creep may be operating over more than 90% of the landscape. The result of this creep activity is a continuing supply of soil material to the stream in the form of encroaching banks and small-scale bank failures. The quantity of soil delivered is quite large, and the supply is continuous from year to year. For example, assuming a creep rate of 10 mm/yr moving mantle material with a dry unit weight of 1,600 kg/m^3 to a stream with a bank approximately 2 m high (conservative estimates for watersheds in pyroclastic materials within the western Cascades Range, Oregon), approximately 64 metric tons/lineal km/yr will be supplied to the channel annually. During high-flow events, this material is carried into the stream by direct water erosion and by undercutting and local bank slumping. Such processes have been demonstrated to be a major contributor to sediment loads of the Eel and Mad rivers in northern

TABLE 2 *Examples of Measured Rates of Natural Creep on Forested Slopes in the Pacific Northwest*

Location	Data source	Parent material	Depth of significant movement (m)	Maximum downslope creep rate — Surface (mm/yr)	Maximum downslope creep rate — Zone of accelerated movement (mm/yr)	Representative creep profile
Coyote Creek South Umpqua River drainage, Cascade Range of Oregon, site C-1	Swanston[a]	Little Butte Volcanic Series: deeply weathered clay-rich andesitic dacitic volcaniclastic rocks	7.3	13.97	10.9	
Blue River drainage— Lookout Creek H. J. Andrews Exp. Forest, Central Cascades of Oregon, site A-1	Swanston[a]	Little Butte Series: (same as above)	5.6	7.9	7.1	
Blue River drainage IBP Experimental Watershed 10, site No. 4	McCorison[b] and Glenn	Little Butte Volcanic Series	0.5	9.0	—	
Baker Creek, Coquille River, Coast Range, Ore., Site B-3	Swanston[a]	Otter Point Formation: highly sheared and altered clay-rich argillite and mudstone	7.3	10.4	10.7	
Bear Creek, Nestucca River, Coast Range, Ore., site N-1	Swanston[a]	Nestucca Formation: deeply weathered pyroclastic rocks and interbedded, shaly siltstones and claystones	15.2	14.9	11.7	
Redwood Creek, Coast Range, northern Calif., site 3-B	Swanston[a]	Kerr Ranch Schist: sheared, deeply weathered clayey schist	2.6	15.2	10.4	

[a]Douglas N. Swanston, unpublished data on file at Forestry Sciences Laboratory, U.S. Department of Agriculture, Forest Service, Pacific Northwest Forest and Range Experiment Station, Corvallis, Ore.
[b]F. Michael McCorison and J. F. Glenn, data on file at Forestry Sciences Laboratory, U.S. Department of Agriculture, Forest Service, Pacific Northwest Forest and Range Experiment Station, Corvallis, Ore.

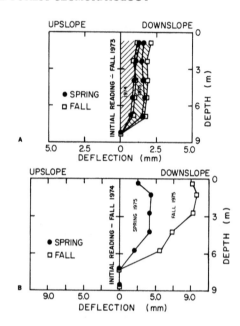

FIGURE 2 *Deformation of inclinometer tubes at two sites in the southern Cascade Range and Coast Range of Oregon. (A) Coyote Creek in the southern Cascade Range showing seasonal variation in movement rate as the result of changing soil-water levels. Note that the difference in readings between spring and fall of each year (dry months) is very small. (B) Baker Creek, Coquille River, Oregon Coast Range, showing constant rate of creep as a result of continual high-water levels.*

California, which have the largest sediment discharges in the world. Anderson (1971) cites estimates that 80 to 85% of the total volume of sediment produced by these two rivers is the result of landslide and streambank erosion. In areas characterized by low-flow conditions supplemented only occasionally by storm flows, creep may fill the channel with soil and debris, and the stream water may be carried by subsurface flow and piping within the channel filling. Only during storm periods is flow great enough to open the channel and remove the debris stored there, resulting in the periodic discharge of excessive sediment loads from affected streams, and occasional torrent flow occurrence where local damming by debris occurs. Such a mechanism dominates in the great majority of smaller watersheds in the creep-dominated areas of the Pacific Northwest.

Impact of Forest Operations

There have been no direct measurements of the impact of harvesting activities on creep rates in the forest environment, mainly because of the long periods of record needed both before and after a disturbance. However, there are a number of indications that creep rates are accelerated by clearcutting and road construction.

Wilson (1970) and others have used inclinometers to verify accelerated creep following modification of slope angle, compaction of fill materials, and redistribution of soil mass at construction sites. The common occurrence of shallow-soil mass movements in these disturbed areas and open tension cracks along roadways at cut and fill slopes suggest that similar features along forest roads are indicators of significantly accelerated creep movement.

On slopes where clearcutting is the principal influence, impact on creep rates may be more subtle, involving modifications of hydrology and root strength. Where creep is a shallow phenomenon (less than 2 to 3 m), the loss of root strength due to clearcutting is likely to be significant. Reduced evapotranspiration following clearcutting (Gray, 1970; Rothacher, 1971) may result in greater duration of the annual period of creep activity and, thereby, increase the annual creep rate.

Brown and Sheu (1975) have developed a mathematical model of creep that accounts for root strength, wind stress on the soil mantle, weight of vegetation, and soil moisture. The model predicts a brief period of increased slope stability following clearcutting due to removal of the weight of vegetation and elimination of wind stress. Thereafter, creep rates are accelerated as soil strength is attenuated by progressive root decay. The net result of the short-term increase and longer-term decrease in slope stability is expected to be an overall increase in creep activity.

SLUMP–EARTHFLOW

Characteristics

In local areas where shear stresses are great enough, discrete failure occurs and slump–earthflow features (Varnes, 1958) are formed. Simple slumping takes place as a rotational movement of a block of earth over a broadly concave slip surface and involves very little breakup of the moving material. Where the moving material slips downslope and is broken up and transported either by a flowage mechanism or by gliding displacement of a series of blocks, the movement is termed slow earthflow (Varnes, 1958). The combined term slump–earthflow is used because many deep-seated mass movements in the Pacific Northwest have slump characteristics in the headwall area and develop earthflow features downslope.

Slump–earthflows have been described by Varnes (1958), Wilson (1970), Colman (1973), Swanson and James (1975), and others. In the Pacific Northwest, these features may range in area from less than 1 hectare to more than several square kilometers. The zone of failure occurs at depths of a few meters up to several tens of meters below the surface. Commonly, there is a slump basin with a headwall scarp at the top of the failure area. Lower ends of earthflows typically run into stream channels. Transfer of earthflow debris to stream channels may take place by shallow, small-scale debris avalanching or by gullying and surface erosion, depending on soil and vegetation conditions. Therefore, the general instability set up by an active slump–earthflow initiates erosion activity by a variety of other processes.

Geologic, vegetative, and hydrologic factors have primary control over slump–earthflow occurrence. Deep, cohesive soils and clay-rich bedrock are especially prone to slump–earthflow failure, particularly where these materials are overlain by hard, competent rock (Wilson, 1970; Swanson and James, 1975). Earthflow movement also appears to be most sensitive to long-term fluctuations in the amount of available soil water (weeks, months, annually) (Wilson, 1970; and others).

Because earthflows are slow moving, deep-seated, poorly drained features, individual storm events probably have much less influence on their movement than on the occurrence of debris avalanches and torrents. Where planes of slump–earthflow failure are more than several meters deep, weight of vegetation and vertical root-anchoring effects are negligible (O'Loughlin, 1974).

Movement Rate and Occurrence

Movement rates of earthflows vary from imperceptibly slow to more than 1 m/day in extreme cases. In parts of the Pacific Northwest, many slump—earthflow areas appear to be presently inactive (Colman, 1973; Swanson and James, 1975). Areas of active movement may be recognized by fresh ground breaks at shear and tension cracks and by tipped and bowed trees. Rates of movement may be monitored directly by repeated surveying of marked points, by inclinometers, and by measuring deflection of roadways and other reference systems. These methods have been used to estimate the rates of earthflow movement shown in Table 3. However, these average rates, stated so simply, are somewhat misleading, because of the great variability of movement rate over both time and space, even for a single slump—earthflow (Colman, 1973; Swanson and James, 1975). Open tension cracks and degree of disturbance of vegetation on slump—earthflows indicate that some part of an earthflow terrane may move rather rapidly, while other areas appear to be temporarily stabilized.

The history of individual slump—earthflows may extend over thousands of years. This is indicated by age estimates based on radiometric dating of included wood, calculation of volume of eroded material, estimated long-term rate of earthflow erosion from site, presence of 7,000-yr-old Mazama ash on preexisting earthflow terrain, and characteristics of drainage development over earthflow surfaces. During this long history, periods of relatively high precipitation or forest removal may increase water content of an earthflow and accelerate movement rate. This downslope movement will decrease the relief at the site until the stability of the mass is increased to a point where the velocity decreases markedly. A period of temporary inactivity will take place until there is reactiviation during a period of high moisture availability, or when the area had been destabilized by stream erosion at the toe of the earthflow.

The areal occurrence of slump—earthflows is mainly determined by bedrock geology. In the Redwood Creek Basin, northern California, Colman (1973) observed that, of the 27.4% of the drainage which is in slumps, earthflows, and older or questionable landslides, a very high percentage of the unstable areas are located in the clay-rich and

TABLE 3   *Observations of Movement Rates of Four Active Earthflows in the Western Cascade Range, Oregon*

Location	Period of record (yr)	Movement rate (cm/yr)	Method of observation
Landes Creek (Sec. 21 T22S R4E)	15	12	Deflection of road
Boone Creek (Sec. 17 T17S R5E)	2	25	Deflection of road
S.W. Cougar Reservoir (Sec. 29 T17S R5E)	2	2.5	Deflection of road
Lookout Creek (Sec. 30 T15S R6E)	1	7	Strain rhombus measurements across active ground breaks

pervasively sheared portions of the Franciscan assemblage of rocks. Areas underlain by schists and other more highly metamorphosed rock are much less prone to deep-seated mass erosion. The areal occurrence of slump—earthflows in volcanic terranes of the Pacific Northwest is also closely linked to bedrock type. In the H. J. Andrews Experimental Forest, western Cascade Range, Oregon, for example, approximately 25.6% of areas underlain by volcaniclastic rocks are included in active and presently inactive slump—earthflows. Less than 1% of areas of basalt and andesite flow rock have undergone slump—earthflow failure.

Impact of Forest Operations

Engineering activities that involve excavation and fills frequently have dramatic impact on slump—earthflow activity (Wilson, 1970). In the forest environment, there are numerous unpublished examples of accelerated or reactivated slump—earthflow movement after forest road construction. Undercutting of toe slopes of earthflows and piling of rock and soil debris on slump blocks are common practices that increase slump—earthflow movement. Stability of such areas is also affected by modification of drainage systems, particularly where road drainage systems route additional water into the slump—earthflow areas. These disturbances may increase movement rates from a few millimeters per year to several tens of centimeters per year or more. Once such areas have been destabilized, they may continue to move at accelerated rates for several years.

Although the impact of clearcutting alone on slump—earthflow movement has not been demonstrated quantitatively, several pieces of evidence suggest that it may be significant. In massive, deep-seated failures, lateral and vertical anchoring of tree-root systems is negligible. However, hydrologic impacts appear to be important. Increased moisture availability due to reduced evapotranspiration will increase the volume of water not utilized by the vegetation. This water is therefore free to pass through the rooting zone to deeper levels of the earthflow. Although the hydrology of slump—earthflows has not yet been investigated, hydrology research on small watersheds suggests that this effect may be substantial. For example, even 9 yr after clearcutting of a small watershed in the H. J. Andrews Experimental Forest, runoff may still exceed by 50 cm the estimated yield for the watershed in a forested condition (Rothacher, 1971; Harr, 1975, pers. comm.). In watersheds with steep slopes and relatively permeable soils, this increased water yield comes as higher base flow during dry months and higher peak and base flow during early and late stages of the wet season (Rothacher, 1971). On poorly drained earthflows, the increased available moisture is likely to be stored in the subsoil for longer periods of time, possibly contributing to increased rate and duration of the wet season earthflow movement. It is not known whether possible clearcutting-related increases in peak discharge of surface and subsurface water influences earthflow movement.

DEBRIS AVALANCHES

Characteristics

Debris avalanches are rapid, shallow-soil mass movements from hillslope areas. Here we use the term "debris avalanche" in a general sense encompassing debris slides, avalanches, and flows, which have been distinguished by Varnes (1958) and others on the basis of increasing water content. From a land-management standpoint, there is little purpose in differentiating failures among these types of shallow hillslope, since the mechanics and controlling and contributing factors are the same, and one frequently leads to another.

Debris avalanches have rather consistent characteristics in the variety of geologic and geomorphic settings extending from northern California to southeast Alaska (Colman, 1973; Morrison, 1975; Dyrness, 1967; Gonsior and Gardner, 1971; Fiksdal, 1974; O'Loughlin, 1972; Bishop and Stevens, 1964; Swanston, 1970). In all these areas, debris avalanches are usually triggered by infrequent, intense storms. For example, in the H. J. Andrews Experimental Forest, Oregon, it has taken storms of a 7-yr return period or greater to initiate debris avalanching in forested areas. Swanston (1969) has correlated storms with a 5-yr return interval with accelerated debris avalanching in coastal Alaska.

Debris avalanches leave scars in the form of spoon-shaped depressions from which less than 10 to more than 10,000 m^3 of soil and organic debris have moved downslope. Average volumes of individual debris avalanches in forested areas in the Pacific Northwest range from about 1,540 to 4,600 m^3.

Several of the factors discussed previously control the occurrence of debris avalanches. Debris-avalanche-prone areas are typified by shallow, noncohesive soils on steep slopes where subsurface water may be concentrated by subtle topography on bedrock or glacial till surfaces. Because debris avalanches are shallow failures, factors such as root strength, anchoring effects, and the transfer of wind stress to the soil mantle, are potentially important influences. Factors that influence antecedent soil moisture conditions and the rate of water supply to the soil by snowmelt and rainfall also have significant control over when and where debris avalanches occur.

Movement Rate and Occurrence

Movement rates of debris avalanches have seldom been measured because of the extreme storm conditions under which they occur. However, based on the few available accounts and the steep slope conditions where debris avalanches occur, their rates of movement probably range as high as 20 m/s.

The rate of occurrence of debris avalanches is controlled by the stability of the landscape and the frequency of storm events severe enough to trigger them. Therefore, the rates of erosion by debris avalanching will vary from one geomorphic–climatic setting to another. Table 4 shows that annual rates of debris-avalanche erosion from forested areas of study sites in Oregon, Washington, and British Columbia range from 11 to 72 m^3/km^2/yr. These estimates are based on surveys and measurements of erosion by each debris avalanche occurring in a particular time period (25 yr or longer) over a large area (12 km^2 or larger).

Impact of Forest Operations

The net impact of engineering activities can be estimated by determining the rate of debris-avalanche erosion in clearcut areas and road rights-of-way and comparing these levels of erosion with the erosion rate for forested areas during the same time period. Such an analysis (Table 4) reveals that clearcutting commonly results in acceleration by a factor of 2 to 4 of debris-avalanche erosion. Roads appear to have a much more profound impact on erosion activity. In the four study areas listed in Table 4, road-related debris-avalanche erosion was increased by factors ranging from 25 to 340 times the rate of debris-avalanche erosion in forested areas.

TABLE 4 *Debris-Avalanche Erosion in Forest, Clearcut, and Roaded Areas*

Site	Period of record (yr)	Area (%)	Area (km²)	Number slides	Debris-aval. erosion (m³/km²/yr)	Rate of debris-avalanche erosion relative to forested areas
Stequaleho Creek, Olympic Peninsula (Fiksdal, 1974)						
Forest	84	79	19.3	25	71.8	X 1.0
Clearcut	6	18	4.4	0	0	0
Road R/W	6	3	0.7	83	11,825	X 165
			24.4	108		
Alder Creek, western Cascade Range, Oregon (Morrison, 1975)						
Forest	25	70.5	12.3	7	45.3	X 1.0
Clearcut	15	26.0	4.5	18	117.1	X 2.6
Road R/W	15	3.5	0.6	75	15,565	X 344
			17.4	100		
Selected Drainages, Coast Mountains, S.W. British Columbia (O'Loughlin, 1972, and pers. comm.)						
Forest	32	88.9	246.1	29	11.2	X 1.0
Clearcut	32	9.5	26.4	18	24.5	X 2.2
Road R/W	32	1.5	4.2	11	282.5[a]	X 25.2
			276.7	58		
H. J. Andrews Experimental Forest, western Cascade Range, Oregon (Swanson and Dyrness, 1975)						
Forest	25	77.5	49.8	31	35.9	X 1.0
Clearcut	25	19.3	12.4	30	132.2	X 3.7
Road R/W	25	3.2	2.0	69	1,772	X 49
			64.2	130		

[a]Calculated from O'Loughlin (1972, and pers. comm.), assuming that area involving road construction in and outside clearcuts is 16% of area clearcut.

The great variability of the impact of roads reflects not only differences in the natural stability of the landscapes, but also differences in road-location design and construction. For example, the Alder Creek and H. J. Andrews Experimental Forest study sites are in similar climatic, geologic, and geomorphic settings, both areas are managed by a single National Forest, and the level of debris-avalanche erosion in forested areas of the two drainage is similar. However, the level of road-related debris-avalanche erosion in the Alder Creek drainage has been nearly eight times greater than in the Andrews Forest. This contrast in road impact appears to result from higher proportions of midslope road mileage and large, unstable hillslopes in the Alder Creek area, which was managed in a less stringent fashion than the experimental forest (Morrison, 1975).

The duration of impacts of clearcutting and road construction on debris-avalanche erosion is not well documented. Ideally, it would be useful to know how the various factors influencing slope stability, such as rooting strength and soil-moisture storage, each vary as a function of time since disturbance. However, such information is known only in a qualitative sense.

The duration of the net effects of harvesting activities can be deduced from historical records such as those for the H. J. Andrews Experimental Forest, where the history of debris avalanches and associated clearcutting and road construction since 1950 has been documented (Swanson, unpublished data). Based on these observations, clearcut slopes appear to undergo a period of increased susceptibility to debris avalanching for about 12 yr after cutting. However, the strength of the conclusions that can be drawn from a record of only 25 yr is limited by the irregular, episodic nature of both storms and harvesting activities. The history of road-related debris avalanches with respect to road age is further complicated by the disproportionately high levels of natural instability in areas where road access has not yet been developed and by the changing standards of road construction and maintenance. These factors contribute to the very irregular patterns of road age at the time of failure shown in Table 5. It is important to note that roads in the two study areas (Table 5) continued to be very active sites of debris-avalanche erosion for more than 16 yr after initial construction of the road. Therefore, the duration of impact of road construction on erosion by debris avalanches may persist more than twice as long as clearcutting impacts.

The relative impacts of clearcutting and road construction on the total level of accelerated erosion are not clearly reflected in the data in Table 4. For example, in the H. J. Andrews Experimental Forest, roads accelerate debris-avalanche erosion to a much greater extent than clearcutting, but road rights-of-way cover much less area of the forest than do clearcut units. When road and clearcutting impacts are weighted by the area influenced by each activity, the two types of forest engineering activities contribute about equally to the total level of accelerated debris-avalanche erosion (Swanson and Dyrness, 1975).

DEBRIS TORRENTS

Characteristics

Debris torrents involve the rapid movement of water-charged soil, rock, and organic material down steep stream channels. Debris torrents are distinguished from debris

TABLE 5 *History of Debris Avalanches with Respect to Road Age*

Road age (yr)	Percent of debris avalanches	
	H. J. Andrews Exp. Forest[a]	Clearwater Natl. Forest[b]
0–5	57	5
6–10	11	50
11–15	20	34
16+	12	11

[a]Swanson, unpublished data, based on a complete 25-yr history of road construction and 73 debris avalanches.
[b]Day and Megahan, 1975, based on debris avalanches occurring in a single large storm during Jan. 1974.

avalanches because the two types of mass movement events occur over different parts of the landscape; consequently, they have differing implications for the land manager.

Debris torrents typically occur in steep intermittent first- and second-order channels. These events are triggered during extreme discharge events by slides from adjacent hillslopes, which enter a channel and move directly downstream, or by the breakup and mobilization of debris accumulations in the channel. The initial slurry of water and associated debris commonly entrains large quantities of additional inorganic and living and dead organic material from the stream bed and banks. Some torrents are triggered by debris avalanches of less than 100 m^3, but ultimately involve 10,000 m^3 of debris entrained along the track of the torrent. As the torrent moves downstream, hundreds of meters of channel may be scoured to bedrock. When a torrent loses momentum, there is deposition of a tangled mass of large organic debris in a matrix of sediment and fine organic material covering areas up to several hectares.

The main factors controlling the occurrence of debris torrents are the quantity and stability of debris in channels, steepness of channel, stability of adjacent hillslopes, and peak discharge characteristics of the channel. The concentration and stability of debris in channels reflects the history of stream flushing and the health and stage of development of the surrounding timber stand (Froehlich, 1973). The stability of adjacent slopes is dependent on a number of factors described in previous sections on other mass erosion processes. The history of storm flows has a controlling influence over the stability of both soils on hillslopes and debris in stream channels.

Movement Rate and Occurrence

Although debris torrents pose very significant environmental hazards in mountainous areas of the Pacific Northwest, they have received little study (Fredriksen, 1963, 1965; Morrison, 1975; Swanson and Lienkaemper, 1975). Velocities of debris torrents, estimated to be up to several tens of meters per second are known only from verbal and a few written accounts. The occurrence of torrents has been systematically documented in only two small areas of the Pacific Northwest, both in the western Cascade Range of Oregon (Morrison, 1975; Swanson, unpublished data). In these studies, rates of occurrence of debris torrents were observed to be 0.005 and 0.008 events/km^2/yr for forested areas (Table 6). Torrent tracks initiated in forest areas ranged in length from 100 to 2,280 m and averaged 610 m of channel length. Debris avalanches have played a dominant role in triggering 83% of all inventoried torrents. Mobilization of stream debris not immediately related to debris avalanches has been a rather minor factor in initiating debris torrents in these Cascade Range streams. Therefore, the susceptibility of an area to debris avalanching is a direct indicator of the potential for debris torrents.

Impact of Forest Operations

Timber-harvesting activities appear to dramatically accelerate the occurrence of debris torrents by increasing the frequency of debris avalanches. Although it has not been demonstrated, it is also possible that increased concentrations of unstable debris in channels during harvesting (Rothacher, 1959; Froehlich, 1973; Swanson and Lienkaemper, 1975) and possible increased peak discharges (Rothacher, 1973; Harr et al., 1975) may accelerate the frequency of debris torrents.

The relative impacts of these factors and of clearcutting and roads may be assessed by using the frequency of occurrence of debris torrents (events/km^2/yr) in forest areas as an estimate of the natural, background level of debris-torrent activity against which to compare the rates of occurrence attributed to roads and clearcutting (Table 6). In the H. J. Andrews Experimental Forest and the Alder Creek study sites, clearcutting appeared to increase the occurrence of debris torrents by 4.5 and 8.8 times; roads were responsible for increases of 42.5 and 133 times.

Although the quantitative reliability of these estimates of harvesting impacts is limited by the small number of events analyzed, there is clear evidence of marked increase in the frequency of debris torrents as a result of clearcutting and road building. The history of debris avalanches in the two study areas clearly indicates that increased debris torrents are primarily a result of two conditions: debris avalanches trigger most debris torrents (Table 6), and the occurrence of debris avalanches is greatly increased by clearcutting and road construction (Table 4).

The data in Table 6 suggest that clearcutting in the H. J. Andrews Experimental Forest may have had a significant impact on frequency of debris torrents. In several cases, increased concentrations of debris in streams appeared to have contributed to debris-torrent occurrence. The extent of management impact on stream debris concentrations is directly related to how carefully engineering activities are designed and carried out. Under different geomorphic conditions and management practices, logging-related increases of debris in streams may have a more dramatic impact on debris-torrent occurrence.

The close relationship between occurrence of debris avalanches and debris torrents also indicates that the duration and the relative impacts of roads and clearcutting on debris torrents and debris avalanches (discussed previously) will be very similar in timing and magnitude.

RELATIONSHIPS AMONG PROCESSES

Creep, slump—earthflows, debris avalanches, and debris torrents function as primary links in the natural transport of soil material to streams in the Pacific Northwest. The importance of the linkage among these processes and the apparent impact of timber harvesting on rate of movement and sediment yield to streams is illustrated by results from ongoing investigations at Coyote Creek in the South Umpqua Experimental Forest and in the H. J. Andrews Experimental Forest, both in Oregon.

Coyote Creek Erosion Research

The Coyote Creek research area is located approximately 65 km southeast of Roseburg, Oregon, in the western Cascade Range. Four research watersheds have been established on deeply weathered volcaniclastic materials of the Little Butte Formation. Since 1966, both streamflow and sediment discharge have been monitored in these watersheds to determine water yield and nutrient outflow under forested and clearcut conditions.[1] Two access roads were constructed across the upper half of watershed 3 in 1971, and the watershed was logged by clearcutting in 1972.

From 1966 to 1970, total bedload export, estimated from volumes of sediment removed from a weir at the mouth of the watershed, was much less than 0.01 m³/m of

---

[1] Study 1602-10, A Study of the Effects of Timber Harvesting on Small Watersheds in the Sugarpine, Douglas-Fir Area of Southwestern Oregon, U.S. Department of Agriculture Forest Service, Forestry Sciences Laboratory, Corvallis, Ore.

TABLE 6 *Characteristics of Debris Torrents with Respect to Debris Avalanches and Landuse Status of Site of Initiation in the H. J. Andrews Experimental Forest (Swanson, Unpublished Data) and Alder Creek Drainage (Morrison, 1975)*

Site	Area of watershed (km²)	Period of record (yr)	Debris torrents triggered by debris avalanches	Debris torrents with no associated debris avalanche	Total No.	Total No./km²/yr	Rate of debris torrent occurrence relative to forested areas
H. J. Andrews Experimental Forest, western Cascade Range, Oregon							
Forest	49.8	25	9	1	10	0.008	× 1.0
Clearcut	12.4	25	5	6	11	0.036	× 4.5
Road	2.0	25	17	—	17	0.340	× 42.0
Total	64.2		31	7	38		
Alder Creek drainage, western Cascade Range, Oregon							
Forest	12.3	90	5	1	6	0.005	× 1.0
Clearcut	4.5	15	2	1	3	0.044	× 8.8
Road	0.6	15	6	—	6	0.667	×133.4
Total	17.4		13	2	15		

TABLE 7 *Estimated Annual Bedload Export from Clearcut Coyote Creek Watersheds 3 and 4*

Year	Precipitation (cm)	Total bedload volume (m)^3	Volume/unit stream length (m^3/m)	Basin condition
1966	120.5	3.9	0.0091	Forested
1967	118.5	0.5	0.0012	Forested
1968	87.6	0.6	0.0014	Forested
1969	110.9	0.2	0.0005	Forested
1970	116.3	1.3	0.0030	Forested
1971	155.7	21.9	0.051	Roads
1972	153.3	69.9[a]	0.163	Clearcut
1973	89.5	3.2	0.0074	Clearcut
1974	156.5	46.4	0.108	Clearcut
1975	122.6	7.7	0.018	Clearcut

[a]Minimum estimate; basin overflowed.

active stream channel/yr (m^3/m/yr) (Table 7; R. L. Fredriksen, pers. comm.). In 1971, sediment yield dramatically increased to approximately 0.05 m^3/m/yr as the result of unusually heavy winter precipitation and increased runoff from the construction of logging access roads. The bedload materials were derived primarily from debris avalanching and rotational slumping along the banks of the stream draining the watershed. In 1972, the first year after the entire watershed was clearcut and also a year of exceptionally heavy rainfall, bedload movement tripled over prelogging and road-building levels to an estimated volume of 0.16 m^3/m/yr. During this period, two new debris avalanches reached the channel from midslope, possibly triggered by reduction of rooting strength of vegetation following the logging process. Part of this overall bedload increase is due to surface erosion of severely scarified soil resulting from piling of slash after logging. However, the greater part can be directly linked to mass movements along the channel.

The major increases in bedload deposition in the weir basins have occurred during storm periods. Reconnaissance of the area and dissection of the weir deposits immediately after a major storm during the winter of 1972 exposed layering of heterogeneous sedimentary materials separated by zones of organic accumulations, which defined short periods or pulses of heavy sediment deposition. Such pulses result from repeated episodes of slumping and debris avalanching into the channel above the weir. A survey of the channel above the weir showed that nine new bank slumps and debris avalanches had occurred as a result of the storm, two of which were large enough to provide the volumes of material necessary to fill the weir basin (approximately 5.4 m^3 each). After 1972, bedload yields have been much lower, but still substantially above prelogging levels. This reflects accelerated bank erosion and removal of stream-stored sediment due to increased peak flows. A detailed survey of the watershed has since revealed over 50 sites of active bank slumping and debris avalanching along the active stream channel, with volumes ranging from 5.6 to 350 m^3. Much of this material moved into the channel, diverting it or causing water to flow beneath the surface. At least 12 debris dams have blocked the channel, leading to temporary storage of from 45 to 1,340 m^3 of alluvium, soil, and organic debris behind each dam. In 1974, the total volume of material available for stream transport was estimated to be 3,100 m³. Of this, 1,090 m³ was stored as slump blocks and 2,010 m³ was stored behind debris dams.

In addition, quantitative creep measurements in the soil and deeply weathered volcaniclastic rocks in the lower half of the watershed indicate creep of the soil in the zone of maximum movement of approximately 10.9 mm/yr, with much of the movement taking place along a narrow zone of weakness at a depth of approximately 8.2 m (Table 2). Since the banks in the watershed average at least 1 m in height, soil material is supplied to the stream channel in the lower part of the watershed at annual rates of 0.02 m³/m of active channel. There are approximately 430 m of active stream channel in watershed 3. Thus, approximately 8.6 m³ of soil are made available for stream transport annually by creep movement. The average annual yield since 1972 has been 31.8 m³/yr. This suggests that about 27% of the annual yield is being supplied by creep processes. The remainder is supplied by active slump–earthflows, surface erosion, and debris avalanching, or derived from stored channel deposits.

H. J. Andrews Experimental Forest

The H. J. Andrews Experimental Forest is located in the western Cascade Range of Oregon in an area characterized by lava flow and volcaniclastic bedrock (Swanson and James, 1975), average annual precipitation of 230 cm, and Douglas fir–western hemlock forest vegetation (Franklin and Dyrness, 1973). Erosion research in the forest has been focussed on two scales: (1) studies in small watersheds (less than 110 hectares) involving continuous monitoring of bedload and suspended sediment outflow since 1956 (Fredriksen, 1963, 1965, 1970), and (2) mapping and historical analysis of slump–earthflows, debris avalanches, and debris torrents throughout the forest (Dyrness, 1967; Swanson and James, 1975).

The history of mass erosion in the small watersheds and the entire experimental forest reveals the close interactions among the various mass erosion processes. As shown diagramatically in Figure 3, areas of creep and slump–earthflow activity may overlap, and these two processes contribute to the instability of areas that ultimately fail by debris avalanching. Debris avalanches, in turn, are dominant initiators of debris torrents.

Creep and slump–earthflow processes are interrelated in at least two senses. Creep deformation is thought to be a common precursor of slump–earthflows, occurring where strain in the form of creep has exceeded the shear strength of soil and rock material. Even where discrete failure has occurred and slump–earthflow movement has begun, superimposed creep deformation is likely to occur in the earthflow material. In one slump–earthflow terrain that is moving at the rate of about 5 cm/yr (Lookout Creek earthflow, Table 3), creep deformation within a single slump block of earthflow material has been measured at 7 mm/yr using an inclinometer tube.

Active slump–earthflow movement, which extends over only 3.3% of the entire forest, appears to have contributed to the instability leading to nearly 40% of the total volume of debris-avalanche erosion from forested areas in the past 25 yr. Slightly more than half of this earthflow generated erosion by debris avalanches occurred in streamside areas where earthflow movement constricted channels, and subsequent streambank cutting has led to rapid, shallow-soil mass movements. Much of the remaining debris-avalanche erosion in undisturbed areas has occurred in steep midslope areas where creep activity is an important destabilizing factor.

These close relationships among mass and fluvial erosion processes suggest that, if one

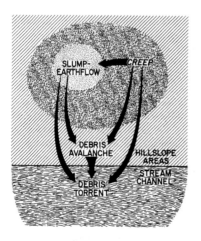

FIGURE 3 *Relationships among mass erosion processes. Arrows point from a process that sets up instability leading to failure by the process at head of the arrow. Width of arrows is a rough indication of degree of influence based on studies in the H. J. Andrews Experimental Forest.*

process is accelerated, the others may also be. Available data are insufficient to demonstrate the extent of impact of timber-harvesting activities on creep and slump–earthflow erosion in the H. J. Andrews Experimental Forest. However, in areas that have been clearcut and where roads have been built, debris-avalanche erosion has been increased by about five times (Swanson and Dyrness, 1975). This might be viewed as a reflection of accelerated creep and slump–earthflow activity in disturbed areas. Increased frequency of occurrence of debris avalanches also has resulted in a dramatic increase in debris torrents (Table 6).

CONCLUSIONS

Creep, slump–earthflows, debris avalanches, and debris torrents function as primary links in the natural transport of soil material to streams in the Pacific Northwest.

In areas characterized by deeply weathered, clay-rich mantle materials, creep movement may range as high as 15 mm/yr. Locally, where strain buildup causes discrete failure and the development of progressive slump–earthflows, transport rates of material to the stream may increase by several orders of magnitude. In areas characterized by steep slopes, shallow, coarse-grained mantle materials and steep, incised drainages, discrete failures producing debris avalanches and debris torrents transport large volumes of material to the stream at rates as high as 20 m/s.

Timber-harvesting operations, particularly clearcutting and road construction, accelerate these processes, the former by destroying the stabilizing influence of vegetation cover and altering the hydrologic regime of the site, the latter by interrupting the balanced strength–stress relationships existing under natural conditions by cut and fill activities, poor construction of fills, and alteration of surface and subsurface water movement.

ACKNOWLEDGMENTS

The work reported in this paper was supported in part by National Science Foundation Grant No. BMS74-20744 to the Coniferous Forest Biome, Ecosystem Analysis

Studies, U.S./International Biological Program. This is contribution no. 216 from the Coniferous Forest Biome.

REFERENCES

Anderson, H. W. 1969. Snowpack management: in *Snow,* seminar of Oregon State University, Oregon Water Resources Research Institute, Corvallis, Ore., p. 27–40.

_____. 1971. Relative contributions of sediment from source areas, and transport processes: in J. T. Krygier and J. D. Hall, eds., *Forest Land Uses and the Stream Environment,* Oregon State University, Corvallis, Ore., p. 55–63.

Barr, D. J., and Swanston, D. N. 1970. Measurement of creep in a shallow, slide-prone till soil: *Amer. Jour. Sci.,* v. 269, p. 467–480.

Bishop, D. M., and Stevens, M. E. 1964. Landslides on logged areas in southeast Alaska: *U.S. Dept. Agr. Forest Serv. Res. Paper NOR-1,* 18 p.

Bjerrum, L. 1967. Progressive failure in slopes of overconsolidated plastic clay and clay shales: *Jour. Soil Mech. Found. Div., Amer. Soc. Civil Engr.,* v. 93, p. 1–49.

Brown, C. B., and Sheu, M. S. 1975. Effects of deforestation on slopes: *Jour. Geotech. Eng. Div., Amer. Soc. Civil Engr.,* v. 101, p. 147–165.

Carson, M. A., and Kirkby, M. J. 1972. *Hillslope Form and Process:* Cambridge University Press, London, 475 p.

Colman, S. M. 1973. The history of mass movement processes in the Redwood Creek basin, Humboldt County, Calif.: M.S. thesis, Pennsylvania State University, University Park, Pa., 151 p.

Croft, A. R., and Adams, J. A. 1950. Landslides and sedimentation in the North Fork of Ogden River, May 1949: *U.S. Dept. Agr. Forest Serv. Res. Paper INT-21,* 4 p.

Culling, W. E. H. 1963. Soil creep and the development of hillside slopes: *Jour. Geol.,* v. 71, p. 127–161.

Day, N. F., and Megahan, W. F. 1975. Landslide occurrence on the Clearwater National Forest, 1974: *Geol. Soc. America Abstr. with Programs,* v. 7, p. 602–603.

Dyrness, C. T. 1967. Mass soil movements in the H. J. Andrews Experimental Forest: *U.S. Dept. Agr. Forest Serv. Res. Paper PNW-42,* 12 p.

Fiksdal, A. J. 1974. A landslide survey of the Stequaleho Creek watershed: Supplement to Final Report FRI-UW-7404, Fisheries Research Institute, University of Washington, Seattle, Wash., 8 p.

Franklin, J. F., and Dyrness, C. T. 1973. Natural vegetation of Oregon and Washington: *U.S. Dept. Agr. Forest Serv. General Tech. Rept. PNW-8,* 417 p.

Fredriksen, R. L. 1963. A case history of a mud and rock slide on an experimental watershed: *U.S. Dept. Agr. Forest Serv. Res. Note PNW-1,* 4 p.

_____. 1965. Christmas storm damage on the H. J. Andrews Experimental Forest: *U.S. Dept. Agr. Forest Serv. Res. Note PNW-29,* 11 p.

_____. 1970. Erosion and sedimentation following road construction and timber harvest on unstable soils in three small western Oregon watersheds: *U.S. Dept. Agr. Forest Serv. Res. Paper PNW-104,* 15 p.

Froehlich, H. A. 1973. Natural and man-caused slash in headwater streams: *Loggers Handbook,* v. 33, 8 p.

Goldstein, M., and Ter-Stepanian, G. 1957. The long-term strength of clays and deep creep of slopes: *Proceedings 4th International Conference on Soil Mechanics and Foundation Engineering,* v. 2, p. 311–314.

Gonsior, M. J., and Gardner, R. B. 1971. Investigation of slope failures in the Idaho Batholith: *U.S. Dept. Agr. Forest Serv. Res. Paper INT-97*, 34 p.

Gray, D. H. 1970. Effects of forest clearcutting on the stability of natural slopes: *Assoc. Eng. Geologists Bull.*, v. 7, p. 45–67.

———. 1973. Effects of forest clearcutting on the stability of natural slopes: results of field studies: Interim report to National Science Foundation for Grant no. GK-24747, 119 p.

———, and Brenner, R. P. 1970. The hydrology and stability of cut-over slopes: *Proceedings Symposium on Interdisciplinary Aspects of Watershed Management*, American Society of Civil Engineers, Bozeman, Mont., p. 295–326.

Haefeli, R. 1965. Creep and progressive failure in snow, soil, rock, and ice: *6th International Conference on Soil Mechanics and Foundation Engineering*, v. 3, p. 134–148.

Harr, R. D. 1975. Forest practices and streamflow in western Oregon: Unpublished paper presented at Symposium on Watershed Management, American Society of Civil Engineers, Logan, Utah, 27 p.

———, Harper, W. C., Krygier, J. T., and Hsiel, F. S. 1975. Changes in storm hydrographs after roadbuilding and clearcutting in the Oregon Coast Range: *Water Resources Res.*, v. 11, p. 436–444.

Kitamura, Y., and Senshi Namba. 1966. A field experiment on the uprooting resistance of tree roots (I): *Proceedings of the 77th Meeting of the Japanese Forestry Society*, pp. 568–570.

———, and Senshi Namba. 1968. A field experiment on the uprooting resistance of tree roots (II): *Proceedings of the 79th Meeting of the Japanese Forestry Society*, p. 360–361.

Kojan, E. 1968. Mechanics and rates of natural soil creep: *Proceedings of 1st Session of International Association of Engineering Geologists*, Prague, p. 122–154.

Manbeian, T. 1973. The influence of soil moisture suction, cyclic wetting and drying, and plant roots on the shear strength of a cohesive soil: Ph.D. thesis, University of California, Berkeley, Calif., 207 p.

Megahan, W. F. 1972. Subsurface flow interception by a logging road in mountains of central Idaho: in *National Symposium on Watersheds in Transition*, Colorado State University, Fort Collins, Colo., p. 350–356.

Morrison, P. H. 1975. Ecological and geomorphological consequences of mass movements in the Alder Creek watershed and implications for forest land management: B.A. thesis, University of Oregon, Eugene, Ore., 102 p.

Nakano, H. 1971. Soil and water conservation functions of forest on mountainous land: *Rept. Forest Influences Devel. Govt. (Japan) Forest Expt. Sta.*, 66 p.

O'Loughlin, C. L. 1972. An investigation of the stability of the steepland forest soils in the Coast Mountains, southwest British Columbia: Ph.D. thesis, University of British Columbia, Vancouver, B.C., 147 p.

———. 1974. The effects of timber removal on the stability of forest soils: *Jour. Hydrology (NZ)*, v. 13, p. 121–134.

Parizek, R. R. 1971. Impact of highways on the hydrogeologic environment: in D. R. Coates, ed., *Environmental Geomorphology*, State University of New York, Binghamton, N.Y., p. 151–199.

Rothacher, J. 1959. How much debris down the drainage: *Timberman*, v. 60, p. 75–76.

———. 1963. Net precipitation under a Douglas-fir forest: *Forest Sci.*, v. 9, p. 423–429.

____. 1971. Regimes of streamflow and their modification by logging: in J. T. Krygier and J. D. Hall, eds., *Forest Land Uses and the Stream Environment,* Oregon State University, Corvallis, Ore., p. 40–54.

____. 1973. Does harvest in west slope Douglas-fir increase peak flow in small forest streams: *U.S. Dept. Agr. Forest Serv. Res. Paper PNW-163,* 13 p.

____, and Glazebrook, T. B. 1968. Flood damage in the national forests of Region 6: U.S. Dept. Agr. Forest Serv., Pacific Northwest Forest and Range Expt. Sta., 20 p.

Saito, M., and Uezawa, H. 1961. Failure of soil due to creep: *Proceedings 5th International Conference on Soil Mechanics and Foundation Engineering,* v. 1, p. 315–318.

Swanson, F. J., and Dyrness, C. T. 1975. Impact of clearcutting and road construction on soil erosion by landslides in the western Cascade Range, Oregon: *Geology,* v. 3, p. 393–396.

____, and James, M. E. 1975. Geology and geomorphology of the H. J. Andrews Experimental Forest, western Cascades, Oregon: *U.S. Dept. Agr. Forest Serv. Res. Paper PNW-188,* 14 p.

____, and Lienkaemper, G. W. 1975. The history and physical effects of large organic debris in western Oregon streams: in *Logging debris in streams,* Oregon State University, Corvallis, Ore., 13 p.

Swanston, D. N. 1969. Mass wasting in coastal Alaska: *U.S. Dept. Agr. Forest Serv. Res. Paper PNW-83,* 15 p.

____. 1970. Mechanics of debris avalanching in shallow till soils of southeast Alaska: *U.S. Dept. Agr. Forest Serv. Res. Paper PNW-103,* 17 p.

Ter-Stepanian, G. 1963. On the long-term stability of slopes: *Norg. Geotek. Inst. Pub.,* v. 52, p. 1–14.

Terzaghi, K. 1953. Some miscellaneous notes on creep: *Proceedings 3rd International Conference on Soil Mechanics and Foundation Engineering,* v. 3, p. 205–206.

Varnes, D. J. 1958. Landslide types and processes: in E. B. Eckel, ed., *Landslides and Engineering Practice, Highway Research Board Spec. Publ. 29,* Washington, D.C., p. 20–47.

Wilson, S. D. 1970. Observational data on ground movements related to slope instability: *Jour. Soil Mech. Found. Div. Amer. Soc. Civil Engr.,* v. 96, p. 1521–1544.

Zarbu, Q., and Menel, V. 1969. *Landslides and Their Control:* Elsevier, New York, 205 p.

11

FORECASTING THE EFFECT OF LANDUSE PLANS ON THE REGIONAL MARKET CONDITIONS OF THE SAND AND GRAVEL BUSINESS

Robert H. Fakundiny

INTRODUCTION

As nonrenewable sand and gravel resources become more scarce, regional landuse planners and the aggregate industry need to develop refined methods of forecasting future market conditions. Implementation of a long-term plan can, because of its own schedule, produce a negative feedback situation in which landuse preempts extraction of sand and gravel reserves that are needed to complete the construction phases of the plan.

It is possible to forecast future market characteristics by using a variation of central-place theory. Marketing dynamics respond to a constantly changing set of demand pressures and transportation constraints that, through time, yield mappable geographic patterns. Initial market-domain patterns in a region are determined by distribution of resources, natural transportation funnels and barriers, and demand. Changes in the distribution of consumption centers, depletion of reserves, and alteration of transportation networks subsequently alter these patterns. These man-induced variations can be controlled through planning. Rochester (New York) area sand and gravel reserves will probably last well into the next century if yearly demand remains constant. However, because demand (consumption) centers will shift with changing development patterns and because production centers will migrate as depletion occurs, the market-domain pattern will vary greatly. Sufficient reserves exist. But the cost in certain areas will rise faster than economic trends would indicate, mainly because hauling distances will be extended. However, with the technique to be described planners should be able to forecast the effect of their plans to some extent so that they can ensure the successful completion of regional development while maintaining a healthy business atmosphere for the aggregate producer.

LANDUSE PLANNING OF SUBURBAN MINERAL RESOURCES

The problem of planning sound landuse allocations while maintaining healthy business conditions is acute because of the great demand that exists for construction aggregate. The sand and gravel industry is, by volume, the largest nonfuel mineral industry in the United States, with a total production value in 1970 of approximately $1.1 billion (Yeend, 1973). Annual increase in demand is estimated to be 3.9 to 4.7%

Published by permission of the Director, New York State Science Service. Journal Series Number 203.

nationally (Cooper, 1970). Because landuse planners may determine, to a large degree, the fate of the local mineral producer, they must understand how landuse changes affect the industry and be able to generate plans that will allow continued production. The manipulation of resource use by means of planning is called by some workers the process of "resource engineering" (Donald R. Coates, pers. comm.).

Until now, most landuse regulation has been established at the county, town, or municipal levels of government in the northeastern United States (LaFleur, 1974). Regulation of larger areas of land under a unifying philosophy developed in response to the need for coordinated large-scale construction projects, such as new cities and national highway networks, and to satisfy the public's wish to preserve specified areas in a natural state. Regional planners must be aware, however, that it is possible for suggested plans to end in economic failure if their implementation places excessive demand upon the local resources needed to complete the project. Such failures, when they occur, are not necessarily indications of the weakness of the regional landuse planning concept. but are more likely the result of a lack of awareness of the limitations of the local mineral resource base (Goldman, 1959; Davis, 1970).

The successful completion of a regional landuse scheme can be inhibited by several inherent characteristics, including incorporation of construction projects that require more raw material than is available (Fakundiny, 1975a, b), premature depletion of reserves by preemptive landuse (Bronitsky and Wallace, 1974a; Dare, 1975), inadaptability to changing transportation conditions (Bronitsky and Wallace, 1974b), and the establishment of transportation barriers (Fakundiny, 1975c). The challenge to planners is to determine what effects a given landuse plan will have upon the sand and gravel industry, and to establish alternative actions or adjustments that will allow the plan to be implemented while maintaining and stimulating a healthy sand and gravel business.

The solution has two aspects: (1) to develop a predictive analytical technique that determines marketing trends through time for a given region and uses these trends to forecast future market conditions, and (2) to convey to the landuse planner and the decision maker the results of the analysis so that they can adapt their long-term plans to accommodate the most advantageous conditions. This can be accomplished by using variations of central-place market theory, which has been known to economic geographers for almost 40 yr. This paper describes an application of the theory to the sand and gravel business in the area around Rochester, New York.

CENTRAL-PLACE MARKET ANALYSIS

Central-place theory of market locations (Hoover, 1948; Lösch, 1954; Isard, 1956; Lefeber, 1958; Moses, 1958; von Böventer, 1964; Devletoglou, 1965; Haggett, 1966; Christaller, 1966; Berry, 1967; King, 1969; Plattner, 1975) provides some of the most powerful empirical tools for determining how marketplace locations shift in response to consumer needs.

Central-place theory sets forth a method for determining the most economical, spatial distribution of market locations in relation to given demand patterns for any particular commodity or set of commodities within a trade area, and is thus the theory of the location, size, and nature of clusters of trading activity (Berry, 1967, p. 3). This concept assumes that many factors are equal, including a featureless landscape, evenly distributed population (demand), no social or political barriers to transportation of consumers or commodities, goods that have freedom to flow in any direction in the market structure, a business whose only motive is economic, freely competitive traders, a standard price for a given commodity, inclusion of the consumer's travel costs to and from the marketplace in the cost of the commodity, and a drop to zero at a given distance from the market place (a zero-demand distance). Zero-demand distance is the point beyond which travel cost exceeds the value of the commodity.

Modern economic geographers realize that the postulates set forth by Christaller (1966) and his contemporaries (Weber, 1929; Lösch, 1954) are too simplistic for direct application to market analysis in areas where social, economic, and physiographic characteristics are complex. To analyze real marketplaces, it becomes necessary to consider such nonuniform conditions as randomly configured transportation networks, which can distort the simple, theoretically most efficient geographic patterns of market areas (Berry, 1967), and political barriers to marketing such as national borders (Isard, 1956). Barriers to transportation should not be confused with economic barriers to entry of new firms into an established market area (Bain, 1956).

Central-place theory was derived from the empirical relationships of marketplaces that shift geographically in response to changes in location and character of demand. Thus, the discipline of geography has principles that can be used for evaluating markets, delineating trading and selling areas, and selecting channels of distribution and locations for wholesale, retail, and service establishments (Applebaum, 1954). Accordingly, this philosophy should be applicable to the study of the market dynamics of the sand and gravel business.

APPLICATION OF CENTRAL-PLACE THEORY TO THE SAND AND GRAVEL BUSINESS

Central-place theory assumes that producers transport their commodity to an advantageous marketplace, and that consumers travel to that marketplace to purchase the commodity, and ultimately transport the commodity to a site of use or consumption. In the usual sand and gravel marketing situation, however, the consumer establishes a site of use (consumption site) to which the producer transports the aggregate from the pit (production site). Except for specially processed aggregate that retains value in a stockpile located away from the pit, the central marketplace is eliminated from the analysis. The theory still holds, however, if it is assumed that the site of the pit acts as the marketplace and that transportation vectors extend from the production site to the consumption site rather than from the consumer's location to the marketplace. The problem is thus redefined to determine the most efficient location for a sand and gravel pit, rather than the most efficient marketplace site.

As in the more sophisticated applications of central-place theory, demand for sand and gravel is not equally distributed throughout a market area, but is concentrated at construction sites. The opening of new pits geographically follows shifts in construction locations. Because of the low-cost, high-bulk character of aggregate, the economic framework of the aggregate industry is tied directly to the cost of transportation (Hudec et al., 1970). Bedrock and surficial deposit landforms determine to a great degree the location and nature of transportation networks, but superimposed on the natural landscape are cultural features and landuses that may alter the character of these transportation channels.

The analysis of the sand and gravel business in western New York uses several empirically derived conditions of the market existing in 1971: (1) marketing patterns of the larger firms seem to apply to the conditions of neighboring small operations and can thus be used to characterize the entire market situation in each trade area; (2) the cost of hauling sand and gravel varies from firm to firm and place to place throughout the region (this is partly because of the differences in business profiles of firms and partly because of a variation in zero-demand distances across the region), and (3) since most firms have only a general notion of the amount of their reserves, it is difficult to establish accurately how much and what kinds of reserves are available without instituting a gigantic field study. Each of the above conditions greatly affects the analysis of the market conditions.

However, if some assumptions and generalizations are made about the prevailing market conditions, the task is greatly simplified. The assumptions used in the study of Rochester are as follows:

1. There is an average sand and gravel product for the area that can be used in calculations. The determination of this average product is made by comparing the weighted averages of the amount sold of processed and high-quality run of bank to run of bank used for fill. It is estimated that 6 million tons (T) of sand and gravel were produced in 1971, half of which was high-quality processed items selling at an average of $2.50 to $3/T and half of which was run of bank selling at an average of $1 to $1.50/T. The average product thus is 6 million T of sand and gravel selling at $2/T (Fakundiny, 1975c), which is approximately the same cost as in other parts of New York (Dunn and Hudec, 1967; Bishko and Wallace, 1972).

2. The average zero-demand distance for sand and gravel in the region is 20 mi. The average hauling cost in 1971 was $0.10/mi. where the material was hauled beyond 5 mi. Within 5 mi of the pit, the haul cost is generally a standard fee because for transport to consumption sites near the pit the loading cost outweighs the actual transport cost. Thus, a commonly used formula for haul charges in the Rochester area is $0.50/T anywhere within the first 5 mi and $0.10/mi/T for every mile thereafter. At 20 mi, the haul cost is $2/T or the same as the material cost. This is a common zero-demand condition in other parts of New York (Dunne and Hudec, 1967; Bishko and Wallace, 1972; Bronitsky and Wallace, 1974b).

3. To determine the amount of reserves without testing each production site, it is necessary to assume that the physical characteristics of the deposit, the operating procedure, and the local demand characteristics will remain stable throughout the development of all remaining reserves.

Once the basic data are collected from field canvassing, a desire-line map is made that shows the location of pits and demand centers, with straight lines connecting each pit to its consumption sites. The resulting diagram will show irregularly spaced rosettes of radiating lines that form multirayed stars (Fig. 1A). A few lines may be longer than the rest because of some unusual transportation easement or an anomalously high demand attraction in that direction. The end of each line is the zero-demand distance in that direction from the pit. If the pit were situated on a featureless plain having equal ease of transportation and equal demand distribution in all directions, the desire-line rosette for any given product would have lines of equal length and spacing. More commonly, the rosettes are lopsided, with rays spaced irregularly and extending farther in some directions than in others.

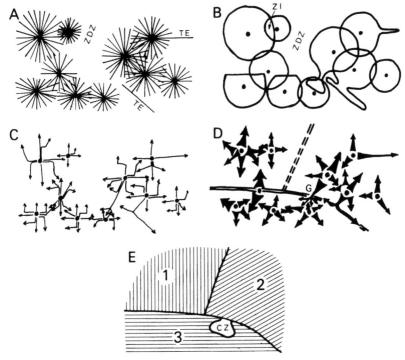

FIGURE 1 *Marketing characteristics analysis technique: (A) desire-line map for hypothetical region: ZDZ, zero-demand zone, TE, desire lines showing unusual transportation ease or unusually high demand pressure; (B) market-area map of region shown in (A) produced by drawing lines around tips of desire-line rosette rays: ZI, zone of indifference; (C) haul-route map of region in (A): pits at dots and consumption sites at arrow heads; (D) funnel map of region in (A): width of arrow tail shows qualitatively the volume transportation along funnel; arrow constricted to gate G; solid double line represents transportation barrier, in this situation a limited-access highway; broken double line represents zero-demand zone boundary; (E) market-domain map of region in (A) showing three domains, two barriers, and a competition zone (CZ).*

By drawing a circular border around the rosette that connects the tips of each ray, a market-area map is produced (Fig. 1B). Each market-area boundary surrounds the area in which the pit, located near the center, does business. Figure 1B is the market-area map of the same hypothetical region. The overlapping areas are zones of indifference (ZI) and represent territory where a consumer is statistically as apt to buy sand and gravel from one producer as from the other, with a probability proportional to the distance his consumption site lies from each (Berry, 1967, p. 40). In the analysis of sand and gravel operations in the Rochester area, it appears that, if several pits are located within 5 mi of each other, they collectively possess a large mutual zone of indifference, especially where many pits are operated by the same company. Where one operation is much larger than its neighbors, its market area overshadows those with lesser production and can be used as the market area for analysis of all other pits adjacent to it. Therefore, study of the larger firms will provide a clear picture of the total sand and gravel market conditions in a region.

By superimposing the market-area map upon a topographic map, it is possible to delineate the barriers to and funnels for transportation. In zones where adjacent market areas do not overlap, such as the zone ZDZ between market areas in the upper, central part of Fig. 1B, a special type of market-area limit is formed, termed herein a "zero-demand zone."

Zones of transportation ease are termed herein "transportation funnels." To delineate transportation funnels a haul-route map is drafted (Fig. 1C). The haul-route map shows avenues of sand and gravel transport along which most of the volume from a pit is carried. These avenues are plotted on topographic maps that show major road networks and that classify the condition of the road. Included on the haul-route map are the data collected in the field from producers and from observation of truck traffic to determine favored transportation routes. Unlike the desire-line map, the haul-route map shows all the bends in the road and produces a truer picture of zero-demand distance by showing road directions and lengths.

Funnels are narrow bands of easy transportation along which the majority of a production site's material flows. The funnel map can be derived from the haul-route map by qualitatively considering the amount of sand and gravel transported along each haul route in a given time period. Figure 1D is a funnel map for the hypothetical region illustrated in Figs. 1A, B, and C. Thus, the funnel map is a way of illustrating the information contained on desire-line, market-area, and haul-route maps.

This information can be synthesized on a market-domain map (Fig. 1E) to show the differences in marketing characteristics across a region. A market domain is a territory that is bounded by transportation barriers and/or zero-demand zones, and is composed of a set of overlapping market areas. All material produced in a market domain is consumed there; production equals consumption and supply equals demand. A special circumstance exists where two adjacent market domains overlap at a gate in a barrier. This area is called a "zone of competition" to distinguish it from zones of indifference between market areas. Zones of competition among market domains are similar to zones of indifference among market areas, except that they are usually larger and define territory where there is abnormal competition between operations in adjacent market domains. Complexity arises in zones of competition because it is more difficult for a producer to know the business conditions of his competitors in another market domain than it is to know the situation of his competition next door. Competition zones, then, are areas where marketing characteristics are most difficult to determine.

Once the current marketing characteristics of a region are determined, including the market domains, barriers, funnels, zero-demand distances, demand, and commodity characteristics, and once these marketing characteristics are analyzed for their interrelationships, the next step is to analyze past situations. Where the changes in marketing characteristics have been consistent and trends have continued in the same direction, a set of long-term market-dynamics criteria can be formulated. These include (1) whether market domains become smaller or larger and whether they become fewer, (2) whether the number of barriers increases, (3) whether the length of funnels increases, (4) whether zero-demand distance changes, and (5) whether these changes are related to changes in the character of the demand and/or to changes in transportation methods and technology. From here, it is simple to forecast future market-domain conditions, if it is assumed that the market dynamics will act as they have in the past and that the same trends will prevail in the future. It is from these assumptions that the predictive capability of the process is derived.

The first set of considerations used in the technique of forecasting future sand and gravel market conditions includes delineation of resources, compilation of reserve inventories, mapping haul routes, determination of funnels and barriers, and, finally, using these to delineate market domains for the contemporary situation. Likewise, this procedure is performed for previous times for which data are available. By comparing the changes from one time to another, market-dynamics trends can be determined. At this point, a proposed landuse plan and its effect upon the transportation network, demand for sand and gravel, and possible conflict of landuse with extraction of reserves are introduced. Funnel- and market-domain maps can then be made for future situations. The funnel map shows where sand and gravel will be produced, and the path it takes to where it is consumed; the market-domain map shows the size of domains, location of barriers, and zones of intensified competition.

Distribution of sand and gravel resources can be determined either by inspecting available surficial deposit maps or by conducting a field mapping program to produce surficial deposit maps. From the producer's point of view, the best deposits of sand and gravel are those that have been naturally processed by wind or water to produce sorted and stratified deposits that can be easily extracted (LaFleur, 1963). Within the glacially derived deposits of the northeastern United States, these include kame and esker complexes, stratified outwash, and glaciolacustrine deltas and beach sands; within postglacial units, these are most commonly sand dunes or stream-deposited sands. Resource distribution maps give the forecaster a basis for predicting the potential that exists for development by indicating where new production sites are likely to be opened in the future. Furthermore, knowledge of the distribution of resources allows the planner to assess where zoning or other landuse decisions will most greatly affect the future availability of sand and gravel.

Natural and man-made barriers and funnels can be easily determined from topographic maps. Barriers commonly include (1) highways with a limited number of crossing points such as underpasses, (2) streams or canals with sparsely distributed bridges, (3) regions of high topographic relief that precludes passable roads, (4) swamps, marshes, and wetlands with few crossing roads, (5) lakes, (6) large areas of restricted access, such as military reserves and urban areas where zoning restricts the size of vehicles allowed to pass over its streets, and (7) preserved lands such as wilderness areas and parks.

Funnels for truck haulage are transportation zones where there is easy hauling, such as (1) valleys with several parallel roads allowing choice of route, (2) high-speed highways, (3) well-maintained roads with few traffic control points, and (4) flat terrain where gas mileage is not reduced by climbing hills. Rail and barge transportation (Dunn and Hudec, 1967) should receive similar analysis, but since these are not currently being considered by Rochester planners for aggregate hauling, they will not be included here.

Market conditions can be determined by polling the producers and consumers about prevailing market situations. This information, along with that derived from site inspections of active pits and construction projects using the material, provide the data needed to characterize the market. These data include amount of reserves, yearly production, grades and uses of aggregate produced, consumption site locations, transportation routes, social, economic, and political influences, and past market conditions and trends.

Some characteristics of market dynamics in the Rochester area are that (1) haul distances (zero-demand distances) are shorter across barriers than along funnels, (2) competition zones tend to cluster near barrier gates and at the ends of funnels, (3) it is easier to establish a new production site than to open a new gate through a barrier to increase the market area of an established site (this is noted by Hoover, 1948, and Isard, 1956, to be a common situation in markets), (4) new barriers create drastic changes in marketing characteristics, (5) depletion by preemptive landuse alters marketing characteristics, (6) major changes in demand patterns alter market-domain configuration, (7) creation of high-demand sites produces competition zones near those barrier gates situated within the zero-demand distance of the new site, (8) market-domain boundaries are constrained by transportation barriers, and (9) market-domain boundaries are extended along funnels.

ROCHESTER EXAMPLE

The Genesee–Finger Lakes Region,[1] formed of eight countries in western New York (Fig. 2), has an area of approximately 4280 mi^2, nearly the size of Connecticut. The region lies adjacent to and south of Lake Ontario, with the majority of the Finger Lakes situated within its southern half. The Genesee River traverses the area, flowing north. Total population of the region in 1970 was 1,075,152, with the highest concentration in the greater metropolitan area of Rochester. The city of Rochester, with approximately 300,000 people, is the third largest city in New York State and houses approximately one third of the region's total population.

The New York State Thruway, a limited-access toll road with a freeway extension into Rochester, is the major highway serving the region. Because it has few entries and exits and only a limited number of crossover points for minor roads and highways, it is a partial barrier to short-haul transportation, for, in some places, one must detour several miles before reaching a crossing to the opposite side. A new highway arterial, the Genesee Expressway, I-390, now in the preliminary stages of development, will connect Dansville in the southern part of the region and other points farther south with the Rochester area.

The interconnected waterway system of the Genesee River, the New York State Barge Canal system, and Lake Ontario provides a network of inexpensive haul routes that has potential as a transportation corridor much beyond its present use and must be considered in any future detailed analyses, since barge haul rates are approximately one tenth the rate for trucks (Dunn and Hudec, 1967; Bronitsky and Wallace, 1974b). Rail transportation should also be considered, because of the established railroad right-of-way network and because the cost of hauling by rail is about one fourth that for trucks (Dunn and Hudec, 1967; Bishko and Wallace, 1972).

Bedrock of the region is composed of relatively flat lying Paleozoic strata of alternating carbonate and clastic sedimentary rocks and is structurally simple compared to other areas of the northeastern United States. In contrast, the overlying glacial,

[1] Genesee–Finger Lakes region will be referred to hereafter as the region.

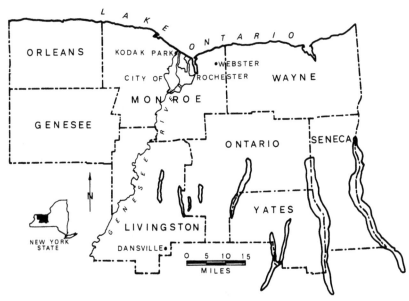

FIGURE 2 *Index map of Genesee–Finger Lakes region.*

glaciolacustrine, glaciofluvial, and postglacial fluvial and aeolian deposits are complex both in geometry and historical development. Bibliographies that will direct the reader to more detailed discussions are found in Muller (1965), Broughton et al. (1966), Rickard and Fisher (1970), the New York State Geological Survey (1972), Fakundiny (1975c), and Muller (1976).

Because sand and gravel represent the third largest mineral resource product produced in the region after salt and limestone, a careful look at future market conditions is necessary, if the mineral resource economic base is to remain a positive factor in the region's future. In 1971, approximately 6 million T of sand and gravel were sold (Fakundiny, 1975c). All aggregate produced from stone quarries, such as crushed stone, is excluded from this analysis and these production figures, although any future analysis of the total aggregate picture must include crushed stone.

Present consumption of sand and gravel is highest in Monroe County with Rochester's Kodak Park and Webster using most of it in large-scale building construction. Expected population growth west and southwest of Rochester will have an impact on future sand and gravel consumption during the construction of two new communities, Gananda in the southwestern part of Wayne County and Riverton in southern Monroe County. Also, at least two predictable new consumers of aggregate will increase aggregate consumption in the region. One is the new steam-generated electric power plants planned to be built along the Lake Ontario and Finger Lakes shorelines; the other is the Genesee Expressway, which will provide easy access to Rochester from Livingston County. The need for low-cost, high-quality road-building aggregate required for the Genesee Expressway will spur prospecting for, and establishment of, new sand and gravel pits in Livingston County.

Total reserves of sand and gravel, regardless of quality or type, are estimated to be approximately 1 billion yd^3 and represent 58% of all original reserves. Monroe County has used 73%, Wayne 36%, Yates 32%, Seneca 84%, Ontario 56%, Livingston 51%, Genesee 59%, and Orleans 38% of all sand and gravel that was originally available on each county's presently mined land (Fakundiny, 1975c). If yearly production continues to be constant, as it has in the last 10 to 15 yr (see U.S. Bureau of Mines Annual Reports, such as Chin, 1973), the projected time of depletion of present reserves will be over 150 yr from now. But, if demand increases at the projected 3.9 to 4.7% (Cooper, 1970), depletion will occur in less than 50 yr.

A field study was conducted in the summer of 1971, and all figures apply to 1971. Costs are in 1971 dollars. This study consisted of visiting the more than 600 sand and gravel production sites (pits) in the region. Where possible, interviews were conducted with owners and/or operators of each site. For all active sites, the operator was interviewed. The data were compiled on a form questionnaire (Fakundiny, 1975c, App. V). Confidential information was used in the analysis technique and is presented only as combined regional figures to preserve the confidentiality of the information received. Among other items, the poll asked for location of pit, zoning restrictions, physiographic setting, geomorphic setting, geologic setting, pit conditions, confidential information used for calculating reserves, types of products, processing, prices, market area, and information related to environmental impact of the operation.

A field reconnaissance investigation of surficial deposits was performed because no complete surficial deposit maps were available. Since the time of the investigation, a new map of part of the area has been published (Muller, 1976). The reconnaissance also determined adequate haul routes and noted which roads were used by trucks to haul sand and gravel. Residents near pits were also interviewed to learn their reaction to the business and their opinions about existing and future zoning. To partially remedy the lack of available surficial geology maps, a geomorphic analysis was made of the 103 U.S. Geological Survey 7½-minute series topographic maps covering the area, using data collected in the field. The resulting interpretation (Fig. 3) shows the distribution of landforms thought to be underlain by stratified sand and gravel deposits. Also shown are the locations of active pits. This map has value not only in showing the distribution of active production sites, but also as a prospecting tool, because it shows the planner and producer where future aggregate sources may be located. However, it must be cautioned that this is not a reserve map.

With these field observations and geomorphic analyses it is possible to reconstruct the market situation in 1971. Figure 4 is the barrier map. Major barriers include Lake Ontario, the Finger Lakes, Genesee River, Oak Orchard Creek in the northwestern corner of the area, marshes, canals, steep slopes such as the hillsides bordering the Finger Lakes valleys, the Thruway, and a military installation in Seneca County. Also shown on the map are crossover points (gates) along barriers. These are bridges over waterways, under- and overpasses along the thruway, and roads through the marshes and swamps. After the gates are plotted, a haul-route map can be made. By incorporating a qualitative estimate of volumes of sand and gravel transported along various haul routes, a funnel map is produced. The funnel map for sand and gravel flow for 1971 (Fig. 5) provides a more detailed measure of the transportation ease and demand than the haul-route map. The

FIGURE 3 *Sand and gravel resources in Genesee–Finger Lakes region: black area underlain by well-sorted and stratified deposits; stipled areas may have naturally refined deposits; circles are active pits. (After Fakundiny, 1975c, plate 2.)*

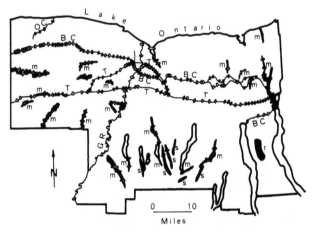

FIGURE 4 *Barrier map of Genesee–Finger Lakes region: GR, Genesee River; OC, Oak Orchard Creek; BC, New York State Barge Canal and remnants of old Erie Canal in the east; T, New York State Thruway and Rochester spur; m, marsh or swamp; s, steep slopes; r, military reserve; oval symbols in barriers are gates.*

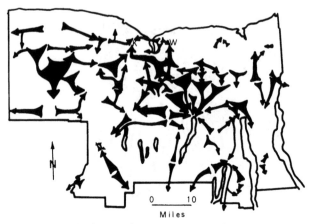

FIGURE 5 *Funnel map of Genesee–Finger Lakes region in 1971: width of arrow tail indicates qualitatively the amount of material transported along arrow (funnel); K, Kodak Park; W, Webster.*

map shows that the thruway acts as a barrier to transportation across it and as a transportation funnel along it.

By combining the information from the desire-line map, the market-area map, and the barrier- and funnel-maps, a market-domain map (Fig. 6) is produced. Figure 6 shows that there are probably 13 different market domains in the region, and also shows where market domains overlap at gates in the barriers. The map shows market-domain boundaries in the southern part of the area where transportation barriers do not exist along

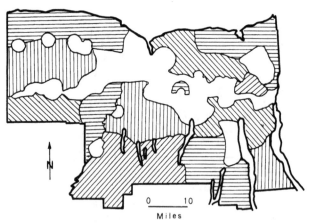

FIGURE 6 *Market-domain map of Genesee–Finger Lakes region in 1971: the area covered by each ruled pattern is a separate market domain; blank areas are competition zones.*

domain borders. Here there are strips of territory without overlapping market areas or zero-demand zones. The long competition zone extending across the region marks the path of the thruway. It is apparent from this diagram that major highways can be formidable barriers to transportation. To test the significance of highway barriers, an investigation was initiated to study the market conditions of the region prior to the construction of the thruway.

Reconstruction of the market-domain situation before the thruway was built requires study of the history of abandoned pits, especially those abandoned during or just after the highway was constructed. By comparing the history of these pits with the history of pits that existed before and continued to operate after the construction, a funnel network map of pre-thruway time can be made (Fig. 7). From this it can be seen that (1) the barge canal was not as substantial a barrier as in 1971, (2) the length of funnels was generally much shorter than at present, and (3) the Genesee River was a barrier then, as now. Funnels were shorter for at least three reasons: (1) there was more aggregate available because fewer pits were depleted, (2) there were fewer zoning restrictions, and (3) the trucks used then were smaller, which made haul costs higher for longer runs. Introduction of the 10-wheel and larger trucks substantially reduced haul costs between the early 1950s and 1971. Before the thruway was built, zero-demand distance was shorter, generally about 10 mi.

From the funnel map it is possible to construct a pre-thruway market-domain map (Fig. 8), which shows there were 13 major domains then also, but they were somewhat larger than at present. Competition zones occupied a smaller portion of the map, and more of the market-domain borders were zones of zero demand.

By comparing the changes in market domains and funnels from pre-thruway time to 1971, several market-dynamics relationships appear. Market domains generally decrease in size through time; funnels become longer through time; zones of competition become larger through time; borders of market domains become more aligned with transportation barriers and less aligned with zero-demand zones; and establishment of limited-access

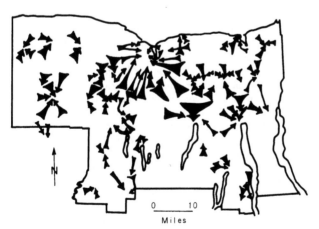

FIGURE 7 *Funnel map of Genesee–Finger Lakes region prior to construction of Thruway.*

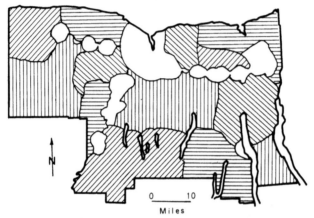

FIGURE 8 *Market-domain map of Genesee–Finger Lakes region prior to construction of Thruway.*

highways with few crossover points produces great changes in the market pattern. Other dynamics that remain constant through time also appear: haul distances across barriers are shorter than along funnels; competition zones tend to cluster near gates; funnel ends tend to concentrate competition; market-domain boundaries are extended by transportation funnels; and depletion of pits alters funnel configuration.

Having established that the construction of limited-access highways with selected overpasses places stress upon market conditions, it is important to determine what effect future highways will have on the market condition even though not many more are likely to be built in the next few decades. The Genesee Expressway is a new highway planned by federal and state transportation departments to extend from Dansville northward to Rochester, as an extension of the highway system leading to New York City across the southern counties of the state. Using very preliminary and extremely rough estimates, it can be assumed that construction will require possibly as much as 4 million yd of state-approved aggregate and as much as 400 million yd of fill. Using the analysis technique described above, the local funnel situation will be approximately as shown in Fig. 9. Most aggregate will move in east–west directions, which will mean heavier traffic on similar-trending roads. Assuming a free competition situation in which most aggregate will come from production sites within 10 mi, it is possible to forecast how this perturbation in the normal development of the area will affect the local market situation.

Figure 10 shows the present reserve conditions in belts within 10 mi and from 10 to 20 mi distant from the proposed highway. There are 207.4 million yd of reserves within 10 mi of the construction, and 112.5 million yd are of state-approved construction quality. High-grade reserves are adequate to supply the 4-million-yd demand for state-approved material. However, only approximately 306 million yd are available for fill within 20 mi of the highway. Clearly, more production sites will have to be developed to satisfy fill requirements. Because of the exclusive zoning within Monroe County, all extra required sand and gravel must come from Genesee, Wyoming, Livingston, and the western half of Ontario counties lying to the south. Comparison of Figs. 3 and 10 reveals the areal

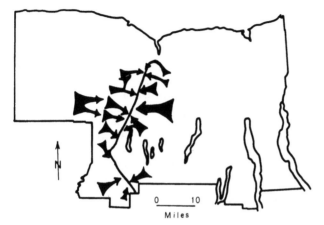

FIGURE 9 *Funnel map of territory around planned Genesee Expressway.*

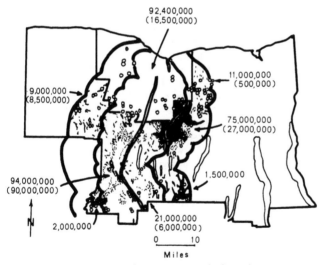

FIGURE 10 *Resources and reserves around planned Genesee Expressway; symbols the same as in Fig. 3. Resources in Monroe County not shown because of prohibitive zoning there; double line is proposed path of the expressway; lines on either side represent 10- and 20-mi haul distances; numbers refer to reserves of sand and gravel in cubic yards; numbers in parentheses are the cubic yards of sand and gravel approved for New York State construction projects.*

distribution of regions that have high likelihood of being underlain by stratified sand and gravel in Monroe County, but which are now lost by preemptive landuse and restrictive zoning.

The analysis is not complete, however, without taking into account other development within the region and assessing the combined effect of the construction of the highway and other construction projects. Figure 11 shows the projected expanded area of high, medium, and low urban development within the greater metropolitan area of Rochester by the year 1990 that was estimated by the New York State Office of Planning Coordination (1971), and the anticipated need for at least four new major steam-generating power plants (New York State Power Pool, 1973). Also shown are two planned residential communities, Gananda in Wayne County and Riverton in Monroe County.

Assuming that demand for aggregate is proportional to overall population, it appears valid to project the increase in demand for sand and gravel by looking at population increase estimates. The population for the region during 1970 was 1,075,152 and is projected to be 1,367,000 in 1990 (New York State Office of Planning Services, 1974). This represents a total increase of almost 30%. Assuming that sand and gravel production will increase from its present 6 million T/yr at the same rate as population, we can expect production of more than 7 million T/yr in 1990. Rough estimates indicate that from 50 to 70 million yd of state-approved sand and gravel and up to 600 million yd of fill will be needed to fulfill the projected special urban and suburban development growth estimated for 1990. This does not deplete all available state-approved reserves; but the need for fill will require total development of all available fill reserves. If these figures are added to the yearly constant demand of 6 to 8 million yd of sand and gravel used for normal development, the total by 1990 will be from 50 to 70 million yd of state-approved sand and gravel, 150 million yd of processed aggregate for other purposes, and 1 billion yd of fill. With the present economic situation and its concomitant slowdown of construction, these figures may be more applicable to the year 2000 than to 1990.

Funnels in 1990 (Fig. 11) will have to be extended by as much as 35 mi. Great pressure will be placed on development of the resources in the central part of the region. It can also be assumed that all reserves for the region within 10 mi of the strip of land between the Genesee River and the Genesee Expressway will be depleted. If crossover points are not carefully chosen for the expressway, development within this strip will feel the high cost of long haul distances. The cost of aggregate may greatly increase the cost of construction within this zone.

Figure 12 shows the market-domain picture for 1990. The most striking feature is the large competition zone across the northern half of the region. Also, market domains have decreased in both number and size.

IMPLICATIONS FOR THE REGION

The field canvass indicates that the region has about 1 billion yd of sand and gravel in reserves. Assuming that the assumptions listed above about development approximate the impending situation, state-approved reserves will suffice until the end of the century, but many more pits will have to be opened to supply road and construction fill and low-quality bulk needs.

Construction of the Genesee Expressway will present some special problems; mainly, if all the available reserves are used from the adjacent areas, there will be added haul costs

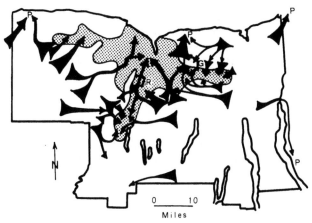

FIGURE 11   *Funnel map of Genesee–Finger Lakes region in 1990: stipled area expected to be under development; P, expected power plant construction; R, Riverton; G, Gananda.*

for sand and gravel needed to develop the region bounded by the Genesee River and the Genesee Expressway. This area will become the most restricted in the region because of the few bridges across the river and because crossover points for the expressway will be more widely spaced than the present road network. Expressway construction will also compete for reserves that are now supplying Rochester's needs. Rochester will have to reach farther into the surrounding countryside to obtain aggregate. The fortunate aspect is that there is ample resource potential, if wise planning of their use is begun now.

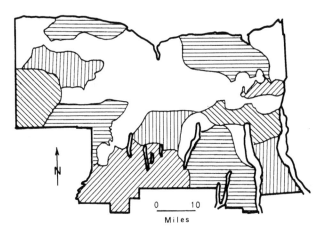

FIGURE 12   *Market-domain map of Genesee–Finger Lakes region in 1990.*

Nevertheless, a careful mapping of the region to delineate the extent of resources and an attempt to define and evaluate reserves are required.

Any plan that considers the future demand for aggregate, the problems of transportation, and the acceptability by the public of opening new production sites must investigate the advantages of hauling aggregate by rail or barge and the need to design extraction procedures in a fashion that allows for adequate reclamation of the land for secondary landuses.

CONCLUSIONS

It is possible to predict the geographically controlled characteristics of the aggregate market condition through a sequence of steps, starting with a field canvass of operators and a reconnaissance of the region's resource potential and physiographic characteristics. In areas where surficial geology maps are inadequate or unavailable, these can be generated by an experienced geomorphologist. The end product is a series of maps illustrating market dynamics and the sequential change in market domains and transportation ease.

The technique can be used by planners to understand what long-term situations will result from their planning schemes and the impact that the plan will have on the public, as well as the aggregate industry. The technique allows them to test the effects of alternative landuse options for a more satisfying time—space development of sand and gravel resources. The method can also be used by the industry to predict where demand centers will be in the future and how they should obtain reserve property to satisfy those demands.

Regardless of who uses the technique, the picture that is emerging for the future of the Rochester area indicates that close cooperation will be needed between planners, who design regional development, the community, who decides zoning regulations, and the industry, which provides building material so that all can mutually benefit.

ACKNOWLEDGMENTS

Paul W. Fickett and David R. Sipperly helped collect field data. The paper was reviewed by James F. Davis, James R. Dunn and the staff of Dunn Geoscience Corporation, Robert G. LaFleur, Robert F. Legget, Fred Pessl, George Toung, and W. A. Wallace. Although the paper reflects many of their excellent and widely varying comments, the emphasis placed on certain topics is my own.

REFERENCES

Applebaum, W. 1954. Marketing geography: in *American Geography: Inventory and Prospect*, Syracuse University, Syracuse, N.Y., p. 245–251.

Bain, J. S. 1956. *Barriers to New Competition:* Harvard University Press, Cambridge, Mass., 329 p.

Berry, B. J. L. 1967. *Geography of Market Centers and Retail Distribution:* Prentice-Hall, Englewood Cliffs, N.J., 146 p.

Bishko, D., and Wallace, W. A. 1972. A planning model for construction minerals: *Management Sci.,* v. 18, no. 10, p. B-502-518.

Bronitsky, L., and Wallace, W. A. 1974a. The impact of urbanization on construction minerals: *OMEGA,* v. 2, no. 6, p. 809–813.

———, and Wallace, W. A. 1974b. The economic impact of urbanization on the mineral aggregate industry: *Econ. Geography,* v. 50, no. 2, p. 130–140.

Broughton, J. G., Fisher, D. W., Isachsen, Y. W., and Richard, L. V. 1966. Geology of New York: a short account: *N.Y. State Museum Sci. Serv. Educ. Leaflet 20,* 49 p.

Chin, E. 1973. The mineral industry of New York: in *Area Reports, Domestic,* U.S. Dept. Interior, Bur. Mines Minerals Yearbook, 1971, v. II, p. 511–523.

Christaller, W. 1966. *Central Places in Southern Germany:* translated from *Die zentralen Orte in Suddeutschland* by C. W. Baskin, Prentice-Hall, Englewood Cliffs, N.J., 230 p.

Cooper, J. D. 1970. Sand and gravel: in Mineral facts and problems, *U.S. Bur. Mines Bull.* 650, p. 1185–1199.

Dare, W. L. 1975. Pre-emptive land-use—its impact on the mineral base: *Virginia Minerals,* v. 21, no. 3, p. 23–26.

Davis, J. F. 1970. Geologic inventory of Central New York Region: Central New York Regional Planning and Development Board, HUD project no. NYP-160, Syracuse, N.Y., 40 p.

Devletoglou, N. E. 1965. A dissenting view of duoply and spatial competition: *Economica,* v. 32, no. 126, p. 140–160.

Dunn, J. R., and Hudec, P. P. 1967. Evaluation of mineral resources of Nassau and Suffolk Counties, New York: N.Y. State Sci. Serv. Geol. Survey, Open File Rept., Albany, N.Y., 94 p.

Fakundiny, R. H. 1975a. New method for predicting the effectiveness of land-use plans upon the sand and gravel business (abs.): in *Geol. Soc. America Abstr. with Programs* (Northeastern Sect., Syracuse, N.Y., 1975, 10th Annual Meeting), v. 7, no. 1, p. 55–56.

———. 1975b. Predicting the effect of future land use upon the sand and gravel business (abs.): in *Geol. Soc. America Abstr. with Programs* (Salt Lake City, Utah, 1975, Annual Meeting), v. 7, no. 7, p. 1071.

———. 1975c. Geologic resources: *Genesee/Finger Lakes Regional Plan. Board Tech. Study Ser., no. 16,* HUD project no. G/FL RPB: 75-NYP-1038-TSR16, 169 p.

Goldman, H. B. 1959. Urbanization and the mineral industry: *Calif. Div. Mines Mineral Inform. Serv.,* v. 12, no. 12, p. 1–5.

Haggett, P. 1966. *Locational Analysis in Human Geography:* St. Martin's, New York, 339 p.

Hoover, E. M. 1948. *The Location of Economic Activity:* McGraw-Hill, New York, 310 p.

Hudec, P. P., Dunn, J. R., and Brown, S. P. 1970. Transportation advantage—a unifying factor in mineral aggregate valuation: in *Proceedings of the Sixth Forum on Geology of Industrial Minerals,* Michigan Dept. Nat. Resources, Misc. 1, p. 87–94.

Isard, W. 1956. *Location and Space-Economy:* Wiley, New York, and The MIT Press, Cambridge, Mass., 350 p.

King, L. J. 1969. *Statistical Analysis in Geography:* Prentice-Hall, Englewood Cliffs, N.J., 288 p.

LaFleur, R. G. 1963. Origin of sand and gravel deposits in New York State: in *First Annual Sand and Gravel Symposium,* Rensselaer Polytechnic Institute, Troy, N.Y., p. 1–27.

――――. 1974. Glacial geology in rural land use planning and zoning: in D. R. Coates, ed., *Glacial Geomorphology,* State University of New York, Binghamton, N.Y., p. 375–388.

Lefeber, L. 1958. *Allocation in Space, Production, Transport, and Industrial Location:* North-Holland, Amsterdam, 151 p.

Lösch, A. 1954. *Economics of Location,* 2d rev. ed.: translated by W. H. Woglom and W. F. Stolper from *Die räumliche Ordnung der Wirtschaft,* (G. Fischer, Jena, 1941), Yale University Press, New Haven, Conn.

Moses, L. N. 1958. Location and the theory of production: *Quart. Jour. Economics,* v. 72, no. 2, p. 259–272.

Muller, E. H. 1965. Bibliography of New York Quaternary: *N.Y. State Museum Sci. Serv. Bull. 398,* 116 p.

――――. 1976. Surficial geology of the Niagara Sheet, New York: *N.Y. State Museum Sci. Serv. Map and Chart Ser. 28* (in press).

New York State Geological Survey. 1972. *A Complete List of Geological Publications of the Geological Survey:* New York State Museum and Science Service, New York State Geological Survey, Albany, N.Y., 26 p.

New York State Office of Planning Coordination. 1971. New York State Development Plan 1: N.Y. State Office Plan. Coord., HUD Proj. N.Y.P. 222, 127 p.

New York State Office of Planning Services. 1974. New York State demographic projections 7–Genesee–Finger Lakes Region (rev. 1974): N.Y. State Office Plan. Services Proj. N.Y.P. 245. (Obtainable from Policy Planning Bureau, New York State Economic Development Board, Alfred E. Smith Building, Box 7027, Albany, N.Y. 12225.)

New York State Power Pool. 1973. 1973 report of the member electric corporations of the New York Power Pool and Empire State Electric Energy Service Corporation to the Public Service Commission pursuant to Article VIII, Section 149-b of the Public Service Law, New York State Power Pool, Albany, N.Y. v. 1: (Obtainable from New York State Public Service Commission, Albany, N.Y. 12208.)

Plattner, S. 1975. Rural market networks: *Sci. Amer.,* v. 232, no. 5, p. 66–79.

Rickard, L. V., and Fisher, D. W. 1970. Geologic map of the state of New York, Niagara and Finger Lakes Sheets: *N.Y. State Museum Sci. Serv. Map and Chart Ser. 15.*

von Böventer, E. 1964. Spatial organization theory as a basis for regional planning: *Amer. Inst. Planners Jour.,* v. 30, p. 90–100.

Weber, A. 1929. Alfred Weber's theory of the location of industries (C. J. Friedrich, ed.): University of Chicago Press, Chicago, 255 p.

Yeend, W. 1973. Sand and gravel: in D. A. Brobst and W. P. Platt, eds., United States mineral resources, *U.S. Geol. Survey Prof. Paper 820,* p. 561–565.

IV
URBANIZATION
EFFECTS

12

THE URBANIZING RIVER: A CASE STUDY IN THE MARYLAND PIEDMONT

Helen L. Fox

INTRODUCTION

The continuing interaction between water flow in its confining channel, changing over time and space and affected by the whims of man and weather, is the raison d'etre of fluvial geomorphology. Its students seek for intellectual satisfaction and for ways to better manage an integral part of our environment. The Patuxent River Basin in Maryland (Fig. 1A) is a rewarding setting for fluvial studies: one river with 2,410 km² in drainage area provides basic data on rural and urbanizing streams; on the effects over time of suburban construction and catastrophic flooding on sediment transport and channel form; and on the impacts of construction and flooding from the Piedmont stream sources through the Maryland Coastal Plain to the river's estuary on the Chesapeake Bay.

The Maryland Department of Natural Resources supported a study in 1972 and 1973 on which this report is based and also a previous study in 1971 from which data are used.

THE PATUXENT RIVER BASIN

The Patuxent River heads at an elevation of 242 m in north-central Maryland near Parr's Ridge, the physiographic boundary between the eastern and western Piedmont divisions. The main stream flows south over the eastern Piedmont and crosses the fall line at an elevation of about 45 m near Laurel, Maryland, between Washington, D.C. and Baltimore, Maryland. The river's course continues south and southeast through the coastal plain of southern Maryland to the Patuxent Estuary and, finally, Chesapeake Bay (Fig. 1B). The major tributary of the stream net, the Little Patuxent River, follows a similar path to the east of the main stream and joins it about 100 km upstream of the estuary mouth.

The lower third of the river forms the Patuxent estuary, and tidal influence extends 96 km upstream to Queen Anne's Bridge, where the main stream drains 950 km². Total drainage area at the estuary mouth is 2,410 km², and the total main stream length is 177 km.

The basin's proximity to the bay and the Atlantic Ocean moderates its humid continental climate. Temperatures average 25°C in summer and 2°C in winter, and about 107 cm of rain falls annually. There is no well-defined rainy season; precipitation intensities and durations vary through the year. From April through September, rain is more likely to fall in brief showers or thundershowers of several hours' length and high intensity. From October through March, frontal storms with longer rain periods (up to several days) of lower intensity are more probable. Tropical storms and hurricanes

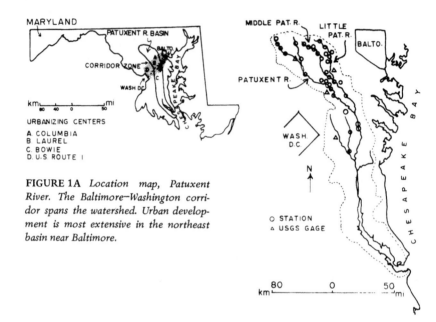

FIGURE 1A *Location map, Patuxent River. The Baltimore–Washington corridor spans the watershed. Urban development is most extensive in the northeast basin near Baltimore.*

FIGURE 1B *Location of study stations. Stations are spread through the freshwater basin and cover a range of landuses on the main stream and tributaries.*

generally occur between August and October once every 4 yr on the average (Dunn and Miller, 1960, p. 292–293). These storms, thundershowers, and severe frontal systems can all bring heavy rains, high winds, and flooding to the study area.

The metamorphic basement rocks of the northern Piedmont quarter of the watershed (655 km^2) are generally more resistant to stream erosion than the unconsolidated sedimentary strata of the coastal plain. Although upland topography is hilly or rolling throughout the watershed, valley-to-ridge relief is 9 to 15 m greater in the Piedmont than in the lower basin. Stream gradients in the Piedmont average 3.4 m/km, compared with 1.9 m/km in the coastal plain. The stream valleys in the upper basin are narrow, with small or discontinuous floodplains a few hundred meters wide at most. Valleys in the coastal plain are several thousand meters wide, frequently with distributary channels and swamps (Figs. 2A and 2B). Floodplains occupy about 4% of the watershed's total area, or 95 km^2.

Streams range from 3 to 100 m wide in the freshwater basin, with banks 2.5 to 3 m high in silty alluvial material containing scattered sand, clay, and gravel lenses. The soils are silty and sandy. Most of the watershed surface is moderately to severely eroded owing to clearing and cultivation, which has occurred since the colonial era.

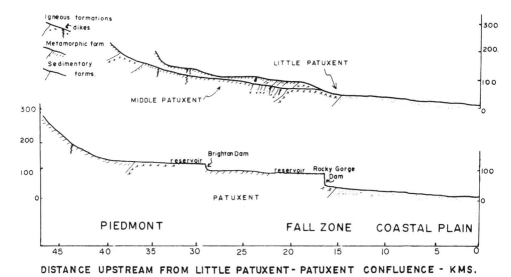

FIGURE 2A *Longitudinal profile and geology of the Patuxent rivers. The headwater channels in the Piedmont country rocks have steeper gradients than the lower streams in the Coastal Plain sediments. (Maryland Geological Survey, 1968.)*

HYDROLOGY

About one third of the precipitation falling on the basin surface enters the channels as storm runoff (The Johns Hopkins University, 1966, p. 34). The U.S. Geological Survey operates six permanent stream gages on the Patuxent and its tributaries (Table 1), including one measuring unregulated flow from a rural catchment at Unity above two water-supply reservoirs, a second below the reservoirs at Rocky Gorge near Laurel, and a third on the unregulated Little Patuxent near Guilford, draining a partly urbanized basin. Flows average about 1 m^3/s each for the rural 90.2 km^2 above the Unity gage and the 98.5 km^2 draining into Guilford, half of which has recently been urbanized. Highest

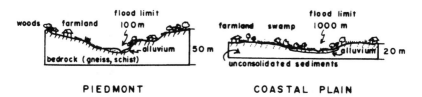

FIGURE 2B *Valley sections, Patuxent basin. Piedmont valleys are generally narrow with small floodplains that funnel high flows quickly downstream compared to the broad flat valleys of the coastal plain.*

TABLE 1 *Gaging Stations in the Patuxent Basin*

Station number (U.S. Geol. Survey)	Station name	Drainage area		Period of record water-years	Nature of record	Duration of daily flow: Discharge equaled or exceeded percent of the time (m³/s)		
		mi²	km²			10%	50%	90%
5910	Patuxent River near Unity	34.8	90.2	1944–present	Long term, continuous	1.8	0.6	0.2
5925	Patuxent River near Laurel (downstream of two reservoirs)	132.0	341.9	1945–present	Long term, continuous	3.2[a] 7.6[b]	0.4[a] 2.8[b]	0.2[a] 1.1[b]
5935	Little Patuxent at Guilford	38.0	98.4	1932–present	Long term, continuous	1.8	0.7	0.3
5945	Western Branch near Largo	30.2	78.2	1950–1967	Long term, continuous	1.6	0.4	0.1
5946[c]	Cocktown Creek near Huntingtown	3.9	9.9	1957–present	Long term, continuous	0.3	0.1	0.01
5948[c]	St. Leonard Creek near St. Leonard	6.7	17.4	1957–present	Long term, continuous	0.4	0.2	0.1

[a]Flow values for period 1945–1953 before installation of T. H. Duckett Reservoir immediately upstream of gage.
[b]Flow values for period 1953–present after installation of reservoir.
[c]Stations not located in the study area.
Source: P. N. Walker, *Flow Characteristics of Maryland Streams,* Md. Geol. Survey, Report of Investigations 16, Baltimore, Md., and U.S. Geol. Survey Water Resources Division, College Park, Md., 1971.

sustained daily flows usually occur in the spring, but floods can take place any time of year. A series of floods in August and September 1971 due to thundershowers and in June 1972 with the onslaught of tropical storm Agnes brought record high flows to the watershed (Table 2).

LANDUSE

About 44% of the basin is presently in forest or swamp cover and will probably remain so for the next 30 or 40 yr. In 1966, agriculture accounted for 41% and urban landuses for 15% of the watershed. By 1980, 27% of the basin will be in agricultural use and 29% will be in urban areas (The Johns Hopkins University, 1966, p. 118). This rapid landuse conversion is centered in the corridor zone between the cities of Washington, D.C., and Baltimore, Maryland (Fig. 1A). The northernmost part of the watershed and the southern third are relatively far removed from the corridor zone, but the central and eastern parts lie within the zone and will be directly affected by its urbanization.

The new town of Columbia, founded in 1966, lies in the Little Patuxent basin. About half this river's Piedmont watershed is urbanized or developing. The rural market centers of Laurel and Bowie in the Patuxent Basin have developed satellite communities and suburbs in the last 20 yr. Scattered subdivisions dot the central two thirds of the entire watershed. A strip of commercial and industrial facilities is centered on Route 1 (Fig. 1A), but the bulk of development is residential U.S. Routes 1, 3-301, 29, 40, and 50 and Interstate Highways 70 and 95 provide residents with ready mobility in any direction.

Such progress brings erosion hazards. The numerous construction sites have exposed and disturbed ground, resulting in sediment damages on downslope land surfaces and in the receiving waterways. The state of Maryland piloted an Erosion and Sediment Control Program on the Patuxent Basin in 1969 and extended the program statewide by law in

TABLE 2 *Magnitude of Floods in the Patuxent Basin, 1971–1972*

| Location of data, date of flood | Peak discharge | | Approximate peak height | | Duration of overbank flow (hours) | Recurrence interval of flow (years)[a] |
	m³/s	m³/km²	Above bed (m)	Above floodplain (m)		
Unity (drainage area 90.2 km²)						
Aug. 3–4, 1971	263	2.9	3.8	2.6	10	Close to 50
Aug. 27, 1971	16	0.2	1.2	Bankfull	–	Less than 2
Sept. 11–12, 1971	620[b]	7.0	5.2	4.0	26	Considerably more than 100
June 21–22, 1972	410	4.5	4.9	3.7	Not known	Over 100
Guilford (drainage area 98.5 km²)						
Aug. 3–4, 1971	85	0.9	3.5	0.9	11	More than 10
Aug. 27, 1971	56	0.6	3.0	0.6	10	Between 2 and 5
Sept. 11–12, 1971	65	0.7	3.2	0.6	20	More than 5
June 21–22, 1972	351[c]	3.6	5.5	3.0	24	Between 50 and 100

[a]Recurrence intervals are by Walker (1971, p. 73 and 78).
[b]Previous highest flow on record at Unity is 303 m³/s on July 21, 1956.
[c]Previous highest flow on record at Guilford is 150 m³/s on Sept. 1, 1952.

1970 (House Bill 1151, 1970). Sediment still remains in some channels and enters streams from open areas, from nonpoint sources, and from uncontrolled sites despite the law. Paving on parking lots and streets, impervious rooftops and patios, new storm drainage systems, and other land-cover changes accompanying development alter the infiltration capacity, runoff patterns, and concentration times in the different subwatersheds involved. Flood peaks tend to increase in height and decrease in lag time in developing basins compared to flows in the same basin under natural or rural cover (Fig. 3). Floodplain management programs in the counties involved should prevent new encroachment and building in the floodplains. Older roads may be threatened by higher flows if building continues in the future, further altering the hydrologic conditions of the watershed.

INTENSIVE STUDY STATIONS

I collected field data for this study at 36 locations in the Piedmont and 11 in the coastal plain, all in the freshwater basin (Table 3). Literature sources provided information on the tidal river and estuary.

The data-collection sites or stations are documented recoverable locations on the tributary and main stream channels of the Patuxent drainage net (Figs. 1, 4A, and 4B). Cross sections, channel parameters, and bed material samples were measured two to four times. Suspended sediment samples were collected during a range of flows at each station between October 1972 and January 1974. Thirty-four of the stations have channel parameters and bed material data from 1971. Station locations were selected to provide a range of subwatershed sizes from 0.52 to 961.2 km^2 in both urban and rural settings with cases of comparable and comparative geologic settings. A station was also chosen to be typical of the channel reach where it was located and to be accessible for measurement in

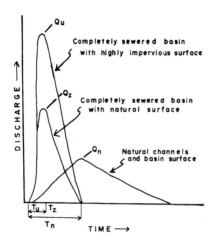

FIGURE 3 *Effect of urban development on flood hydrographs. Peak discharges (Q) are higher and occur sooner after runoff starts (T) in basins after they have been developed or sewered.*

TABLE 3   *Stations Classified by Landuse, Geology, and Reach*

Landuse[a]	Total number of stations	Geology		Reach, number of sections[b]		
		Piedmont	Coastal Plain	Pool	Riffle	Uniform
Rural	25	20	5	17	19	1
Urban	22	16	6	18	25	3
Total	47	36	11	35	44	4

[a]Urban stations are those located in urban, suburban, or developing environments, and stations with 30% or more of their basins so developed. Rural stations have wooded or agricultural environments and less than 30% of their basins affected by development.
[b]Two or more cross sections were measured at a number of stations.

varying flows. Urban stations (26 sites from 0.52 to 380 km) are those in urban, built-up, or developing areas, or sites with 30% or more of their contributing watersheds so urbanized. Twenty-eight rural stations draining 1.3 to 961.1 km^2 occur in rural, wooded, or swampy surroundings and have less than 30% of their basins in urbanized landuses.

The reach of individual stations is defined as "pool" or "riffle" when possible. Pools are defined in relation to riffles in their immediate vicinity: the pool has greater depth and lower velocity for a given flow than its corresponding riffle. In the coastal plain, many main-stream reaches are uniform in average flows. Even in headwater reaches, the pool and riffle sequences washes out in high flows. The terms are used in this discussion as they apply to the reaches present in average flow conditions.

The field data taken at each site provide a sediment transport rate and a number of channel cross sections for that point on the channel, with material describing the stream bed and floodplain in the immediate vicinity (Fig. 4A). The shape and size of the channel cross section are expressed by the form or width—depth ratio and the cross-sectional area measured from an arbitrary point. As successive measurements are made, the changes in the section's shape and size are expressed by a monthly rate of change. The rates are based on each section's original measurements and expressed as a percentage of that original value, so that an areal enlargement of 2 m^2 occurring over 2 months in a section that originally measured 4 m^2 is expressed as a 25% monthly rate. Overall alterations taking place in the channel are expressed by adding the rates of size and shape change to form a change index, which is also expressed as a monthly rate of alteration from the original measures for convenient comparison between different sections. In general, a positive change in the width—depth ratio indicates a widening or shallowing of the section, and a negative change, a deepening of the section. A positive rate of areal change indicates scour or enlargement of the section, and a decrease of area from bed deposition or bar building results in a negative rate of areal change. The change index is formed from absolute magnitudes and shows channel alteration without regard to the kind of process causing it.

MEASUREMENTS AT A STUDY STATION

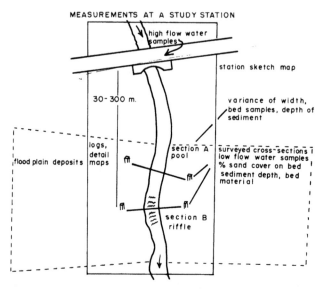

FIGURE 4A *Measurements at a study station. A variety of data were collected near the cross sections to detail their local environments. The section locations are documented for future studies.*

CATASTROPHIC FLOODING

Storms of 1971–1972

Four major floods occurred in the Patuxent Basin in 1971 and 1972 (Table 2), making it possible to determine the response of the channels to repeated flooding. Detailed field observations were made at 31 stations in the Piedmont, where data on channel and bed parameters were available for comparison from 1971 just prior to the flooding.

The first flood occurred after sporadic thunderstorms deposited 140 to 190 mm of rain on the northern part of the watershed from August 1 to 4, 1971. Peak flow at Unity was 263 m³/s, near the estimated 50-yr event, and overbank flooding lasted over 10 h. Water was overbank. for 11 h at Guilford, but the flood itself was not so severe in the Little Patuxent Basin, being about the size of the 10-yr event. On August 27, tropical storm Doria moved northward along the Atlantic Coast, bringing 25 to 127 mm of rain to the area. The waters were bank-full at Unity, and overbank flooding at Guilford lasted almost 10 h. Intense thundershowers on September 11 and 12 produced 60 to 115 mm of rainfall and led to record flows with over-100-yrs recurrence intervals on the Patuxent. Flows were overbank at Unity for 26 h with a peak discharge of 620 m³/s. Guilford recorded 20 h of overbank flow peaking at 65 m³/s.

Despite the previous summer's floods, the June 1972 precipitation was unprecedented in the watershed due to tropical storm Agnes, which brought 305 mm of rain to the area

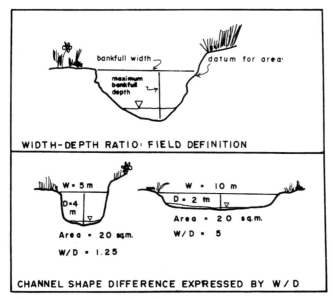

FIGURE 4B *Width–depth ratio. This parameter was defined using the bank top with the lower elevation to define "bankfull." Shape differences are easily expressed by the ratio.*

between June 21 and 23. Between the afternoon of June 21 and the morning of June 22, 75% of the total storm precipitation fell on the Upper Patuxent Basin. The Unity stream gage was not operating automatically, but field investigation fixed the peak discharge at about 410 m³/s (Taylor, 1972). The maximum flood of record occurred at Guilford, with a peak flow of 351 m³/s and 24 h of overbank flow. Details of the floods and the station hydrographs are given in Table 2 and Fig. 5. In this discussion of data collected before and after the floods, the following questions were considered:

1. How were the channels, their bed materials, and channel-bar configurations altered during the floods?

2. Are these patterns and types of channel, bed, and bar alterations common to both urban and rural stations?

3. Is more damage of some kind sustained by urban or by rural reaches in a discernible pattern?

4. How much sediment deposition does major flooding produce on the river flood-plains and in the estuary?

Changes within the Channel

The most striking effect of the floods was channel widening throughout the basin. There was an almost linear relation between widening of the channel bed and drainage area as observed after the September flood (Fig. 6A). Channel widening at the top of the banks also increased downstream, although without a linear relation to drainage area (Fig.

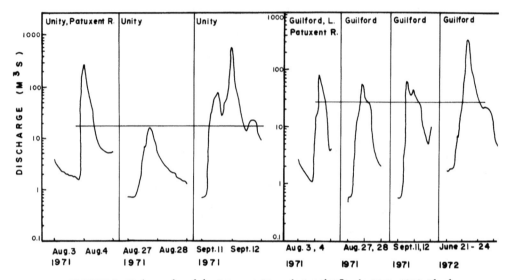

FIGURE 5   *Hydrographs of the Patuxent River during the floods, 1971–1972. The four floods were unprecedented in the basin for their magnitude and their close occurrence. The September 1971 flood was the record flood on the Patuxent, as was the June 1972 storm on the Little Patuxent.*

6B). In general, the banks were widened more at the top than at the bottom, resulting in a more dish-shaped channel section than had existed previously. Both kinds of bank widening occurred again in June 1972, although to a lesser degree. Presumably the September 1971 flood enlarged a channel originally suited to much smaller bankfull capacity to allow its passage, whereas flows in June 1972 did less damage because the channels were already enlarged to carry the volumes of flow. The flooded cross-sectional areas of the channels were directly proportional to the drainage area contributing to the flows, with no significant difference between urban and rural sites.

Width–depth ratios expressing channel shape averaged 5.7 in 1971 prior to the storms, with very little dependence on drainage area or landuse difference. After the floods, the average ratio was 8.4 (Table 4), indicating a general widening of the channels rather than scouring. On the average, rural channels increased their width–depth ratios by 47.5% and urban channels by 70.5%.

Transport competence increased markedly during the high flows, with boulders 300 by 300 by 200 mm being moved over the floodplains as well as on the stream beds (Gupta and Fox, 1974, p. 503). The streams sorted bed materials in association with pools and riffles. The riffles were coarsened by the addition of cobbles and pebbles, and swept clear of sand and finer particles. These fines were deposited downstream in pools, on floodplains, and in the reservoirs, or transported into the Coastal Plain river reaches.

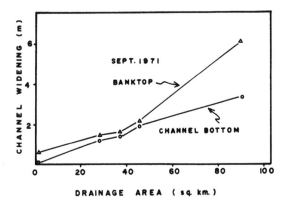

FIGURE 6A *Channel widening in floods. The September 1971 flood widened the channels more at the top of the cross section near the bank edge than at the botton near the bed.*

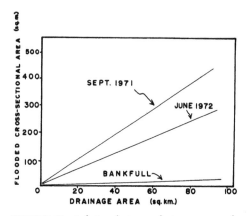

FIGURE 6B *Relation between drainage area and channel area during the floods. The September 1971 flood widened channels that were much too small to accommodate its flow; the flood of June 1972 flowed in a channel that was already enlarged. Both flows were so much larger than the normal bankfull flow that they used the valley sections as floodways. (Gupta and Fox, 1974.)*

TABLE 4 *Measurements of Channel Change*

Parameter	Time of observations	Conditions	No. of stations	Mean value of parameters	Standard deviation
Width–depth	1971	All sites	32	5.7[a]	3.6
ratio (w/d)		Rural	22	5.4	2.7
		Urban	10	6.3	5.2
	1972,	All sites	25	8.4	5.5
	after Agnes	Rural	14	7.3	2.9
		Urban	11	9.7	7.6
Change in	1971–1972;	All sites	25	57.6[b]	67.9
width–depth	after Agnes	Rural	14	47.5	37.5
ratio, $\phi\,(w/d)$		Urban	11	70.5	94.3
Width–depth	1972, fall	All sites	54	8.1[a]	4.5
ratio (w/d)		Rural	28	8.2	3.2
		Urban	26	8.1	5.6
	1973,	All sites	54	9.1[a]	5.5
	end of study	Rural	28	8.6	3.4
		Urban	26	9.7	7.1
Change in	1972–1973	All sites	54	2.4[b]	5.1
width–depth		Rural	28	0.8	1.6
ratio, $\phi\,(w/d)$		Urban	26	4.1	6.8
Change in	1972–1973	All sites	54	1.2[c]	5.8
cross-sectional		Rural	28	0.1	2.0
area (ϕA)		Urban	26	2.4	8.1
Increase in		Rural	15	1.6	1.3
cross-sectional		Urban	16	5.6	8.6
area (ϕA)					
Decrease in		Rural	13	1.6	1.2
cross-sectional		Urban	10	2.8	3.1
area (ϕA)					
Change index:	1972–1973	All sites	54	5.8[d]	7.5
first to last		Rural	28	2.8	1.9
measurements		Urban	26	9.0	9.7
$\phi\,(w/d) + \phi A = CI$					
Change index:	1972–1973	All sites	54	15.4	22.0
winter–spring		Rural	28	5.4	5.7
measurements		Urban	26	25.9	27.6
Change index:		All sites	54	13.1	32.0
summer–fall		Rural	28	8.6	22.6
measurements		Urban	26	17.9	39.6

[a] w/d mean values and standard deviations expressed in meters/meters.
[b] $\phi(w/d)$ mean values and standard deviations expressed in %/month.
[c] ϕA mean values and standard deviations expressed in %/month.
[d] CI mean values and standard deviations expressed in %/month.

Channel bars eroded to a fraction of their original size and after each flood were left composed of coarser materials than before, with sands and finer particles removed. The location of pools, riffles, and bars did not change significantly, even though the flow swept out the channels in high stages, suggesting that these features are associated with the channel configuration in low- and medium-flow ranges rather than with the catastrophic event (Gupta and Fox, 1974, p. 503–504).

Although the bed materials in all the reaches observed were somewhat coarsened, urban reaches contained more sand and small pebbles immediately after the storms than did rural ones. Sandy dunes as high as 460 mm choked some urban tributaries, and bars of coarse sand formed in the larger urban channels of the Little Patuxent in Columbia as the receding water lost transport capacity to carry the excess loads of sediment from open construction sites upstream. Sandy bars were virtually absent in the rural Piedmont reaches after the floods; only pebble and cobble cores remained where before there had been sandier bars.

Damage to the River Floodplains

The long periods of overbank flow destroyed floodplain vegetation within 30 m of the channel banks. Near the stream channels, trees 7.6 m high were uprooted, broken, or bent downstream by the force of the flow, and larger trees 12 to 24 m high on the bank edge toppled owing to bank undercutting and slumping. Lower-story shrubs were uprooted and broken, and an assortment of more pliant shrubs, smaller plants, and herbaceous and grassy annuals was carried away or flattened out in place. Twigs, leaf litter, and debris were swept away and a new increment strewn over the ground as the waters receded.

The bottomland vegetation and undergrowth functioned as sediment and debris traps as the river rose out of its normal channel and used the valley section as a floodway. Overbank deposition was most conspicuous where the water assumed a straight course over the floodplain across bends. The most common deposits were coarse sands and pebbles 2.5 to 5.0 cm in size, but textures ranged from fine sand through cobble sizes at different locations. Elliptical splays up to 500 m^2 in area and 0.6 m thick were left on the floodplain in exceptional instances. However, such deposits were isolated, and the surface cover was generally discontinuous and thin, averaging 0.6 cm distributed over the entire floodplain surface in spite of the size of individual deposits. Floodplain scour was limited even more to isolated cases of channels cut behind boulders protruding into the flow or root exposure near the channel edge. Scouring occurred mainly in narrow valleys where the overbank flow had considerable depth and velocity.

The river floodplain recovered rapidly after the storms, and most evidence of the flooding was concealed very effectively within 4 to 5 months after Agnes. Naturally, the larger trees could not be replaced by mature ones, but the bottomland understory species produced a dense growth through the summer. Grassy and herbaceous annuals grew back thickly and masked both deposits and scours, making their positive identification difficult after as little as 2 months. The pebble and cobble splays, so striking immediately after the storms, were rapidly absorbed into the landscape.

Assuming an average depth of 0.6 cm over the total floodplain area of 1.1×10^9 m^2 for the entire basin (2,410 km^2), the volume of flood deposits was about 6.5×10^5 m^3. This is about the same as the volume of deposition estimated to have accumulated during 100 yr of agricultural activity in the basin, at an annual deposition rate of 0.06 cm (Table

TABLE 5 *Estimated Storage in the Floodplain, Estuary, and Channels, Patuxent River Basin*

Storage location and origin of deposits	Volume (m^3)	Volume (metric tons)	Total thickness (m)	Rate of deposition (cm/yr)	Remarks
Floodplain: Area 1.1 × 10⁸ m²					
Unknown	2.9×10^8	3.8×10^8	2.4 (Piedmont) 2.8 (Coastal Plain)	0.06 (?)	Total storage in all floodplain, age of deposits unknown
Floods 1971–1972	6.7×10^5	8.7×10^5	0.006	—	0.6 cm deposited on total floodplain
Last 100 years	6.7×10^5	8.7×10^5	0.06	0.06	Total over period, on 10% of total floodplain
	6.7×10^3	8.7×10^3	—	0.06	Yearly rate
Total urbanization	7.8×10^5	1.0×10^6	0.007	0.45	Basin totally urbanized in 1½ yr, all material transported by streams deposited on floodplain
Patuxent Estuary: Area 9.2 × 10⁷ m²					
Agriculture (?)	2.5×10^6	3.3×10^6	—	2.7	Yearly rate
1859–1944	2.1×10^7	2.8×10^8	1.5–3.0	—	Total in 85 years[a]
1966 annual rate agriculture and urbanization	1.2×10^5	1.7×10^5	—	0.12	[a]
Agnes (30 × average annual sediment yield)	3.9×10^6	5.1×10^6	0.04	—	Estimate after Schubel (1974)
Total urbanization	7.8×10^5	1.0×10^6	0.009	0.6	All material transported by streams deposited in estuary
Channel					
Channel storage 1966	4.2×10^5	5.3×10^5			15% basin urbanized[a]
Channel storage 2010	8.7×10^5	1.1×10^6			41% basin urbanized[a]
Total urbanization	1.9×10^6	2.5×10^6			100% basin urbanized in 1½ yr

[a]Data from references, The Johns Hopkins University, 1966.

5) typical of Piedmont areas, with the deposits concentrated on that tenth of the floodplain surface nearest the channel itself. Such increments are not large in comparison with the total volume of floodplain deposits, some 2.8×10^8 m^3, that have accumulated over an unknown span of time (Table 5).

Effect of Flooding on the Patuxent Estuary

The estuary below the tidal river is the final sink for sediment entering from the sources in the basin. Studies of the effects of Agnes on the Chesapeake Bay show that the Susquehanna River contributed most of the sediment entering the bay during the storm period, and that very little escaped from the tributary estuaries (Chesapeake Research Consortium, 1974). Schubel (1974) estimated the sediment loads on the Susquehanna were 30 times its average annual load during the storm period. No comparable measures were made on the Patuxent estuary at this time, but an average annual load of 1.7×10^5 metric tons of sediment was determined as the river's contribution to the estuary in 1966 (The Johns Hopkins University, 1966, p. 67). Assuming that the Patuxent carried about the same relative amount of sediment as did the Susquehanna, one then finds that 5.0×10^6 metric tons of material entered the estuary during the Agnes flood (Table 5). This would amount to nearly 40 mm of deposition spread over the estuary's channel bottom area of 9.2×10^7 km, with a density of 1.3×10^3 kg/m^3 (Leopold, Wolman, and Miller, 1964, p. 505).

Seismic bottom profiles run in 1969 (Stiles and Weisnet, 1970) revealed a tongue of mud and silty deposits in the estuary bottom up to 5.3 m deep. Much of this layer could have built up in the measured period between 1859 and 1944, when sediment collected at an average annual rate of 2.7 cm (The Johns Hopkins University, 1966, p. 68). Agriculture was widely practiced then without the benefit of erosion control or conservation efforts, and the existing two water supply reservoirs now on the main Patuxent channel had not yet been built. Some 2.8×10^8 metric tons of sediment accumulated in this period on the estuary floor. Two years of normal deposition during this period would have exceeded the entire contribution made to the estuary during Agnes, the 100-yr storm, indicating that in terms of sediment deposition the effects of the storms on the estuary were relatively small compared to past accumulations.

IMPACTS OF URBANIZATION

Introduction

Sedimentation of channels, changes in their form and size, and alterations of their natural patterns occur as previously rural streams adjust to their new environments during urbanization. The data collected at the study stations during the period after the great floods were analyzed with the following questions in mind:

1. What differences are there between rural and urban stations that are not related to the major floods, but more likely due to urbanization?

2. Do these differences persist downstream away from a local urban center or construction site? If so, how far below the source area are the effects noticeable?

3. How much sediment is present in urban stream channels compared to rural ones?

4. Where is the sediment produced during urbanization finally deposited?

5. What is the impact of urbanization on the rivers compared to the impact of the floods? Are such effects short term and localized, or will they alter the entire river system in time?

Local Changes in Pools and Riffles

The width (w) and cross-sectional area (A) of the streams at 41 stations on the channels were related to the discharges at a constant frequency[1] to define a downstream hydraulic geometry for the watershed (Table 6). A statistical analysis of the data from pools and riffles in both urban and rural stream reaches shows that the width and area of a section are significantly related to the discharge at that section. However, none of the urban–rural pairs of equations, such as those for pools in both settings in the Piedmont, were significantly different from each other at the 5% level. The development of the hydraulic geometry relations (Leopold and Maddock, 1953), using a number of stations of different sizes, apparently smooths out differences between them that are not very strongly developed; thus, this kind of analysis is not sensitive to the impacts of urbanization in their early phases.

The regression relations themselves suggest some differences in the response of pools and riffles to changing landuse. The intercepts of the equations define the initial values of the width or area at a discharge of 0.03 m^3/s (Q15 for a small headwater channel draining about 0.9 km^2). The exponents give the rate of change in the width or area with discharge as the discharge increases downstream (Table 6). Urban pools have smaller intercepts and larger exponents than do rural pools, whereas urban riffles resemble their rural counterparts more closely. Due to their lower velocities, pools in an urbanizing channel collect sediment and debris more readily than do riffles during the earlier phases of development. For a while the pools undergo more modification than do the riffles, but eventually the riffles begin to scour. In time, the differences between the two reach types are reduced and the channels becomre more or less uniform, rather than a sequence of shallows and deeps. The resulting regression equations for urban reaches would have similar parameters for both pools and riffles and would be significantly different from the equations for rural reaches. The Patuxent streams seem to be in the first phase of the sequence during which the urban pools are filling.

Local Changes in Channel Form and Size

During the 18 months following Agnes, low to medium flows predominated in the watershed. Two-year floods were the highest ones recorded at either the Unity or the Guilford gage. During this period, the rivers had opportunity to begin their recovery after the catastrophic storm flows. Low flows deposited fine sediment and sand opposite eroded banks and between channel bars and the banks, reducing the size of the postflood channels. Fine particles lodging in the coarsened stream beds built up a fine matrix between larger particles. Although some of the fines came from surface erosion during rainstorms, much was contributed from eroded banks that slumped into the channels during the great floods and later disintegrated. Silt and sands deposited on the bars increased their areas at the expense of the channel cross-sectional areas and decreased stream depths. As the bar cores built up with fines, small plants frequently rooted on them, further consolidating them.

[1] The discharge equaled or exceeded 15% of the time, Q15.

TABLE 6 *Linear Regression Equations of Pools and Riffles, Hydraulic Geometry Parameters*[a]

Geology	Landuse	Reach type	Regression equation	Correlation coefficient	Exponent
Piedmont and	Urban	Pool	$w = 0.4 + 0.59Q$	0.93	+0.59
Coastal Plain	Rural	Pool	$w = 0.62 + 0.41Q$	0.92	+0.41
	Urban	Riffle	$w = 0.51 + 0.43Q$	0.93	+0.43
	Rural	Riffle	$w = 0.54 + 0.41Q$	0.88	+0.41
	Urban	Pool	$A = 0.039 + 1.00Q$	0.97	+1.00
	Rural	Pool	$A = 0.11 + 0.78Q$	0.90	+0.78
	Urban	Riffle	$A = 0.036 + 0.78Q$	0.94	+0.78
	Rural	Riffle	$A = 0.039 + 0.78Q$	0.95	+0.78
Piedmont	Urban	Pool	$w = 0.38 + 0.66Q$	0.94	+0.66
	Rural	Pool	$w = 0.62 + 0.44Q$	0.84	+0.44
	Urban	Riffle	$w = 0.51 + 0.43Q$	0.94	+0.43
	Urban	Pool	$A = 0.028 + 1.07Q$	0.96	+1.07
	Rural	Pool	$A = 0.124 + 0.78Q$	0.81	+0.78
	Urban	Riffle	$A = 0.035 + 0.98Q$	0.97	+0.98
	Rural	Riffle	$A = 0.079 + 0.56Q$	0.91	+0.56

Regression equation: $X = A + B(Q)$, where

X = dependent variable: w = width; A = cross-sectional area
A = intercept of regression line at $Q = 0$
B = exponent of Q in hydraulic geometry relation $x = Q^b$ (Leopold and Maddock, 1953); slope of the regression line
Q = stream discharge in ft^3/s; independent variable

[a]Original field data processed by computer and statistical analyses. References for general methods: Cooley and Lohnes, 1971; Miller and Freud, 1965; Yevjevich, 1972.

This reacquisition of fines raised the channel-bed level and caused minor bank erosion in some places. Channel shapes shallowed as a consequence and widened slightly, and the average width–depth ratio increased from its postflood value of 8.5 to 9.1 in rural stream reaches by January 1974 (Table 4).

Channel adjustment by bar construction and shoaling also occurred in urban stream reaches, but with the addition of sediment loads averaging 340 metric tons/km$_2$/yr or about fifteen times the loads typical of rural channels (22 metric tons/km^2). Flow regimes in urban channels also differ from those in rural basins, especially in small tributary streams where a single development tract may cover a large percentage of the stream's drainage area. Consequently, the magnitudes of the channel alterations and the rates at which they occurred were very different between the two kinds of reaches. The average monthly rate of change in the width–depth ratio was 0.8%[1] for rural stations, with a standard deviation of 1.24% (Table 4). Urban stations averaged 4.1% change monthly in width–depth ratios during the same period, with a standard deviation of 6.6%. The highest individual rate of change for 28 width–depth ratios on rural stations was near

[1] Rates of change are measured relative to each station's original width–depth ratio, area, etc., and expressed as a percent of change occurring per month.

4.5%/month, whereas over a quarter of the 26 urban sites changed at greater rates, up to 27.4%/month.

The average cross-sectional area increased for rural sites by 0.1%/month (standard deviation 2.0%), probably owing to the redistribution of slumped bank material from consolidated blocks to layers of fines incorporated into the bed and bars over areas downstream. Urban channels also increased their cross-sectional areas by 2.4%/month (standard deviation 8.1%), mainly by bank scouring. Severe bank erosion continued after the floods in a number of urban stations. The monthly change index, combining magnitudes of both shape and areal changes, averaged 2.8% for rural stations compared to 9.0% for urban stations, nearly a third as large as the rural value. Standard deviations of the change index were 5.7% in rural stations and 9.7% in urban ones, following the consistent pattern of greater magnitudes of change combined with greater variability among stations for the urban measurements.

The largest change index recorded for a rural site was nearly 5%/month; therefore, values of the index up to 5% could be due to channel recovery after the floods. However, more than half the total number of urban stations altered size and shape at rates in excess of 5%/month, ranging up to 34%/month. There is some indication of possible seasonal differences in the alteration of the streams affected by different landuses as well. Although urban channels had higher change indexes than rural ones for any season during the study, the largest amount of change in urban sites occurred during the winter and spring, whereas the greatest changes in rural sites took place during the summer and fall. The different rainfall patterns during the time periods may produce different rates of erosion and deposition in streams with varying landuse and hydrologic conditions, but longer studies are necessary to determine if these observations were only chance or not.

Channel cross-sectional changes are most pronounced near construction sites in urbanizing basins on the smaller tributary streams directly draining the work areas. After

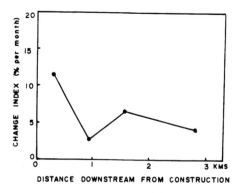

FIGURE 7A *Change index variation downstream of sediment sources. The channel shape and form changes are local and fall to rural levels after a kilometer or so of river flow downstream from a construction area. Near the construction site and in urban centers change index values are high.*

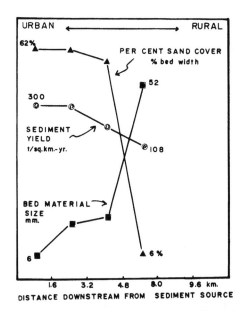

FIGURE 7B *Downstream   sediment   dispersal.
Sediment yield and bed material parameters remain typical of urban sites or about 5.6 km below
a source area in contrast to the change index.*

1.5 km or so of river flow downstream, channel change values fall to lower levels more typical of rural watersheds (Fig. 7A). In extended suburban developments surrounding Columbia and Laurel, most of the tributary channels are affected, as are the main stream reaches flowing through the cities. The magnitude of channel alterations falls off rapidly below the urban and construction centers, and channel size and shape alterations are localized in the immediate vicinity of their causes (Table 10).

This tendency of stream channels to widen, whether slowly in rural areas or more rapidly in urban ones, cannot continue indefinitely. Streams thrown out of equilibrium by floods and landuse changes, such as the Patuxent rivers have been, follow patterns of channel readjustment to their changed environments. In rural channels, downcutting and scouring eventually balance bank erosion and bottom deposition as the fluvial system works to equalize its sediment input with output. The total environment determining the streams' sediment and water budgets was not really altered by the floods, which only affected the channels themselves and their immediate floodplains, comprising about 4% of the total basin area. The rural streams resulting after the postflood adjustment has taken place will be very similar to those that existed before the floods, as flow regimes are the same as they were prior to the storms. Sediment loads reaching the channels from agricultural sources and scattered small building tracts are not excessive. The postflood adjustment is currently taking place in the rural reaches at low rates, with scouring and filling processes nearly balanced. The smaller headwater streams have already returned to their general preflood characteristics.

Channels in the urbanizing parts of the watershed have a continually changing environment that provides different sediment and runoff inputs to the streams with each new building tract. Urban channel development is difficult to predict; it is affected by changing basin characteristics, as well as by variable climatic and weather conditions. Such streams have been observed to undergo cycles of drastic alteration, with periods of excessive sediment deposition followed by severe bank erosion and channel scouring (Wolman, 1967). Eventually, these streams also reach an equilibrium, usually only after active construction on their contributing watersheds has ceased. The resulting waterways bear little resemblance to the original channels.

Sedimentation in Urban and Rural Streams

Streams carry sediment from a number of natural and man-made sources, including upland surfaces, eroding stream banks, cultivated fields, roadcuts, gullies, and construction sites. The last source usually produces larger amounts of sediment than any of the others, resulting in sediment accumulation and channel choking in urban reaches. The Little Patuxent River, with more extensive development than the other rivers in the basin, has higher sediment yields throughout its length than comparable rural reaches in the other major streams (Table 7). The annual sediment yield at Guilford on the Little Patuxent is 236 metric tons per square kilometer (t/km^2) annually, 324% greater than that at Unity on the Patuxent, whose rural watershed produces 56 t/km^2 annually. During one storm on June 22, 1973, nearly bankfull flows carried an average 21.5 metric tons of sediment from each square kilometer of basin past the gage at Guilford in 4.45 h. This yield is almost equal to the average total sediment yield for a whole year from the rural headwater streams, which produce 21.8 $t/km^2/yr$. Basins with active construction sites have sediment yields some 15 times greater, averaging 337 $t/km^2/yr$. Urbanized basins in which the building is completed produce an average of 37 $t/km^2/yr$. The sediment yields of the larger study sites, encompassing a range of basin sizes and a variety of landuse combinations (Table 7), still hold this basic pattern. Stations with the highest yields are near construction sites, and those with the lowest yield values are mainly agricultural. The range of sediment yield values for different landuses in the Patuxent Basin agrees with previously published data for similar watersheds in Maryland and Virginia (Wark and Keller, 1963; Wolman and Schick, 1967; Vice, Guy, and Ferguson, 1969).

Sediment carried by runoff waters moves in a series of steps in different flows downstream to final depositions on the floodplain or in the estuary. Temporary sediment storage occurs on slopes downhill of the source area, on floodplains, and in stream channels as bed fills or bars. The differences in sediment yields in successive downriver reaches are due to either sediment deposition and storage or the dilution of heavily laden flows by the incoming clearer flows of rural tributaries, or a combination of the two. Near Columbia and in the Coastal Plain, the downstream decrease in sediment yields cannot be totally accounted for by flow dilution within the reach, and sediment deposition and temporary storage occur (Table 8).

Although all sizes of particles may be eroded from the basin surface, clay, silt, and sand usually reach the channels directly while larger particles are more likely to be trapped on the floodplain or slopes (Vice, Guy, and Ferguson, 1969). Silt and finer

TABLE 7 *Sediment Yields in the Patuxent River Basin*

Station	Yield (metric tons of sediment/km² /yr)	Station	Yield (metric tons of sediment/km² /yr)
Patuxent River, Piedmont		Middle Patuxent, Piedmont (con't.)	
Rural, 1.3 km²	21.8	Route 108, rural	150.8
Annapolis Rock, rural	116.1	Simpsonville, rural with construction	198.9
Unity, rural	55.5	Route 29, rural with construction	122.3
Roxbury Mills, rural with construction	299.2		
Patuxent River, fall zone		Little Patuxent, Piedmont	
Laurel, Rte 1,[a] urban, regulated	410.9	Golf course, rural with temporary construction	20.1
Patuxent River, coastal plain		Route 40, rural	16.6
Brock Bridge,[a] urban, regulated	200.5	Annapolis Road, urbanizing	65.5
		Wilde Lake, urban	93.1
Priest Bridge, rural[a]	84.7	Urban, 1.3 km²	36.6
Queen Anne's Bridge,[a] rural	101.3	Columbia Station, urban	50.9
		Owen Brown, urban	59.6
Leeland, rural with construction	140.0	Guilford, urbanizing	235.7
		Savage Gauge, urban with construction	246.5
Middle Patuxent, Piedmont		Dorsey Run, urbanizing	343.4
		Little Patuxent, coastal plain	107.4
Construction, 1.3 km²	337.3	Symond's Bridge, rural	252.4
Inwood, rural with construction	908.8	Fort Meade, rural with construction	
Folly Quarter, rural with local construction	183.0	Patuxent Village, rural	73.7
Route 98, rural	134.4	Route 424, rural	87.6

[a]Areas and sediment yields corrected for storage of sediment from headwater reaches in the two Washington Suburban Sanitary Commission water-supply reservoirs on the main stream of the Patuxent River between Laurel and Unity.

TABLE 8 *Reaches with Decreasing Sediment Yield Due to Dilution and Deposition*

Reach	% Change of area	% Change of yield	Suspected process
Patuxent River			
Annapolis Rock to Unity	+141.6	−52.2	Dilution
Laurel Rte. 1 to Brock Bridge	+6.3	−51.2	Deposition > dilution
Brock Bridge to Priest Bridge	+133.5	−57.6	Dilution
Middle Patuxent River			
Inwood to Folly Quarter	+1,070.0	−79.8	Dilution
Folly Quarter to Rte. 98	+21.3	−26.5	Dilution and deposition
Simpsonville to Rte. 29	+8.2	−38.4	Deposition > dilution
Little Patuxent River			
Golf course to Rte. 40	+79.4	−17.1	Dilution
Wilde Lake Station to	+8.6	−45.4	Deposition > dilution
Columbia Station			
Savage Gauge to Symond's Bridge	+27.1	−56.4	Deposition > dilution
Fort Meade to Patuxent Village	+10.9	−70.7	Deposition > dilution
Rte. 424 to Priest Bridge	+119.3	−2.9	Dilution

particles entering the flow are carried downstream, while sand and small pebbles are deposited nearer their sources in the channel bed and bars. These deposits migrate downstream at different intervals in flows of varying magnitude (Table 9).

Rural stream-bed materials in the Piedmont range from boulders to sand, with cobbles and pebbles predominating. More sand and small pebbles occur in urban channels than in rural ones of similar size and geology. The difference in bed materials is especially striking in streams with construction in progress on their watersheds. Sand and granules dominate most channel beds in the Coastal Plain river reaches, regardless of landuse.

Changes in Stream-Bed Materials

The floods of 1971–1972 coarsened stream-bed and bar materials throughout the watershed, more so in rural reaches than in urban ones. Rural Piedmont streams had recovered their preflood characteristics by the fall of 1972 and remained stable during the study period. Changes of bed materials at urban sites that occurred during the study period are due to the impact of urbanization rather than to the floods, as the streams had already had time to recover from these.

Once the rural streams had regained their preflood bed materials, they remained stable in composition. Striking changes of bed materials took place at several urbanizing stations draining 1.2 to 52.0 km^2 in Columbia. At three separate locations, sand steadily invaded pebble–cobble stream beds through reaches 30 to 150 m long, blanketing from 9% to 38% of the stream bed width per month in the different locations. On the average, sand covers 57% of the total bed widths of urbanizing channels in Columbia and Laurel and below highway construction sites. This percentage is 2.5 times the average value for rural streams, in which 23% of the beds are covered with sand in those locations where sand is found in the bed. Coastal Plain streams have 80% sandy bed covers due to geology and hydraulic factors rather than to urbanization.

Loose sandy deposits are deepest in channels near construction, averaging 0.4 m deep. Urban areas developed 2 to 10 yr before the study have thinner deposits about 0.1 m deep. The average sand layer for all urban channels was 0.2 m, 1.8 times the average layer depth of about 0.1 m found in the rural stream reaches. However, for every three sites with a measurable sand layer in urban reaches, only one rural site had a measurable sand layer; thus, the deposits in urban reaches are not only wider and deeper, but are also more widespread than in rural reaches. Overall, there is 13.5 times as much sand stored in Piedmont urban stream channels in the Patuxent basins as in rural ones. Although the natural channel beds in the Coastal Plain mask incoming sand from construction sites, the comparative amounts stored in rural and urbanizing reaches is probably about the same order of magnitude as in the Piedmont reaches. The bulk of this material in urban stream channels is in temporary storage, and observations suggest that most sand moves through a reach within 1 yr on its way downstream (Table 9).

Changes in Channel Bars

Bars are another form of channel storage. They differ from stream-bed covers described above in their more concentrated physical form and their more sporadic movement downstream. Rural channels frequently have pebble or cobble bars along meander bends, near bridge constrictions, in riffles and flood channels, and below eroding banks. These features were coarsened by the floods, but regained their preflood compositions in several months with subsequent low and medium flows, and afterward remained quite stable. Vegetation often roots on their exposed surfaces and increases their stability. Floods with recurrence intervals over 50 yrs may erode them badly, but smaller floods do only minor damage.

Sandy bars are more typical of urban streams in the Piedmont, and these features are more numerous and more unstable than rural bars. Of the rural stations in the study, 60% have bars as compared with 90% of the urban stations. These urban bars accumulate and lose material rapidly. A large one near construction on the Little Patuxent headwaters

TABLE 9 *Lifetimes of Sedimentary Deposits*

Deposits	Location	Lifetimes at one point (yr)
Cobbles to sand in overbank layers and piles	Floodplains, urban and rural	>50; long term
Sand to boulders, loose particles	Channel beds, rural	<1 to 50
Vegetated point bars, sand to cobbles	Meander bends, rural	>50; long term
Unvegetated granule, pebble, or cobble bars	Channel beds, rural	5–50
Sand to pebbles, loose particles	Channel beds, urban	1–2
Sand waves and wedges	Channel beds, urban	<1
Vegetated sand and pebble bars	Channel beds, urban	10–50
Unvegetated sand and granule bars covered in high flows	Channel beds, urban	2–10
Unvegetated sand bars covered in medium flows	Channel beds, urban	1–2
Unvegetated sand bars, covered by low flows	Channel beds, urban	<1

collected 13.3 metric tons of sand during the 14-month study and stabilized with vegetation as well. Other urban streams near Columbia and Laurel repeatedly built up bars and washed them away, only to rebuild them again in a few weeks. The smaller sandy bars subject to such flushing do not support vegetation. Since urban bars are 1.5 times as numerous as rural ones, their relative storage capacity combined with that of channel beds suggests that there is 15 times as much sediment present in urban river reaches as in rural ones.

High sediment yields and extensive sandy bed covers persist for about 5.6 km downstream of a construction site or urban center (Table 10). After this distance, the streams bear more resemblance to rural than to urban channels, as the incoming sediment is dispersed in larger channels among the natural bed cover (Fig. 7B). After the rivers enter the Coastal Plain province, the sandy sediment blends into the sandy beds and becomes indistinguishable from them. The low gradients of the Coastal Plain reaches promote deposition, and the broad swampy floodplains act as natural sediment traps during overbank flow.

Channel and Estuarine Storage

Detailed studies on Baisman Run, a small Piedmont stream lying to the east of the Patuxent Basin in Baltimore County, Maryland, show that the channel stores about four times the basin's annual sediment yield (Wolman, 1974, pers. comm.). Using this factor for the Patuxent Basin, which has similar physical characteristices, one finds that each rural square kilometer contains 85.2 metric tons of sediment in channel storage, and each urban square kilometer contains 1,280 metric tons of deposits. With 15% of the watershed urbanized in 1966, the channels would have contained 4.2×10^5 m^3 or 5.3×10^5 metric tons of sediment. If the basin is 41% urbanized in the year 2000, the sediment then held in channel storage would increase to 8.7×10^5 m^3 or 1.1×10^6 metric tons, only 20% of the sediment load that is estimated to have reached the estuary during Agnes (Table 5). If the entire basin were urbanized in an average 1.5 yr, a grand total of 1.9×10^6 m^3 or 2.5×10^6 metric tons of sediment would be held in channel storage as a result. These figures are clearly speculative, but they suggest that a good deal, perhaps 70%, of the sediment derived from construction remains temporarily in channel storage en route to the estuary, where most of it is finally deposited. The actual time of transport depends on the sediment particle size, the flows available for its transport, and the forms of temporary storage the particles encounter on their journey downstream.

Sand deposited on the floodplain during a storm may remain in place many years until another flood removes it, whereas sand in the channel below it proceeds down-

TABLE 10 *Downstream Variation of Bed and Sediment Parameters*

Distance downstream of sediment source (km)	Size of average bed material (mm)	Average percent sand cover of bed (% of total width)	Average sand layer depth (cm)	Average suspended sediment yield (t/km²/yr)
0.0–1.6	5.9	61.9	11.2	300
1.6–3.2	14.5	61.3	17.6	302
3.2–4.8	16.1	58.5	8.5	181
4.8–8.0	51.7	5.6	Trace	108

stream relatively rapidly (Table 9). When the sediment from construction sites reaches the estuary, only a small amount of average total deposition may result from the first increments of silt and clays. As the sandy material held in channel storage moves by stages down into the estuary, its total volume of 1.0×10^6 m^3 would produce some 0.03 m of deposition if it were spread evenly over the estuary bottom. This does not generally happen, and the incoming sediment concentrates in harbors, embayments, and shoals, where it is a nuisance to navigation and gives the impression of being part of a larger influx than it really is. Nonetheless, when the total amounts of material are compared to other deposits in the estuary, it is evident that the contributions due to urbanization are not really very large (Table 5).

CONCLUSIONS

The Patuxent River Basin is ideal for comparing the effects of catastrophic flooding and suburban development on a single watershed by using data on subwatersheds with different landuses within the larger one. The floods widened channels throughout the basin, damaging urban streams half again as much as rural ones. Stream beds and bars were coarsened in rural channels as the floodwaters removed sand and fine particles, leaving only pebbles and cobbles. Urban stream channels were not cleared of sands and developed new bars and dunes. The banks in urban reaches were more severely eroded than in rural ones, but overbank deposition, although discontinuous, was widespread in occurrence throughout the basin. Particles ranging from boulder-sized to sand grains were scattered over the valley floors. Although a number of sites had striking overbank deposits, the average depth of deposition over the entire floodplain surface was about 0.6 cm. Approximately the same amount of material would have been deposited on the floodplain surface during the last 100 yr. The average total depth of floodplain deposits is 2.6 m, accumulated over an unknown period of time.

During the Agnes flood, an estimated 3.9×10^6 m^3 of sediment entered the estuary, increasing the depth of bottom deposits there by 4 cm. This increment is hardly significant compared to deposits 1.5 to 3.0 m deep that accumulated between 1859 and 1944 on the estuary bed. Although the major floods caused severe channel damages on a local and temporary scale, natural channels quickly recovered. The volumes of sediment they distributed on the floodplains and the estuary are not significant compared to the total amounts of material already present. Rural streams had largely recovered from the floods in 4 to 5 months. A lush regrowth of vegetation masked overbank deposits and smoothed over the scarred channel banks. Silt and sand filled in the cobbles and pebbles left in the channel beds. Measurable cross-sectional parameters still reflect the floods' occurrence and will do so for some time, but the general appearance of the stream courses no longer attests to the very recent passage of two catastrophic floods.

Subwatersheds with suburban development in the Patuxent Basin are markedly different in physical characteristics and in behavior from rural watersheds. Urbanizing basins yield 340 metric tons of sediment/km^2/yr, compared to 22 metric tons produced by the same area in a rural district. Such excessive sediment loads choke local streams with deep sandy fills and bars for about 5.6 km downstream of the sediment source area. Urban channels hold in storage about 15 times as much sediment as rural channels, and deposition and scour are more severe. The size and shape of urban channels change at rates at least three times greater than those found in comparable rural stations for about

1.5 km below construction areas. In time, the processes of deposition and scour will stabilize urban channels after widespread construction has ceased and the sediment source areas have stabilized. Until then, bars built up by excess loads wash away unpredictably in storm flows only to rebuild again. In contrast, the basin's rural channels are largely stable, having adjusted to their flood damages and presently showing a balance between scour and fill. Their channels and channel features remain the same through the normal ranges of flows and are similar to preflood conditions. The future Patuxent will be a collage of streams, with rural headwaters much the same as they have been for the last several hundred years. Urbanized reaches in the central corridor zone will be prone to rapid changes in channel geometry from excessive deposition and scour. Below them, the Coastal Plain rivers, flowing beyond the most damaging impacts of development, will maintain most of their present characteristics and seem relatively untouched by man's activity above them.

REFERENCES

Chesapeake Research Consortium. 1974. Symposium on the effects of tropical storm Agnes on the Chesapeake Bay, May 1974: The Johns Hopkins University, University of Maryland, Smithsonian Institution, and Virginia Institute of Marine Science, 32 p.

Cooley, W. W., and Lohnes, R. R. 1971. *Multivariate Data Analysis:* Wiley, New York, 364 p.

Dunn, G. E., and Miller, B. I. 1960. *Atlantic Hurricanes:* Louisiana University Press, Baton Rouge, La., 326 p.

Gupta, A., and Fox, H. 1974. Effects of high magnitude floods on channel form: a case study in the Maryland Piedmont: *Water Resources Res.,* v. 10, no. 3, p. 499–503.

The Johns Hopkins University, Water Sciences and Management Seminar, Department of Geography and Environmental Engineering. 1966. *Report on the Patuxent River Basin:* The Johns Hopkins University, Baltimore, Md., 228 p.

Laws of the State of Maryland. 1970. Chapter 245, House Bill 1151, vol. 1: King Bros., Baltimore, Md., 1582 p.

Leopold, L. B., and Maddock, T., Jr. 1953. The hydraulic geometry of stream channels and some physiographic implications: *U.S. Geol. Survey Prof. Paper 252,* 56 p.

———, Wolman, M. G., and Miller, J. P. 1964. *Fluvial Processes in Geomorphology,* W. H. Freeman, San Francisco, 522 p.

Maryland Geological Survey. 1968. *Geological Map of Maryland:* Williams and Heinz Map Corp., Washington, D.C.

Miller, I., and Freud, J. E. 1965. *Probability and Statistics for Engineers:* Prentice-Hall, Englewood Cliffs, N.J., 432 p.

Schubel, J. R. 1974. Effects of Agnes on the suspended sediment of the upper Bay: *Chesapeake Research Consortium Symposium Abstracts,* May 1974, p. 8.

Stiles, T. N., and Weisnet, D. R. 1970. Isopachous mapping of the lower Patuxent Estuary by continuous seismic profiling techniques: Informal Rept. IR-70-37, Naval Oceanographic Office, Washington, D.C., 23 p.

Taylor, K. P. 1972. Summary of peak discharges: Maryland, Delaware, and Washington, D.C., for the flood of June 1972: U.S. Geological Survey, Water Resources Division, Open File Report, Parkville, Md., 13 p.

Vice, R. B., Guy, H. P., and Ferguson, G. E. 1969. Sediment movement in an area of highway construction, Scott Run Basin, Fairfax County, Va., 1961–1964: *U.S. Geol. Survey Water-Supply Paper 1591-E,* 41 p.

Walker, P. N. 1971. Flow characteristics of Maryland Streams: *U.S. Geol. Survey and Md. Geol. Survey Report of Investigations 16,* Baltimore, Md., 160 p.

Wark, J. W., and Keller, F. J. 1963. Preliminary study of sediment sources and transport in the Potomac River basin: *U.S. Dept. Interior, U.S. Geol. Survey, and Interstate Comm. on the Potomac River Basin Tech. Rept. 1963-11,* p. 27.

Wolman, M. G. 1967. A cycle of sedimentation and erosion in urban river channels: *Geog. Annaler 49A,* v. 24, pp. 385–395.

_____, and Schick, A. P. 1967. Effects of construction on fluvial sediment, urban and suburban areas of Maryland: *Water Resources Res.,* v. 3, no. 2, pp. 451–464.

Yevjevich, V. 1972. *Probability and Statistics in Hydrology:* Water Resources Publ., Fort Collins, Colo., 302 p.

13

GEOMORPHOLOGY AND ENGINEERING
CONTROL OF LANDSLIDES

F. Beach Leighton

INTRODUCTION

Geomorphology and engineering are dependent upon each other in the solution of landslide problems. The geomorphologist can provide answers to key engineering questions such as the following:

1. What has been the history of slope stability in the area of concern?

2. What is the approximate most representative cross section of the landslide(s)?

3. What is the best volumetric estimate of slide debris, pending subsurface exploration?

4. Is the landslide likely to increase in size? If so, where and to what extent?

5. Can the earth materials in the landslide be removed readily and used as select or acceptable fill elsewhere?

6. What other problems are associated with the landslide (e.g., seismic and groundwater)?

Each landslide problem requires a different blend of the two disciplines. Guidelines for differentiating the areas of emphasis between geomorphic–geologic and engineering aspects of landslide investigations are presented in Table 1.

Geomorphology, more than any other branch of geology, emphasizes the tools of photogeology and landform analysis (geomorphic interpretation), both important in the engineering control of landslides.[1] Moreover, the approach of the geomorphologist in deciphering events in the history of landforms and anticipating future events is commonly the direction that the engineer needs for design purposes.

The geomorphologist's role in the application of geomorphic principles to the engineering control of landslides, whether corrective or preventative, can be divided into three components:

1. Geomorphic identification: where is the landslide problem?

2. Geomorphic interpretation: why is the landslide there?

3. Engineering control: what is the best landslide remedy?

These three components are treated in summary form in the following sections.

---

[1] Landslides form a continuum from sliding to flowage of viscous fluids, and from large-scale bedrock movements to small-scale dislodgement of fragments and accelerated erosion. Only those slides whose sole is on bedrock are considered here.

TABLE 1 *Differentiation of Geomorphic–Geologic and Engineering Areas of Emphasis in Landslide Investigation*

Landslide problems	Geomorphic–geologic aspects	Civil engineering aspects (chiefly soil engineering)	Common ground (with special areas shown that require consultation)
Landslides on natural slopes	1. *Slides:* Chiefly bedrock types; areas of emphasis include description (including three-dimensional geometry and physiochemical characteristics), classification, origin, history, rates of movement and/or probability of future movement, seismic and geohydrologic parameters; development alternatives. 2. *Natural slopes to be developed:* Description including three-dimensional geometry and seismic–geohydrologic parameters; restricted use areas, recommendations for future development; special removal and proper drainage above natural slopes; necessary setbacks of structures above natural slopes; special planting programs.	1. *Slides:* Analyses of potential arcuate soil and fill failures; analyses of potential planar slides; areas of emphasis include soil mechanics, rock mechanics, design of remedial structures including subdrains; permeability and pore pressure tests; development alternatives. 2. *Slopes to be left natural:* Foundation investigations for stilt-type construction; stability analyses; buttress design; need for subdrainage at contact of natural slope and fill below; design and supervision of installation of subdrains; design for landuse potential.	1. Subsurface explorations; determination of groundwater levels; control and corrections; inspections for proper removal and benching; inspections for hazards during removal and their correction. 2. Haul roads for grading purposes that create instability and divert runoff; inspection of daylight areas during adjacent grading; setbacks of structures; restricted use areas.
Cut-slope stability	1. Types of *earth material,* their configuration, lithology, and geologic structure. 2. *Cut-slope angle and benches in bedrock cuts* on basis of above constitution and presence of groundwater problems; determination of geologic–seismic parameters and preparation of cross sections for calculation of buttresses by civil engineer.	1. *Soil engineering properties* of earth materials (sampling, testing, and analysis). 2. *Applied engineering procedures* (e.g., the design of restraining structures—buttresses, retaining walls, etc.; final determination of slope angle in unconsolidated essentially homogeneous earth materials (e.g., topsoil, slopewash, old fill); terracing and drainage in these materials; stability analyses of bedrock slopes, based upon geologic, soil, and seismic parameters.	1. Subsurface program of exploration and testing: choice of equipment, access of rigs, location of probes, depths of probes, sampling, and measurements of groundwater levels. 2. Stability equivalents; cuts that expose unconsolidated earth materials; inspections to determine whether grading follows combined recommendations; unforeseen problems that occur during grading; construction of temporary reservoir sites; bedrock drainage; inspection of excavated keyways, benches, and footings during excavation for restraining structures.

3. *Excavation difficulty* (analysis from subsurface exploration, seismic studies, groundwater information, etc.).	3. *Excavation problems* (difficulty of ripping from borehole information, etc.).	3. Consultation.
4. *Necessity of setbacks* for structures above and below cuts, compacted soil blankets for cut-lot pads, slough barriers and/or special planting programs; construction of haul roads for grading purposes; construction of sewer trenches; safety of temporary cuts for construction purposes.	4. *Design specifications and construction*; supervision of effective soil blankets, slough barriers, and retaining walls.	4. Problems arising because of relationships to adjoining properties, particularly removal of support from adjacent slopes.
5. *Recommendations for elimination, reorientation, or repositioning* of cuts, and/or reduction of cut heights; areas to be left natural ground.	5. *Recommendations for redesign of cut slopes* on basis of stability analyses; economics of retaining walls vs. buttress vs. redesigned cut slope, etc.	5. Consultation.
6. *Necessity of subdrains* to relieve groundwater problems.	6. *Design specifications and supervision* of installation of subdrains.	6. Location and length of subdrains.
Fill-slope stability		
1. *Geologic description* of materials available for fill.	1. *Soil engineering properties* of materials available for fill; feasibility, design, placement, and compaction control.	1. Consultation.
2. *Volumetric estimates* of surficial materials subject to removal prior to placement of fill.	2. *Final determination of materials subject to removal* prior to placement of fills where problem is bearing capacity.	2. Inspection of canyon cleanout during benching needed to remove unsuitable materials.
3. *Natural problems* in areas to be loaded with fill (e.g., groundwater, slides, faults, dip slopes, potentially compressible materials, etc.).	3. *Design, feasibility, preparation of ground* to be loaded with fill; provision for setbacks of structures; overfilling and excavation of natural ground and perched soil zones; reduction in slope angles and height; buttressing, special planting programs.	3. Elimination, reduction of height and repositioning on basis of underlying geologic problems; construction of reservoir sites and haul roads for grading purposes.

continued

TABLE 1 (continued)

Landslide problems	Geomorphic–geologic aspects	Civil engineering aspects (chiefly soil engineering)	Common ground (with special areas shown that require consultation)
	4. *Necessity of subdrains* to relieve groundwater problems.	4. *Design specifications* and supervision of installation of subdrains; recommendations for subdrains on basis of protecting fills.	4. Location and length of subdrains.
	5. *Configuration of old fills*, well sites, seepage pits and sumps, mine workings, dump areas, etc., from surface mapping.	5. *Stability evaluation* and field handling of old uncompacted fills and other unsuitable man-made fills; conversion of sliver fills to stability fills, fill problems related to adjoining properties.	5. Subsurface exploration of old fills and other man-made features that need delineation; preparation of cross sections to evaluate lateral stability of fills; problems related to surcharged cut slopes and adjoining properties.
Surcharges[a]	1. *Geologic description* of underlying earth materials, natural problems in areas to be loaded with fill (e.g., groundwater, faults, dip slopes, potentially compressible materials, etc.).	1. *Feasibility*: quantitative analysis of load effects on basis of geologic data; design of slopes, including fill setbacks.	1. Subsurface exploration, elimination, reorientation, buttressing of surcharged slope; surface drainage; stripping at daylight lines.
	2. *Necessity of subdrains* to relieve groundwater problems.	2. *Design specifications* and supervision of installation of subdrains; recommendations for subdrains on basis of protecting fills.	2. Location and length of subdrains.

[a]*Surcharges* refer to overloading a slope (by human or natural agencies), usually by means of artificial fills.

CAPSULE OF GEOMORPHOLOGIST ROLE IN LANDSLIDE CONTROL

Geomorphic Identification

The old cookbook recipe for rabbit stew that began with "First catch the rabbit" can be applied similarly to the first step in the engineering control of a landslide: "First catch the landslides." Geomorphology is the principal discipline that provides the tools to "catch" the landslides that may require engineering control.

A firsthand acquaintance with the geomorphic setting is a helpful first step in the identification of landslides. Armed with available maps and aerial photos, the geomorphologist benefits from an initial field reconnaissance in areas where landslides are unmapped. An appreciation of the tendency toward landsliding is gained from this appraisal of the natural setting. This is a useful approach, even where landslides have occurred after urban development.

Stereoscopic study of time-lapse aerial photographs has proved to be an invaluable aid, not only in the detection of landslides, but in unraveling their history and mode of formation. These photographs are available for much of California over a time span that extends back to 1928. Boundaries of landslides can be plotted directly on the aerial photographs, and recent photographs compared to the older sets to map their progressive development.

A scale of approximately 1 : 12,000 has been found most useful for photomapping. However, photoanalysis with both small-scale photographs (1:80,000±) and large-scale photographs (1 : 12,000) permits both an overall perspective and more detailed mapping. In addition, different sets of years (and scales) enable more detailed analysis of perplexing features under different conditions of time, viewing direction, light, and, in some cases, stage of slide activity.

Slides identified can generally be classified by the degree of positive identification of slide existence, e.g., *definite, probable,* and *questionable.* Landslides denoted as questionable include those with somewhat anomalous topography that could have formed by differential weathering and erosion or by man-made activities, as well as by landsliding. Some landslides may have been so modified by man-made grading and building, masked by vegetation, or buried by surficial deposits that their delineation is difficult or impossible from aerial photographs.

Active landslides can commonly be differentiated from inactive slides, although in some areas a classification of *active, recently active,* and *inactive* may have to be used. It is commonly important to distinguish whether sliding has been as a unit or in separate blocks, because one portion of a slide may show activity or more recent activation than other parts. To identify slides that have occurred since the date of the most recent photographs, field observations may be necessary.

Geomorphic Interpretation

Given a landslide, the most vital geomorphic question is, why is it a landslide? Factors that must be interpreted to answer this question include slope geometry, lithology, geologic structure, stage of geomorphic development, climatic regimen, and association with other geomorphic processes in the area.

Geomorphic variables related to landslides have to be viewed in three dimensions. The chief variables are *slope geometry, lithology* and *geologic structure.* Each case history presented in the next section demonstrates the interaction of these three variables and, more importantly, the savings that can result from geomorphic interpretation of these variables. *Geohydrology* and *seismicity* are also important variables, but are not as significant in the case histories presented.

Obviously, some unique weakness must be present in the slope that has slid. Exploration of this reasoning by the geomorphologist enables more judicious positioning of subsurface probes in this and nearby slopes, leading to clues of where additional landslides might be expected.

Interpretation of the type of landslide is commonly a key to the type of weakness present. For example, a block glide is common on dip slopes, and slumps are common on neutral or anti-dip slopes and in nonbedded materials. Earth- and mudflows are common on clay-rich slopes and those blanketed by thick colluvium.

On the basis of photogeology, slopes can generally be rated by the geomorphologist for landuse planning purposes as those having (1) little or no tendency to slide, (2) a tendency to develop local and surficial failures, and (3) a potential for slope failure of a magnitude that is difficult or unfeasible to prevent or repair (Leighton, 1976).

Engineering Control of Landslides

Geomorphologists must be acquainted with the alternative methods of engineering control in order to provide appropriate geologic data, determine the feasibility of each alternative, and make sound recommendations based upon this information. Commonly, where multiple alternatives of correction are available, combinations of these alternatives can be used.

Designed masses of compacted fill with high shear strength and a system of subdrains are generally used to support sections of a slide block or sections of a potential slide block. They are divided into three types:

1. *Earth buttresses* generally support the entire unstable block, as in Fig. 1.

2. *Shear keys* are essentially prisms of compacted fill placed to support only sections of a slide block (Fig. 2).

3. *Replacement fills* are used to entirely replace unstable areas and therefore do not render support to landslides, existing or potential.

Other means of engineering control of landslides include (1) avoidance of the slide area, (2) entire removal of the unstable area, (3) construction of retaining walls and similar retention devices, and (4) slope flattening, lowering, reshaping, or a combination of these.

Generally speaking, buttress fills have replaced retaining walls and other retention devices in California because of their lower initial cost, maintenance expense, and aesthetic potential. The advantages of buttresses over slope flattening are that (1) some areas can be stablized that cannot be flattened, (2) they free more land for other uses, as well as avoiding condemnation and purchase of valuable upslope property, (3) they reduce the net cost of stabilization and maintenance, (4) they reduce the slope breadth and the scarred appearance of the flattened cut face, and (5) they engender intensive geotechnical and engineering design in advance of grading operations, thereby saving both time and money during construction.

Buttresses can be appraised by the geomorphologist as economically feasible, economically marginal, or economically unfeasible. Unless the cost of a buttress can be spread over a large segment of the project, it is commonly unfeasible. A major buttress for two residential lots generally costs more than the profit margin on the two lots. A

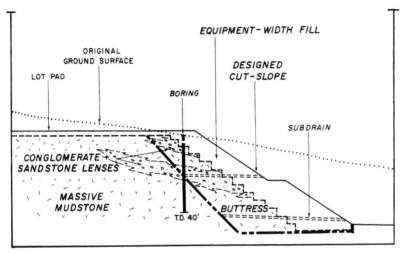

FIGURE 1 *Type of designed buttress at Porter Ranch that was alleviated by combination of geomorphic interpretation and drilling program. Equipment-width fill was placed on many of the excavated slopes with the intent of inhibiting accelerated erosion.*

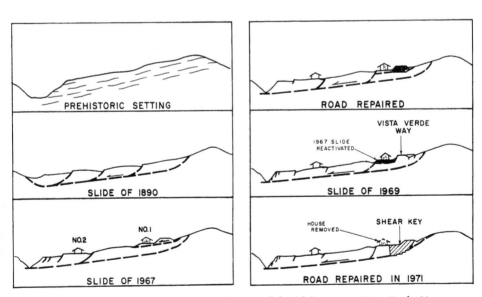

FIGURE 2 *Schematic reconstruction of principal landslide events, Vista Verde Way landslide, San Mateo County, Calif.*

279

common rule of thumb is that the basal width of a buttress with a 2 : 1 outer fill slope extending to the top of the buttressed slope must be one half to three quarters of the height of the slope to be buttressed. Seismic loading increases this ratio to two thirds to one.

For buttress design purposes, the geomorphologist can supply the engineer with (1) three-dimensional geometry of the slide or potential slide area, (2) geohydrology and seismic conditions in the area, and (3) longitudinal cross sections in the direction of slide movement as well as directly downslope. Maps and cross sections should extend beyond the boundaries of the slide and usually to the crest of slope. Influential man-made and geologic structures should be incorporated on the cross sections.

CASE HISTORIES

San Juan Creek Landslide, Orange County, California

The San Juan Creek landslide developed in pre-Holocene time and is located on the south side of San Juan Creek near where it enters the Pacific Ocean between Dana Point and Capistrano Beach. It lies at the south border of San Juan Capistrano. Whereas the interest of stratigraphers and petroleum geologists has centered on the turbidite sandstone sequence of the Capistrano Formation, the interest of the geomorphologist and engineering geologist has centered on the weak siltstone units that contribute to numerous landslides, including this one (Asquith and Leighton, 1972). The San Juan Creek landslide (Fig. 3) was originally identified by Vedder et al. (1957). Further fieldwork was conducted in 1970 by the author and Peter Tresselt for purposes of determining the feasibility of development. Following geologic mapping and stereoscopic study of aerial photographs taken in 1929, 1938, and 1947, a drilling and trenching program was undertaken. This investigation indicates that the landslide is approximately the same age as one of the nearby slides dated by radiocarbon methods of Stout (1969) as 17,180± yrs B.P., a time when sea level was much lower than present and the stream valley slopes were oversteepened.

The property studied is a part of a relatively flat river terrace approximately 30 m above the present river bed of San Juan Creek. Terrace deposits occupy a portion of the property studied and lie unconformably upon an old erosional surface of the marine Capistrano Formation (Miocene–Pliocene).

Geomorphic interpretation combined with drilling data show that the slide occurred prior to the present erosion cycle and that there is no evidence of reactivation during the present erosion cycle. One boring on an adjacent property by Moore and Taber revealed that the basal rupture surface of the slide lies over 30 m beneath the Holocene alluvium of San Juan Creek. Geomorphology played an important role in working out the following sequence of events and the conclusion that the property is reasonably safe from large-scale landsliding.

Three stages in the evolution of the landslide are illustrated by Fig. 1. Stage I shows sliding of the siltstone facies of the Capistrano Formation. In stage II, the landslide is naturally buttressed by deposition of alluvium as San Juan Creek aggrades its channel. Stage III illustrates the effects of relative uplift, which caused renewed degradation by

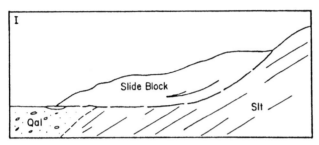

STAGE I. Landslide stage. Original valley cutting removes support, causing sliding onto stream deposits.

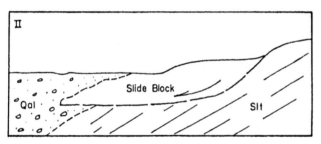

STAGE II. Valley filling stage. Sea level rises, valley fills to cover toe of slide.

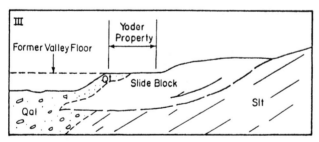

STAGE III. Valley cutting leaves old stream deposit as terrace on top of slide block. Slide is still adequately buttressed by stream deposits

FIGURE 3 Three stages in the evolution of the San Juan Creek landslike. Yoder property studied in detail (Yoder) was found to be developable, because the landslide is adequately buttressed by the alluvium of San Juan Creek.

San Juan Creek. At the present, the slide block is sufficiently buttressed so that no further reactivation can be anticipated due to removal of toe support.

Porter Ranch,
Los Angeles, California

Porter Ranch, of roughly 4,300 acres, lies on the south flank of the Santa Susana Mountains in the northwest portion of the San Fernando Valley. It has been the site of intensive residential development since 1965. Gently rolling to steep hillside terrain has been converted by massive grading to terraced subdivisions with major canyons retained in their natural state as recreational areas.

The Saugus Formation (Plio-Pleistocene) is the principal bedrock unit in the area. It underlies the south-facing slopes illustrated in Fig. 1. This unit exhibits rapid lithologic changes between siltstone, mudstone, sandstone, and conglomerate. In the tract illustrated, the formation consists of massive mudstone, with scattered lenses of friable conglomerate and sandstone. These lenses are subject to accelerated erosion but not landsliding.

Geomorphic interpretation provided a major key in distinguishing the slopes susceptible to landsliding. Landslide-prone slopes revealed a history of Holocene landsliding, whereas, in the case at hand, no history of landsliding was recorded. However, adverse bedding attitudes had been recorded in sparse outcrops and shallow trenches in an earlier geologic survey, leading to the recommendation that large-scale buttresses of the type shown in Fig. 1 be designed. If built, these buttresses would have made development uneconomical. In view of the geomorphic findings, it seemed wise for the developer to invest in a large number of borings placed in each proposed cut slope. Downhole descent and measurements by the geologist during a drilling program revealed the massive nature of the mudstone; the only rock defects recorded were associated with the irregular contacts of the conglomerate-sandstone lenses. As a result, the designed buttresses were avoided, saving over $500,000 in grading costs.

Abdale Landslide, Newhall–Saugus Area,
Los Angeles County, California

The Newhall–Saugus area has a long history not only of urban landslides, but also of expansive and collapsible soils, high groundwater levels, and high seismicity. The landslide shown in Fig. 4 began in late 1967 and reached its climax during the wet season of 1969. Seven homes, constructed between 1965 and 1966, were damaged beyond repair. Costs exceeded $1 million, including the expenses of litigation.

The slide involved interbedded siltstone and sandstone of the Mint Canyon Formation (Miocene) and surficial earth units consisting of colluvium, creep debris, prehistoric slide debris, and artificial fill (Fig. 5). The basal rupture zone of the slide ranged from 1 to 5 m in thickness and contained gnarled and twisted slickensided surfaces. The zone was heavily fractured, open structured, and clay rich with a high montmorillonitic content. The basal rupture surface served as a perched water table.

To define the spatial aspects of the slide and determine its mode of origin, 17 borings were drilled with a bucket auger rig sufficiently large in diameter (54 to 62 cm) to permit descent and geologic logging within the hole. A portable hillside rig was used to drill on side slopes and in some backyard areas. In addition, numerous shallow holes were hand dug to permit differentiation of fill, creep and slide debris, and bedrock. Geologic logging

A

B

FIGURE 4 *(A) High-angle oblique of the Abdale landslide, Newhall–Saugus area, Los Angeles County, Calif., looking southwest. Photo taken June 20, 1968, by Department of Los Angeles County Engineer at time landslide was discovered. Slide measured 100 m in length, 80 m in width, and 21 m in maximum thickness. Failing slope is composite fill and cut slope 21 m high. (B) Photograph of Abdale Landslide, Newhall-Saugus Area, Los Angeles County, California. This landslide reached its zenith during the wet season of 1969 at the time this photograph was taken.*

on a rope ladder within the drill holes was mechanically difficult because of the small diameter of the hillside holes (54 cm), the lack of oxygen, which required the use of breathing air equipment, caving intervals and seeping groundwater, and actual shearing of holes left open (but covered) for later study.

The unusual and complex character of the slide can be explained geomorphologically by the following simplified sequence, worked out by the author during a postmortem investigation in 1968–1969:

1. A prehistoric earthflow, greatly modified by erosion and not obvious on some aerial photographs, occupied this slope and extended headward of the damaging slide. Its earlier movement buried natural colluvium and slope debris shed along a vertical cliff near the crest of this major topographic ridge.

2. Grading of the tract removed support from the prehistoric slide and, in combination with the placement of a fill surcharge, gradually overcame the low shear strength of the old rupture zone.

3. A slow glacier-like movement was initiated about a year later, triggered by water of diverse origin that filled the rupture zone and reduced the shear strength.

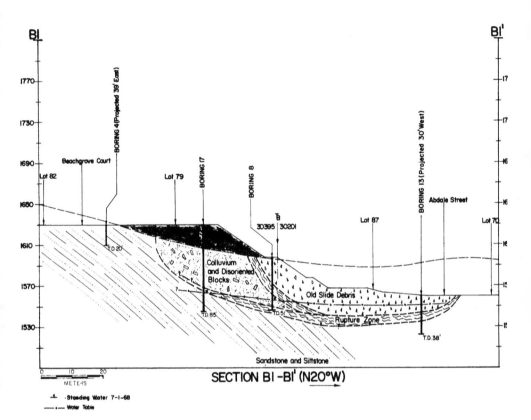

FIGURE 5 *Cross section of Abdale landslide, Newhall–Saugus area, Los Angeles County, Calif. North is to the right.*

The chief geomorphic lessons from this landslide as related to engineering control were as follows:

1. The prehistoric landslide might have been detected before development had all geomorphic tools been used. This could have resulted in the slope being left in a natural state or stabilized adequately.

2. Geomorphic observations during the postmortem study reduced costs of subsurface exploration by reducing the number of trenches and borings needed and by making it possible to position them to greater advantage.

3. Despite recommendations to the contrary, lime stabilization by grouting was undertaken to stabilize the slide; it was unsuccessful partly because stabilization was begun surficially before the true geometry of the slide was known. Redesign of the slope, including an earth buttress, was necessary to assure stabilization. The slope stands as open ground today, mute testimony to failure in landslide diagnosis and inadequate implementation of stabilization measures.

284

Vista Verde Way Landslide,
County of San Mateo, California

The Vista Verde Way landslide, a composite block glide approximately 580 m in length, is located south of Los Trancos Woods and Portola Valley, near the junction of Old Spanish Trail and Vista Verde Way (Mader and Crowder). It occurs on the western flank of Coal Mine Ridge, a ridge bounded on the east by the San Andreas Fault zone and on the west by the Pilarcitos Fault zone. This area is underlain by sandstone, siltstone, and conglomerate of the Santa Clara Formation (Lower Pleistocene). Seams of clay and coal are scattered throughout these bedrock units.

The history of the Vista Verde Way slide is reconstructed schematically in Fig. 2. In 1890, the landslide underwent major movement, creating some of the ponds and the hummocky topography between segmented blocks of the slide. Between 1956 and 1963, Vista Verde Way was constructed across the head of the slide and two residences were constructed on two segmented levels downslope of Vista Verde Way. In 1967, Vista Verde Way dropped approximately 10 vertical ft and was shifted downslope as a result of reactivation of the upper portion of the slide. The residence immediately downslope was destroyed, but the lower residence was not damaged. Vista Verde Way was reconstructed at a cost of $60,000 between late 1967 or 1968, but slid again in 1969.

The basic factors responsible for sliding have been (1) thin plastic clay beds and seams of low shear strength that became the basal rupture surfaces, (2) an unfavorable geologic structure consisting of a northwest plunging syncline, (3) removal of toe support ancestrally by stream erosion and probable faulting, (4) perched groundwater, and (5) loading the head of the 1967 slide with fill.

To reconstruct damaged Vista Verde Way, the upper portion of the landslide was stabilized by a shear key. Two principal engineering problems involved in the reconstruction were temporary stability in cut slopes of the shear key and adequate subdrainage. Engineering design was able to proceed without delay as a result of the geologic control obtained prior to reconstruction, but at a cost of over $200,000. Had geomorphic interpretation been completed prior to construction of Vista Verde Way, it is likely that the road could have circumvented the slide or at least been constructed safely and relatively inexpensively.

SUMMARY AND MYTHS TO DISPEL

How important is geomorphology to the engineering control of landslides? The preceding case histories offer documented examples of geomorphic contributions to the identification and interpretation of actual landslide areas and suspected landslide areas. The geomorphologist provides valuable guideposts to the best landslide remedy, as well as basic data for design purposes. The chief lesson that stands out from the analysis of the case histories is that geomorphology can be extremely effective in cost and hazard reduction for the engineer. Overdesign and underdesign of corrective and preventive measures can be avoided. When diagnosed in advance of construction, many landslides can be essentially totally prevented or their effects neutralized. In many cases landslide conditions in urban areas and along our roadways could have been detected before development, thus preventing injury and huge dollar losses.

Emphasis is still shifting in engineering control from corrective measures to preventive measures for obvious socioeconomic reasons, but the change in most cases is still painfully slow as lessons of yesterday are rediscovered today. The geomorphologist has a responsibility for educating the general public and separating fact from fancy about landslides. Myths the geomorphologist can help to dispel in the engineering control of landslides include the following:

1. *The locations of most landslides and associated stability problems are unpredictable.* Geomorphic studies, aided by subsurface investigations can identify existing and potential landslide conditions in advance of reactivation or initial movement. Postmortems afford the best evidence of this.

2. *The empirical approach of stability analysis by the engineer is the only sound method of solving slope stability problems.* This method omits much of the critical evidence of slope behavior that can be obtained from interpretive geomorphology. Many slopes evaluated by the engineer as having a safety factor greater than 1.5 have failed where geomorphic analysis was lacking.

3. *Weather or some other triggering device such as groundwater from an earthquake or sonic boom is usually the principal cause of the landslide.* Adverse geologic materials and geologic structures are the basic culprits, aided by the effects of groundwater circulation. Triggering events will have little or no effect on most stable areas. These triggering events are more like the "straw that broke the camel's back" than like the "squeezed trigger of a loaded gun."

4. *Detailed geology should be left to the grading stage when exposures are more abundant and complete.* This approach eases the geologic task and facilitates meeting preliminary time schedules, but it upsets construction budgets, develops crises during grading, and creates blind alleys in design and construction.

5. *Our most stable building sites have been utilized in most areas.* Early sites were without benefit of geologic-soils investigations. Many of these sites are related to existing or potential natural hazards, such as landslides. In addition, mass grading without adequate geologic-soils control has produced instability in many areas. These older sites commonly suffer damage in time.

6. *All landslides per se are bad news.* Some slides are ancient, immobilized, and well consolidated. Others form natural shoulders, benches, and more gentle inclines that are favorable building sites. Others have created important natural resource and recreational areas. Others occur during grading and may aid in crushing and transporting material downslope.

CONCLUSIONS

Geomorphology plays an important role in the engineering control of landslides. Photogeology and landform analysis enable the geomorphologist to identify landslide problems, interpret them, and offer guidelines and alternatives for their remedy. Slope geometry, lithology, and geologic structure are the most important variables in the case histories presented, including the San Juan Creek landslide of Orange County, California, the Abdale landslide of northern Los Angeles County, and the Vista Verde Way landslide of San Mateo County, California. In each of these case histories the principles of geomorphology proved extremely effective in cost and hazard reduction for the engineer.

REFERENCES

Asquith, D. O., and Leighton, F. B. 1972. Landslides in the Capistrano Formation and their engineering and environmental significance: *Geologic Guidebook to the Northern Peninsular Ranges, Orange and Riverside Counties, California,* p. 69–77.

Leighton, F. B. 1976. Urban landslides: Targets for land-use planning in California: *Geol. Soc. America Spec. Paper 174,* D. R. Coates, ed. (in press).

Stout, M. L. 1969. Radiocarbon dating of landslides in Southern California and engineering geology implications: *Geol. Soc. America Spec. Paper 123,* 7 p.

Vedder, J. G., Yerkes, R. F., and Schoellhamer, J. E. 1957. Geologic map of the San Joaquin Hills–San Juan Capistrano area, Orange County, California: *U.S. Geol. Survey Oil and Gas Inv. Map OM 193.*

14

SCIENTIFIC AND ENGINEERING PARAMETERS IN PLANNING AND DEVELOPMENT OF A LANDFILL SITE IN PENNSYLVANIA

Richard M. Foose and Paul W. Hess

INTRODUCTION

Man has written a mute history of the evolution of his culture with his solid waste. Much of what we know of prehistoric man and his association with his biological and physical surroundings has been revealed through archaeological studies of his early living places. By sorting through his refuse heaps in caves and covered cities, we have pieced together a picture of his changing environment. We know the species of animal that man first domesticated; the grains he first grew, harvested, and stored; the hand tools he first fashioned; the containers his hands first shaped. Man's trash pile is a monument to his "advance."

As man advanced in technology, he also started to produce and dump into his environment materials never before known on earth and, in many instances, incompatible with natural processes. Whereas most of his earlier refuse had been plant or animal in structure and, therefore, amenable to natural decay and recycling, much of modern-day refuse is nonbiodegradable—steel, plastics, ceramics, glass, toxic chemicals, pesticides, etc. Now man produces many solid wastes that are almost impossible to "get rid of."

The problem of removing the unwanted and unusuable materials from the usable has become larger and more complex simply because the world's population has become larger, and in the more advanced societies each member of that population individually produces more waste than before. Apparently, several options remain for us to economically recover energy and reusable materials from our wastes.

Bellamy (1969) and Han et al. (1969) conducted several studies on the conversion of municipal waste to single-cell protein for human consumption. Fitzpatrick (1973) stated that there has been an increasing awareness of the potential value of the energy in solid waste, and the energy crisis has demanded that the energy in refuse be extracted for use *in an economic way.* Many early experiments in recovering resources from mixed municipal refuse, particularly in the period from the late 1940s through the early 1960s, were basically concerned with composting, and all have failed. They have failed primarily because of the decline in the proportion of compostable organics in municipal refuse.

Recycling and reuse schemes for total recycling include some that are extremely complex (Stern, 1971). Some schemes have involved the collection of concentrated pollutants into larger piles. Locally, spent nuclear fuels have been concentrated and allowed to pollute groundwater. Often the systems designed to reduce the problem have demanded more energy than could be extracted from the materials recycled; hence, there was no net gain (Abert, 1975). In addition, political exploitation of some situations or

289

bowing to pressure groups has often confused real issues and defeated attempts to apply technical logic and sound scientific and engineering planning.

In the meantime, the problem of solid-waste disposal remains and grows more critical (Avers, 1975). It is imperative to review all feasible methods for reducing the bulk of our refuse and for reusing what we can. However, this must be done with a scientific approach, with engineering logic and methods, and with patience. And the problems must be solved at the least cost to society.

Society usually protects itself through its governments. In the United States, attempts have been made to put into laws and regulations some sense of direction for the handling of wastes produced by our affluent way of life. On October 20, 1965, the Solid Waste Disposal Act was passed by Congress to authorize a research and development program with respect to solid-waste disposal. The act was amended by the Resource Recovery Act of 1970, and the two pieces of legislation set into action a series of events. In response to the act, each state was mandated to develop its own legislation for the control and disposal of solid wastes.

In July 31, 1968, the commonwealth of Pennsylvania passed the Pennsylvania Solid Waste Management Act, which provided for the planning and regulation of solid-waste storage, collection, transportation, processing, and disposal systems, and required municipalities to submit plans conforming to the act's mandates. The act was amended in January 1970 and August 1972, and was made effective by a series of regulations amended through October 1973. Standards were set for planning, design, and operation of any solid-waste processing or disposal facility or area of a solid-waste management system, including but not limited to sanitary landfills, incinerators, compost plants, transfer stations, and solid-waste salvage operations. Regional solid-waste management plans were written and placed into operation.

As in most cases, the laws and regulations were guidelines for what to do or not to do. The development of specific techniques and methods for handling, disposing of, and managing solid wastes was left to local governing bodies. In response to these needs, the board of supervisors of Derry Township, Pennsylvania, appointed a landfill team and charged them with the responsibility of developing and maintaining a system that would meet legal requirements for the disposal of township wastes. Unfortunately, township wastes were being disposed of already at a site that could best be called the "town dump." Few, if any, of the main criteria of the American Public Works Association for sanitary landfills were being observed:

1. Vector breeding and sustenance must be prevented.
2. Air pollution by dust, smoke, and odor must be controlled.
3. Fire hazards must be avoided.
4. Pollution of surface and groundwater must be precluded.
5. All nuisances must be controlled.

The landfill team included technical expertise in the persons of an engineer, a biologist–chemist, and a geologist. Through their efforts, the town dump was converted to a sanitary landfill with an approved permit from the Pennsylvania Department of Environmental Resources. The trench method of refuse disposal was employed beginning in 1970, and it continued until exhaustion of the 34-acre site in 1975. The technical considerations involved in the creation of the sanitary landfill have been described by Foose (1972). Some earlier data gathered and interpreted by the landfill team at that time are included in this paper, in which we discuss the following:

1. The geologic parameters associated with sanitary landfill siting and development.

2. The critical data collected and interpreted during the 5-yr period (1970–1975) of operation of the trench-type landfill in Derry Township.

3. The design and instrumentation planning for the extension of the existing landfill by its conversion to an area-type landfill, for which a permit was granted by the Pennsylvania Department of Environmental Resources in September 1975.

GEOLOGIC PARAMETERS
IN LANDFILL DEVELOPMENT

Although some important nongeologic parameters need to be carefully considered in planning for a sanitary landfill site, most of the critical ones are geologic. These are set forth below and discussed briefly so as to clarify their relationship to each other. Obviously, the importance of each would vary from one site to another. But none may be overlooked because, in every case, the land—its surface and subsurface—is to be evaluated for its ability to withstand or respond to the impact of man's action, that of dumping solid refuse.

Atmospheric Parameters

Wind. Prevailing wind direction, average wind velocity, and the incidence of wind gusts may be important in site selection with respect to the potential for blowing paper litter.

Precipitation and Evaporation. More important to the geologic regime is the average yearly precipitation, its distribution in time, and the average evaporative rate; all these are important with respect to the potential for runoff and erosion, for influent seepage and the level of the groundwater surface, and for the rate of biodegradation of the refuse and the potential volume of leachate production. These data should be collected and interpreted prior to site selection. Needless to say, the longer the data time base, the better the possibility is for effective evaluation.

Surface Parameters

Topography and Slope. If the topography is very uneven or if the slopes are steep (more than 8 to 10%), a potential site has a greater possibility of being unacceptable. At least, it would be more difficult to work with because of the increased likelihood of erosion and the greater potential for slope failure due to increased weight (from water saturation) and pore pressure and the decreased coefficient of friction between particles. With the availability of large earth-moving equipment, slopes and topography may, of course, be changed.

By maintaining boundary slopes in the landfill as steep as possible, the total area available for refuse emplacement may be increased.

Vegetative Cover. The extent and kind of vegetative cover has a direct effect upon the amount and rate of runoff, transpiration, and influent seepage. This is true both with respect to the original vegetative cover of the chosen site and also of the developing or developed site and its planted cover. Erosion control and slope stability both may be influenced and, to a considerable extent, controlled by the planted vegetative cover.

Drainage and Flooding. The rate and path of runoff, particularly at times of high precipitation, are important. Drainage must be good enough so that water is not ponded on the landfill and influent seepage is not excessive; otherwise, the groundwater surface may rise too high and/or slope failure could occur because of increased weight and pore pressure. However, the runoff channels should be graded and controlled so that excessive erosion is prevented. Generally, a site with a potential for flooding is unacceptable both with respect to considerations of ground- and surface-water quality and to the maintenance of site integrity (erosion prevention).

Soil

Physical Character of the Soil. The extent of weathering, particle size, and the permeability of the soil profile are all important with respect to the soil's ability to allow water to move downward through it or to be forced to drain across its surface.

Thickness and Volume of Soil. The thickness of the soil is important with respect to the ease with which it may be moved and the depth to which it may be excavated for the development of trenches for refuse emplacement. Perhaps most important is the volume (tonnage) of soil *on the site* that is available for cover material to seal off the daily refuse. If it becomes necessary to import soil for the purpose, that could represent a major added cost to the operation.

The final cover material must be such that there will be no escape of odor or exposure of buried refuse and no easy access by rodents, but will provide for good support of vegetative cover, good drainage, minor influent seepage, and a gradual escape of generated gases, as well as "adjusting" without cracking to any changes in surface elevation due to compaction. These qualifications would generally be met by a silty clay or a mixture of very fine sand and clay. Obviously, there is a need to study the physical characteristics of the soil on the proposed site or of the soil that would be imported for use on the site.

Surface Water

Runoff from the landfill and also groundwater effluent as seeps or springs will discharge into the nearest stream. It is important to establish base-line data for that stream or streams with respect both to volume of discharge and quality of water on a seasonal time base. The minimal flow of the stream or streams involved on or near a landfill and the range of concentration of pH, B.O.D., C.O.D., total dissolved solids, and selected chemical ions from the contributing source (seep, spring, or runoff) will dictate whether the quality of water in the stream is less than or in excess of permitted standards. If the quality does not meet legal standards, the "leachate" must be contained and controlled on the site or treated to upgrade it to "clear stream" standards before discharge. Obviously, a large stream with a large discharge could accept some poor-quality water (leachate) and, by dilution, still meet standards. A small stream, or a stream with a seasonally small discharge, could not.

Groundwater

Much of the water falling as rain on the landfill will soak in and percolate downward through the refuse, chemically altering the refuse and taking selected ions (SO_4, Cl, Fe, Cu, Pb, etc.) into solution, thus producing leachate. If the groundwater surface is below the bottom of the landfill, then all the refuse will be subjected to some oxidation and to the maximum chemical change. Also, there will be an opportunity for the downward percolating leachate to be "cleaned" by its chemical interaction with the soil and rock above the groundwater surface. Hence, it is critical to monitor the depth to the groundwater surface and the seasonal fluctuation of groundwater levels with respect to planning the maximum depth of refuse emplacement.

The other groundwater parameters that must be monitored are the following:

1. The elevation and configuration of the groundwater surface and, hence, its gradient and direction of movement.

2. Knowledge of the structures in the unweathered bedrock (bedding, joints, cleavage, and faults) and their influence upon the direction of groundwater flow.

3. Rate of groundwater flow.

4. Quality of the groundwater at multiple monitoring locations within and adjacent to the landfill site.

5. Location of leachate emission to the landfill surface, its quantity and quality.

Regarding the emission of leachate, it must either be treated before discharge to a stream or it must be controlled (retained) on the site. Collection and retention of leachate may involve simple or complicated and extensive engineering design and construction. Three major possibilities exist:

1. If groundwater flow is directed to a single discharge point, the leachate may be retained in a constructed pond or ponds. Treatment prior to discharge or recirculation of the leachate on the site then may take place.

2. If groundwater flow emanates at the surface at multiple points, it could be collected in multiple ponds. However, it might better be collected by pumping from wells, thereby eliminating groundwater effluence as seeps or springs, and then directed to a single pond for storage treatment or on-site recirculation.

3. In the initial construction stage of the landfill, an impermeable "liner" of clay, plastic, asphalt, or other material might be installed so that all downward percolating water is forced to flow in one direction and may be collected at a single emission point, there to be retained in a pond for treatment or on-site recirculation. The effective installation of a liner that will have long life is not easy. The use of natural clays, if carefully compacted, probably provides the very best liner. Artificial materials may deteriorate with time or may react chemically with the leachate so that deterioration ensues. This has been true of some plastics. In addition, the danger of ripping the liner during refuse emplacement is high unless there is a layer of protective material, such as sand, placed above the liner.

Other Critical Parameters

Whereas the parameters discussed above are dominantly geologic or geomorphic and involve, in most cases, measurements and observations that should be initiated prior to landfill development (during the site selection and design stages), there are other critical parameters that need careful attention *after* landfill operations have begun and probably even after they have been completed. These have to do with the biological and chemical changes that take place within the refuse itself. The most important results of the biochemical alteration of refuse materials are

1. Production of leachate.

2. Production of gases.

3. Changes in volume of the refuse and of the landfill.

The production of leachate has already been discussed.

Gas generation by biochemical change has been shown to be dominantly methane (CH_4). Depending both upon the nature of the refuse and on rates of oxidation and bacterial action, there may be large or small quantities of methane produced. If the cover material has a permeability conducive to continual gas escape, there would be no problem as long as the landfill site were not to be developed for housing or another purpose involving much human use. On the other hand, if entrapment of gas were caused by very low permeability cover, explosive situations could be created. Hence, there is a need to develop a gas-monitoring system that would document the rates of gas production and dissipation and, also, provide the basis for estimation of the stage of "maturation" of the landfill, i.e., the evolutionary stage of biodegradation of the refuse within the landfill. An important scientific objective for all landfill operations should be to collect data of this kind over a long enough time base so that useful interpretations could be made regarding the evolutionary life cycle of a landfill (from its inception to its "death," when continuing biochemical changes would be minimal and no longer pose any concern regarding alteration of the environment).

Associated with measurements of rates of gas generation due to internal biochemical changes are two other parameters, which, if monitored concurrently, would also help to provide an understanding of the evolutionary stage, or "maturation," of the landfill. These are as follows:

1. Changes in chemistry of the soil water, i.e., of the percolating water in the vadose zone above the groundwater surface. Such changes can be monitored at differing depths by the use of lysimeters emplaced in the landfill.

2. Changes in the thermal regime, i.e., precise temperature measurements at differing depths within the landfill, as determined by thermistors. It is to be expected that the biochemical changes within the refuse would alter (increase) the ambient thermal conditions.

Finally, changes in the volume of the landfill would occur as a consequence of the following:

1. Reduction of the original volume of refuse due to biochemical changes, resulting in shrinkage.

2. Compaction due to the total load of refuse and of the final cover material, and, in the case of the Derry Township landfill, due to an additional 20 ft of area fill on top of the original trench fill area.

Steel plates set on concrete pads at the landfill surface would serve as bench marks for repetitive level surveys by which absolute amounts and rates of settlement would be documented. The *successful* use of completed landfill sites for other purposes might well depend on knowing that

1. Production of leachate was minimal.

2. Gas generation had ceased or was minimal.

3. Land surface was not subject to further settlement.

DERRY TOWNSHIP LANDFILL SITE

General Setting

Location and Topography.   The landfill site is approximately 2 mi northeast of the town of Hershey and 0.5 mi west of Palmyra (Fig. 1). The site includes approximately 34

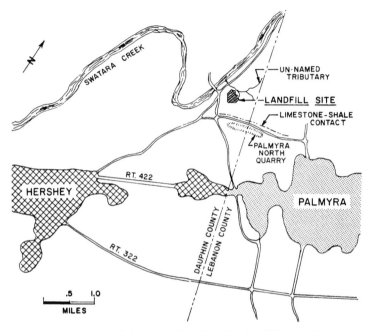

FIGURE 1 *Location of the Derry Township, Pa., landfill site with respect to the towns of Hershey and Palmyra.*

acres, mostly on the nearly flat top of a large hill that slopes gently both east and west (Fig. 2). The easternmost part of the site slopes steeply downward to a small north-flowing tributary to Swatara Creek (Fig. 1). Total relief is about 100 ft.

General Geology. The site is underlain by the Martinsburg shale (Ordovician), a gray to tan, soft, thin-bedded shale with occasional cherty layers and quartz veins but of generally uniform appearance and composition. The bedding planes of the shale strike N.75 E. and dip southward at angles ranging between 40 and 60 degrees. Two steeply dipping joint sets trend northwest and northeast. The bedding and joint surfaces help to control the direction of groundwater movement through the shale. Approximately 500 ft south of the site there is a contact between the shale and the Hershey Limestone, and about an equal distance farther south is a large flooded quarry in the limestone (Fig. 1). The limestone has some interconnected solution openings that allow easy groundwater movement. Fortunately, the rate of movement of groundwater through the shale is so slow, the bedding planes are so steep, and the distance to the limestone is so great that there is very little danger of poor-quality groundwater (leachate) ever reaching the limestone.

Both the shale and limestone formations are on the overturned limb of a very large anticline, which is part of the Appalachian Mountains fold system.

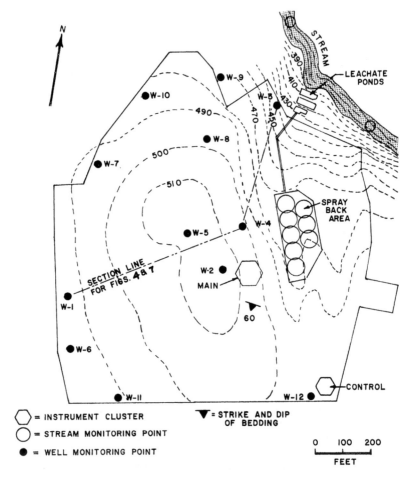

FIGURE 2 *Derry Township landfill site showing location of monitoring stations, leach-ate retention ponds, and spray-back area for leachate recirculation.*

TRENCH-FILL PHASE:
DERRY TOWNSHIP LANDFILL

The conversion of an existing "town dump" in Derry Township to a trench-type sanitary landfill in 1970 has already been described by Foose (1972). During the 5-yr period to 1975 the 34-acre site was fully developed by the filling of 22 trenches, thus completing the life of the landfill by the trench-fill method. Each of the east-west trenches was approximately 10 ft deep, 24 ft wide at its bottom, and separated by a wedge-shaped rib of weathered Martinsburg shale 3 ft wide at the top. With 5 yr of experience stemming from the site development and a large body of systematically collected data at the site, it might be useful to summarize the most important aspects of the landfill development, directly relating these to the geologic and geomorphic parameters discussed earlier.

Topography, Drainage, and Slope Stability

Early in the life of the landfill it became apparent that careful geomorphic attention had to be given to the shaping of the landfill surface by the creation and maintenance of gentle slopes so that too rapid runoff would not occur; otherwise, serious erosion would ensue from sheetwash, particularly prior to the establishment of any vegetative cover. A low berm, about 2 ft high, was developed at the frontal edge of the landfill and, behind it, a broad and shallow channelway with a low gradient to drain off surface water. The channelway was lined with clay so as to reduce influent seepage to a minimum. See the design in Fig. 3. This was particularly important because of the need to maintain the stability of the frontal slope of the landfill at approximately 38 degrees. This could have been jeopardized by influent water seepage because of increased weight and decreased coefficient of friction between soil particles. Periodic soil-moisture measurements were made of the frontal slope area, which showed variations within the narrow range of 15 to 17% to a depth of 2 ft. These are low values and are conducive to maximum slope stability. From time to time it was necessary to repair the berm and the channelway and, occasionally, to fill in small gullies caused by fast runoff at that place where the channelway left the landfill site to join the small stream nearby.

The planting of vegetative cover had the obvious effect of reducing minor rilling and gullying of the surface by runoff, and the less obvious effect of decreasing influent seepage by evapotranspiration.

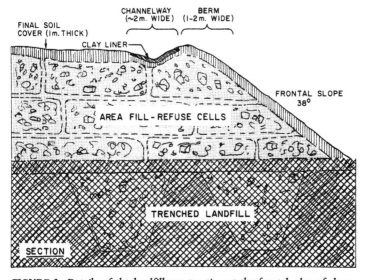

FIGURE 3 *Details of the landfill construction at the frontal edge of slope.*

As noted above, careful attention was given to maintaining the stability of the frontal slope so as to prevent slumping and exposure of the refuse within. This involved periodic examination of the slope, the berm, and the channelway for the presence of tensional cracks. None has ever formed.

Soil

The initial exploration of the site in 1970 established that the depth of weathering was great enough so that adequate cover material would be available for the full life of the trench-fill phase of the landfill (Fig. 4). The top 3 to 4 ft of soil and weathered shale always was stripped off and stockpiled in advance of trench cutting. The weathered shale excavated from the trenches was also stockpiled, and these materials were used as needed for cover material within the trenches and for the final landfill cover. Large stockpiles still exist upon completion of the trenches; these will ensure adequate cover material for much, but perhaps not all, of the area-fill phase of the landfill, which began in mid-1975.

During the early stages of landfill operation it was observed empirically that the cover material, when compacted by the normal use of the landfill equipment, achieved a density and permeability that kept influent seepage at a level so that no abnormal fluctuation of the groundwater surface occurred.

Surface Water

Since September 1970, the discharge of the small stream located along the northeast boundary of the landfill (Fig. 1) has been measured two to four times a year at stations

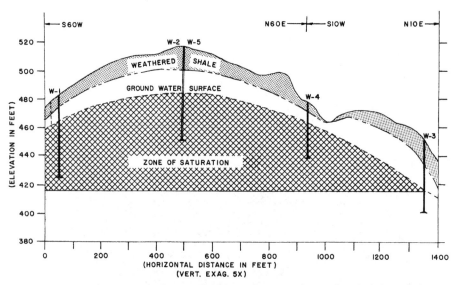

FIGURE 4 *Cross section showing thickness of extensively weathered shale and the average level of the groundwater surface. Line of section shown on Fig. 2. (After Foose, 1972.)*

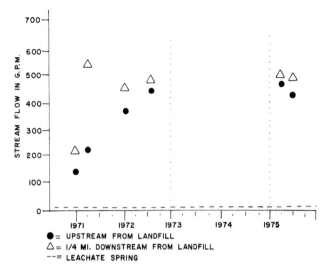

FIGURE 5 *Five-year (1970–1975) discharge record of the small stream near the landfill and of the leachate spring.*

both upstream and downstream from the site. The discharge values for the 5-yr period (1970–1975) are shown in Fig. 5. They range from 121 gallons per minute (gal/min) to 472 gal/min and show typically low flow during summer and fall and the highest flows during winter and spring. Similar data for discharge from the leachate spring (Fig. 5) reveal much smaller variations and an average flow of 3.3 gal/min. Hence, discharge by the small stream ranges between 25 and 150 times greater than that from the leachate spring. This could be important if, for any reason, there were to be small amounts of leachate released to the stream because of the extent to which dilution would occur.

Water-quality data for the stream collected on the same time interval as that for stream discharge reveal that the stream is relatively unaffected by its proximity to the landfill. Figure 6 shows the relative water quality (by measurement of specific conductance, B.O.D., chloride, and iron) in well 5 (Fig. 2) located in the middle of the landfill and also of the landfill stream. These data show that during the 5-yr period (1970–1975) there has been minimal "leakage" to the stream from the leachate ponds and only a very small contribution, if any, of leachate to the stream through the mechanism of groundwater underflow.

Groundwater

Inasmuch as the geologic parameters associated with groundwater are the most important, because of the potential for leachate generation, the earliest consideration of the Derry Township landfill site involved the drilling of a series of five holes. These established the depth to the groundwater surface, the thickness of the weathered shale above fresh unweathered bedrock, and, by subsequent monitoring within these holes, the time and amount of fluctuation of the groundwater surface. Figure 7 shows a northeast–

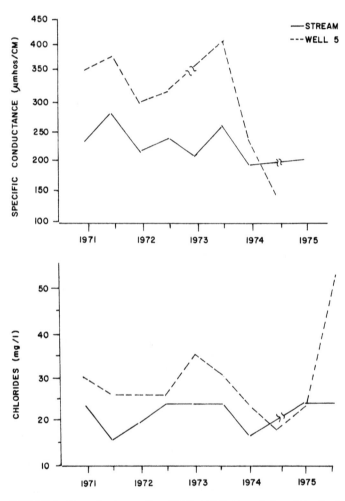

FIGURE 6 *Groundwater quality (1970–1975) in well 5 and in the landfill stream as shown by specific conductance, B.O.D., chloride, and iron.*

southwest cross section through the landfill site that was drawn on the basis of the data observed in the earliest drill holes. The section clearly shows that a large volume of weathered shale existed, and that the groundwater surface was approximately 35 to 40 ft below the original surface of the site. In the 1971–1975 period, an additional eight wells have been drilled to provide adequate monitoring for all the site (Fig. 2). Water levels in all the wells have been monitored every 2 weeks to the present time. A Leupold–Stevens automatic recorder was installed in well W-2 (Fig. 2) in 1973; since then, it has provided a continuous hydrograph of the groundwater surface. The 5-yr (1970–1975) data reveal a

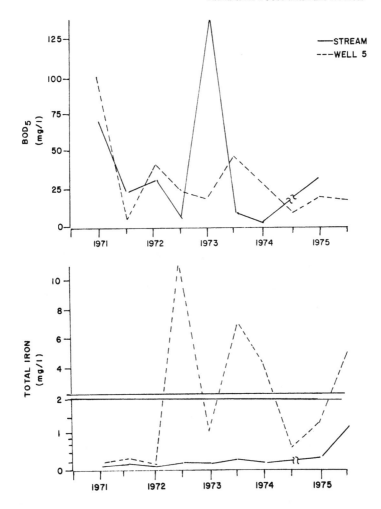

maximum range of water surface fluctuation of 7.4 ft and an average seasonal range (lowest in summer, highest in spring) of 3.7 ft (Fig. 8). These data firmly corroborate the initial conclusion, based on similar measurements in other wells within the Martinsburg shale in the general vicinity of the site prior to 1970, that it would be feasible to design a trench-type landfill on the site that would meet the Pennsylvania requirement that the bottoms of the trenches would be at a height above the groundwater surface at least equal to the thickness of the refuse that would be emplaced in the trenches. See Fig. 7 for an explanation of this relationship. Thus, there has been an average section of partly weathered shale 15 to 20 ft thick below the filled trenches through which groundwater and leachate have been able to percolate before reaching the groundwater surface. Regeneration of leachate has occurred in that vertical section.

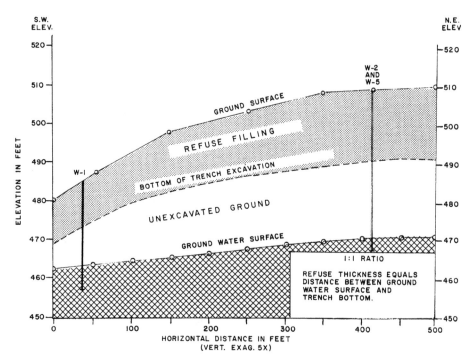

FIGURE 7   *Cross section showing the relationship between the level of the groundwater surface and depth of trench excavation. (After Foose, 1972.)*

Water Quality.   One or more water samples have been collected from every monitoring well on the landfill site every 3 months during the 5-yr period (1970–1975). The data show that only wells 6 and 7 have revealed any significant progressive degradation of quality with time, and for well 6 the quality has improved in the past year. In all wells the value of specific measured parameters does vary from one time of analysis to another, probably in response to such factors as variations in amount of precipitation and influent seepage, times of completion of trench filling close to specific monitoring wells, and the stage of biodegradation of the buried refuse, but the variations are small. Table 1 shows the range of values for the key parameters of pH, B.O.D., C.O.D., and total iron and chloride in wells 6 and 7. The completion of refuse filling in the trenches closest to well 6 occurred in 1973, and the highest values of B.O.D., C.O.D., iron, and chloride were all recorded in March 1974. Since that time, the leachate has improved. The completion of refuse filling in the trenches closest to well 7 occurred in 1974, and the highest values of B.O.D., C.O.D., iron, and chloride were recorded in February 1975. It appears that biodegradation of the refuse may reach its peak approximately 9 to 15 months after refuse emplacement.

Groundwater Flow.   In March 1971, fluorescein was charged to the groundwater in well W-2. Its identification in the water of well W-3, approximately 1,000 ft to the

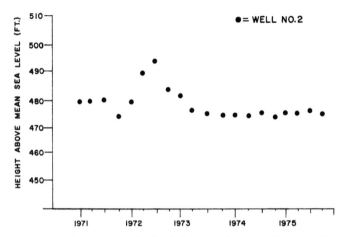

FIGURE 8 *Fluctuation in level of the groundwater surface in well 2 during the period 1970–1975.*

northeast, 9 days later led to the conclusion that the dominant direction of groundwater movement within the landfill site was to the northeast at a maximum rate of slightly more than 100 ft/day. Movement is strongly controlled by the bedding and joint plane attitudes within the bedrock and by their planar intersections. It is also strongly influenced by the depth of weathering, and moves at a more rapid rate in the weathered zone above fresh bedrock.

In 1973, another test of similar nature corroborated the direction and rate of groundwater flow.

Control of Leachate. Except for several instances of small seepage along the west side of the landfill site following times of very high precipitation, groundwater and leachate have emanated at the surface *only* at the location of the leachate spring (Fig. 2)

TABLE 1 *Quality of Water in Monitoring Wells 6 and 7, 1971–1975[a]*

Parameter measured:	PH		B.O.D. (mg/liter)		C.O.D. (mg/liter)		Total iron (mg/liter)		Chloride (mg/liter)	
Well number:	6	7	6	7	6	7	6	7	6	7
Sept. 1971	6.90	7.3	11.5	6.0	90	20	1.35	0.85	26.5	22.7
Mar. 1972	6.80	7.2	32.2	42.0	0	0	0.11	0.11	11.4	22.7
Aug. 1972	6.80	7.4	31.2	28.0	20	125	11.5	6.10	41.7	22.7
Feb. 1973	6.60	7.4	96.5	49.2	300	0	35.5	4.50	64.4	26.5
July 1973	6.30	7.08	156.5	43.9	415	24	71.0	5.50	79.5	18.9
Mar. 1974	6.39	6.6	167.7	27.1	720	20	128.0	0.6o	90.9	49.2
Sept. 1974	6.45	6.32	150.0	32.0	610	10	13.5	0.58	87.1	64.4
Feb. 1975	6.60	6.65	103.1	159.7	250	260	38.0	48.0	64.4	53.03
July 1975	6.55	6.48	47.8	150.8	23	340	27.0	27.25	75.8	37.9

[a]Measurements were made every 3 months. These data are from 6-month intervals.

on the northeast side of the landfill during the entire 5-yr period of trench-filling operations at the landfill site. Hence, groundwater and leachate have been "controlled" by being collected in constructed retention ponds adjacent to the leachate spring. During the 1970–1973 period, the original pond was augmented by two others.

These have provided enough surface area so that the iron in solution has been completely oxidized and precipitated as an iron oxide flocculent and, except on occasions of unusually high precipitation, the evaporative loss from the ponds has exceeded the flow of the leachate spring.

AREA-FILL PHASE:
DERRY TOWNSHIP LANDFILL

At the beginning of the trench-fill phase of landfill development in 1970, the projected life for the site was 7 yr. Continued growth of nearby communities and increased per capita production of refuse made it apparent in 1974 that only another year remained before all trenches would be filled. The landfill team proposed to extend the life of the site by converting it from a trench-fill to an area-fill site and designed a plan that would raise the final elevation of much of the site by 20 ft. This would extend the life of the site an additional 7 to 9 yr. Every geological parameter was carefully studied to determine whether such a conversion from trench fill to area fill could successfully be made, and a proposal was submitted to the Department of Environmental Resources for approval. Their permit for carrying out the conversion was granted in September 1975.

Figure 9 illustrates the general method of operation during the trench-fill phase of landfill development (1970–1975), and Fig. 10 shows the system now being employed as the site is being converted to and developed as an area type landfill.

Regenerative Capacity of the Shale

In an experiment in 1971 (Foose, 1972, p. 13), weathered shale fragments ranging in size from fractions of an inch to more than 1 in. were gently packed into a plastic tube 6 ft long with a diameter of 3 in. Raw leachate that had been collected from the toe of the landfill at its point of origin was poured into the top of the column, and the effluent was collected at the bottom. The leachate was dramatically modified from an ugly red liquid of poor quality to a nearly colorless liquid with essentially no iron, and with its hardness greatly reduced by a combination of adsorption by the clay minerals exposed on the surfaces of the weathered shale fragments and ion exchange between the leachate and the clay minerals themselves.

Further experimentation of similar nature in 1974 was designed to evaluate the capability for "cleansing" the leachate by unweathered shale versus weathered shale. If the experimental results were favorable, a spray irrigation treatment system could be proposed to handle the expected increase in the volume of leachate that would accompany raising the elevation of the landfill surface and increasing its mass. The results of these experiments are summarized in Table 2. They show that, as the raw oxidizing leachate percolated through the shale, the following levels were reduced: specific conductance, total solids, turbidity (Jackson turbidity units, JTU), total organic carbon, total hardness, and total iron. The chloride level was the exception; this increased with the weathered shales more than with the unweathered. This was probably caused by shale dust percolating through with the leachate and then being analyzed.

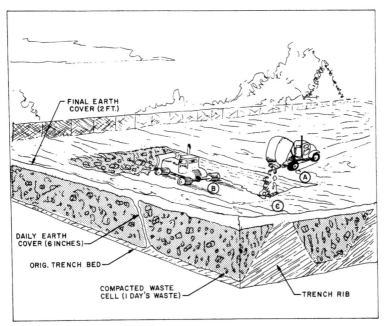

FIGURE 9 *Trench-type landfill operation.*

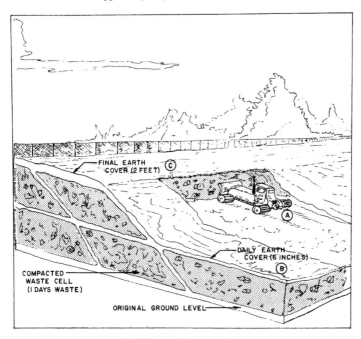

FIGURE 10 *Area-type landfill operation.*

TABLE 2 *"Cleansing" of Leachate by Percolation Through Shale*

	Leachate in:			Leachate out:		
	Unweathered shale	Weathered shale		Unweathered shale	Weathered shale	
pH	6.9	6.7	6.8	7.8	7.6	7.9
Sp. conductance (μS/Cm)	1,187	1,433	1,183	1,010	1,157	877
Total solids (mg/liter)	825	1,051	908	549	744	547
Turbidity (JTU)	431	584		26	7	
Total organic carbon (mg/liter)	73	72		38	47	
Total hardness (mg/liter)	529	670	420	412	528	286
Chlorides (mg/liter)	105	28	81	107	78	88
Total iron (mg/liter)	32	57	55	0.2	0	0.04

Leachate Recycling

The significance of these experiments is that the natural regenerative capacity of both weathered and unweathered Martinsburg shale is high. The success of the experiments led to the conclusion that an increased volume of leachate could be treated on the site by designing a mobile spray system that would involve the pumping of leachate from the storage ponds back onto the landfill surface. The leachate entering the ponds will be monitored to determine concentration levels of pollutants entering the ponds. When the levels of pollutants dictate, the leachate will be recycled onto the surface of the old landfill for treatment. The spray irrigation type of leachate treatment will allow for a loss of volume through evaporation while spraying, thus reducing the total volume of recirculating fluid, and it will substantially reduce organics and inorganics in the leachate, through the biological filter. There is sufficient acreage available for rotation of the areas sprayed upon so that fresh shale will be available for the reactive process during the life of the landfill. Spray application at the rate of approximately 0.5 in./acre/week would be more than sufficient to handle all the leachate now produced.

Figure 11 shows the concept and general scheme for the installation of the spray-back system at the landfill in the spring of 1976, and Fig. 2 shows its location.

Lysimeter Monitoring. Suction lysimeters have been emplaced at several depths below the surface in four different parts of the landfill to provide for the collection of percolating water from the soil and, thus, to analyze it for key parameters, such as pH, B.O.D., C.O.D., iron, and chloride. The lysimeter systems have been installed within the spray-back area (where it is expected to provide information on the regenerative action of the shale), and in a control plot where no spray back will be carried on at any time. In addition, lysimeter systems have been installed both within a refuse-filled trench and within a shale barrier that separates two trenches so as to provide the basis for comparison of soil-water chemistry in the two different regimes. The data from the four different lysimeter installations, collected at regular intervals, should provide additional insight into the rate of biodegradation of the refuse and the maturation of the landfill.

Figure 12 shows the details of a lysimeter installation. Soil water "percolates" into the porous porcelain cup and may be collected at any time by activation of the portable vacuum pump. Figures 13 and 14 show lysimeter installations as part of an instrument

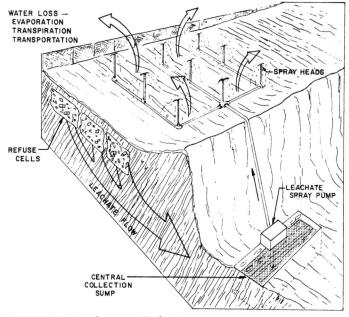

FIGURE 11 *Leachate spray-back system.*

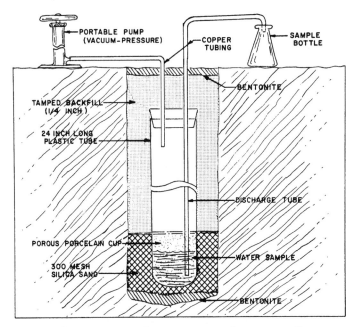

FIGURE 12 *Cross section of a pressure-vacuum lysimeter installation.*

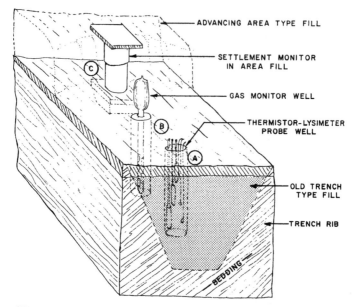

FIGURE 13 *Instrument cluster installed within or at the surface of a trench.*

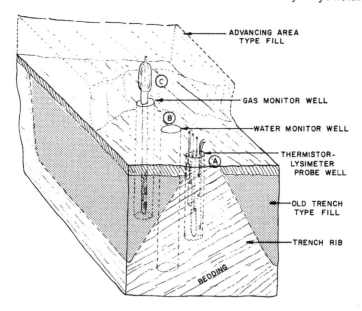

FIGURE 14 *Instrument cluster installed within a trench rib.*

cluster within a trench and an intervening rock rib. The lysimeter installation would, in both cases, be within the same vertical column as the thermistor probes.

Gas Monitoring

As discussed earlier, methane is the most common gaseous product of biochemical decomposition of refuse within a landfill. If the compacted refuse and the cover material above each refuse cell are permeable, the methane and any other gases produced will escape. However, the possibility of local entrapment clearly exists in every landfill. In an effort to learn more about the rate and volume of gas production, four gas-monitoring locations have been installed at the Derry Township landfill. Two will monitor methane gas production from one of the trenches. One is in the middle of a trench and the other at the end of the trench where it exits onto the hillside. A third monitor has been placed into the solid bedrock that constitutes the barrier between two trenches, and a fourth will be placed in the new area fill.

Figure 15 shows a typical gas-monitoring installation. Bentonite clay is used to seal off the area of the installation. Any installation within an existing trench involves the placement of perforated polyvinyl chloride (PVC) pipe within a porous stone stack (of 2-B grade size) *below* the bentonite clay layer. *Above* the bentonite clay layer, solid wall PVC pipe will be extended periodically as the different lifts are constructed. In this way, methane gas generation from within the existing trench will be monitored at the surface throughout the remaining life of the landfill.

In the area-fill monitoring installation, a layer of bentonite clay is placed at the top of the refuse in the existing trenches, and the existing cover is replaced and compacted in an

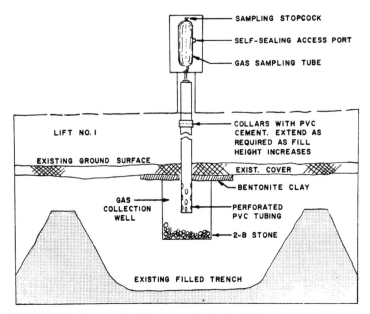

FIGURE 15 *Cross section of a gas-monitoring installation.*

attempt to prevent any gas generated within the trenches from moving directly into the proposed area fill. The grading in the trench with bentonite liners is toward the center of the fill. The stone stack would be extended periodically as the different lifts are constructed. The perforated PVC pipe is inserted in the top of the stone stack so that the gas sample may be collected. As the stone stack is extended to accommodate another lift, the perforated PVC pipe will be removed from the stone stack, the stone stack extended, and the perforated PVC pipe reinserted in the top of the stone stack. In this way, gas generation within all lifts of the proposed area fill will be monitored throughout the remaining life of the landfill.

Figures 13 and 14 show gas-monitoring installations as part of an instrument cluster within a trench and within an intervening trench rib.

An important objective of the gas-monitoring program is to relate the volume and rate of gas production to the age of the disposed refuse so that it may be possible in the future to use such data as a guide to the stage of maturation of the landfill.

Temperature Monitoring

In addition to gas production, the temperature regime within the landfill also may be altered by the decomposition of refuse. It may be possible to document the stage of refuse decomposition by routine temperature monitoring in a manner similar to that of gas monitoring.

At the Derry Township landfill site, temperature is being measured at three depths (3, 6, and 9 ft below the surface) and at the same locations as the lysimeter installations (Figs. 13 and 14). At each location the thermistors have leads that come to the surface. These are available for attachment to a direct temperature-monitoring device. The ultimate number of thermistor locations will be dictated by the success of the first set installed.

Settlement Monitoring

The addition of 20 ft of fill on top of the existing landfill site with its alternating filled trenches and intervening rock ribs will, of course, result in further compaction of the filled trenches, thereby producing differential settlement at the surface.

Two settlement monitoring plates are being installed, one on top of a completed trench and the other, at a later date, on top of the completed 20-ft-high area fill.

The settlement plate used to monitor settlement of the existing trench consists of a 1-in galvanized steel pipe welded to a steel plate, which, in turn, is fastened to a mass of concrete. Four- and 6-in galvanized steel pipe sleeves are used to permit the 1-in pipe to move freely as settlement occurs (Fig. 16). As settlement occurs, the frictional resistance of the surrounding soil against the 6-in pipe will hold this pipe in place. The 5-ft length of 4-in pipe will serve as an additional sleeve to keep soil or refuse from impeding the free movement of the 1-in pipe (Fig. 16).

The settlement plate used to monitor settlement of the completed 20-ft-high area fill will be simply a concrete mass. It will be placed immediately after the 20 ft height has been completed.

Precise leveling surveys will be made periodically to determine the amount and rate of settlement at both monitoring stations. The location of the settlement monitor above a trench is shown in Fig. 13.

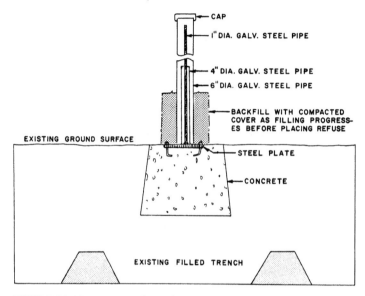

FIGURE 16 *Cross section of a settlement monitor.*

ACKNOWLEDGMENTS

We thank Don Hornung, of the Department of Environmental Affairs of Hershey Foods Corporation, for his processing of many of the data and for his carrying out the experiments dealing with leachate regeneration. We appreciate the contribution of Dr. Grover H. Emrich, geologist with A. W. Martin Associates, Inc., on the use of liners in landfills. The illustrations have been prepared by the Engineering Department of Hershey Foods Corporation. Inevitably, there have been many contributions from professional colleagues stemming from discussions in field and office. Chief among these are William Christensen, Hershey Foods Corporation, and Kenneth Williams, LeVan, Inc. Finally, we recognize with appreciation the hard work of those who, on a daily basis, and often under adverse physical conditions, supervised the growth of the landfill at the site: John Leppard and Harold Landvater and, during the past year, John Weigel, III, the Derry Township Manager.

SUMMARY

The Derry Township landfill near Hershey, Pennsylvania, was converted in 1970 from a town dump to a trench-type landfill following careful evaluation of geologic data concerned with groundwater elevation, direction and rate of groundwater flow, ground- and surface-water quality, depth of weathering of the shale bedrock, availability of adequate cover material, topography, drainage, and slope stability.

During the 5-yr life (1970–1975) of the trench-type landfill, routine monitoring of groundwater levels, water quality, groundwater flow, stream discharge, and slope stability has shown that the landfill operations have been carried out so that no significant alteration of the environment has resulted. Leachate produced by water percolating through the landfill flows at a spring at a rate of less than 5 gal/min. The leachate has been controlled by retention in constructed ponds and by evaporation.

Upon completion of the trench-type landfill, operations have been converted to an area-type landfill, with plans to raise the final surface elevation on the site by 20 ft, thereby extending the life of the landfill 7 to 9 yr.

To cope with the anticipated increase in volume of leachate, a mobile spray-back system has been designed to recirculate leachate through the landfill. Experiments have shown that evaporative losses will reduce the total volume of fluid. In addition, a combination of adsorption on clay mineral surfaces and ion exchange with clay minerals in the shale will "regenerate" the leachate by raising its pH and by reducing turbidity, dissolved solids, B.O.D., C.O.D., and iron.

In addition to continued monitoring of the geologic parameters named above, new monitoring systems have been installed for measuring soil-water chemistry, gas generation, internal temperature, and heat flow within the landfill, and surface settlement as a consequence of compaction.

The newly designed area-type landfill will involve raising the elevation of the original site by 20 ft. Associated with the conversion, which was begun in 1975, are a number of important parameters that are being measured for the first time at newly created monitoring stations: soil-water chemistry, temperature, gas production, and surface settlement.

Although each potential landfill site is different and will have its own set of problems, many will be similar to those described in this chapter. It is hoped this may serve as a guide to others by identifying the critical geologic and geomorphic parameters and discussing effective methods for solving problems.

REFERENCES

Abert, J. G. 1975. Resource Recovery: The economics and the risks: *Prof. Engr.*, Nov. 1975, p. 29–31.

Avers, C. C. 1975. Nashville confronts technical, economic problems in refuse-to-energy: *Prof. Engr.*, Nov. 1975, p. 32.

Bellamy, W. D. 1969. Cellulose as a source of single-cell-proteins: a preliminary evaluation: *General Electric Report No. 69-C-335*, 5 p.

Fitzpatrick, J. V. 1973. Energy recovery from municipal solid waste: present status and future prospects: Plastics Waste Management Committee Meeting, Society of Plastics Industry, Inc., Cincinnati, Ohio, 11 p.

Foose, R. M. 1972. From town dump to sanitary landfill: a case history: *Bull. Assoc. Eng. Geologists*, v. 9, no. 1, p. 1–16.

Han, Y. W., Dunlap, C. E., and Callihan, C. D. 1969. Single-cell-protein from cellulosic wastes: Louisiana State University, Baton Rouge, La., 20 p.

Stern, C. J. 1971. Technical and economic evaluation report: Total Recycling Systems, Inc., Narberth, Pa., 21 p.

V
GEOMORPHIC SYNTHESIS

15

THE ROLE OF GEOMORPHOLOGY
IN PLANNING

Robert F. Legget[1]

INTRODUCTION

Les Éboulements is the name, in Canada's second language, of a pleasant village in the province of Quebec on the north shore of the River St. Lawrence, close to the widening out of the river into the great gulf. As may be seen from Fig. 1, the village has a somewhat unusual location but this adds to the charm of its locale. The tongue of land on which the village is situated will be found, upon closer examination, to be the flat conical delta of an ancient massive landslide of the sensitive marine (Leda) clay so characteristic of the St. Lawrence Valley. Once disturbed, these clays can flow like a viscous liquid, but after such disturbances, with the release of the excess pore water, they change to a stable insensitive material on which buildings can safely be erected. All this is reflected in the name of the village, probably unique, since *éboulement* is one of the French words for landslides.

The date of this major landslide has not yet been determined with any accuracy, but it probably antedates the first white settlements on the St. Lawrence. The founders of the village had no geotechnical or geomorphological assistance to help them with the start of

FIGURE 1   *Village of Les Éboulements on the St. Lawrence River, Que., Can. (Photograph by C. B. Crawford.)*

[1] Formerly Director, Division of Building Research, National Research Council of Canada.

their building, but their sound common sense led them to follow a perfectly safe course. How strange and significant it is to think that the planners and builders of a modern housing project, about 80 mi northwest of Les Éboulements, who could have availed themselves of modern geological and geotechnical studies had they wished, placed their buildings on undisturbed Leda clay in a location so critical that on May 4, 1971, a massive landslide—that of St. Jean-Viannet—took place, carrying 40 new homes to destruction and 31 persons to their deaths.

Les Éboulements may therefore fittingly serve as a symbol for this chapter, representing the safe and efficient use of a landform. It is not alone in its distinction. In the canton of Glarus, Switzerland, on the road from Glarus to Altdorf, one passes the picturesque village of Urnerboden in an incomparable mountain setting, the great peak of Claridenstock forming a splendid backdrop. If one can divert attention from the scenery around and look at the village, its unusual position will be noticed—the few houses clustered on a valley-floor ridge that might at first be taken for a glacial feature, but which is actually another landslide debris cone, now safe as a building site but once the result of a major landslide and rockfall from the steep slope to the east. Around the world there must be many other human settlements founded quite securely on ancient landslides, even though they are not so well known as the locations of towns and settlements that have been destroyed or damaged by falls of soil and rock. The Frank "slide" always comes first to mind in North America, but in Europe the destruction of Taurentunum in the year 563 is well recorded, the first of a succession of such disasters.

Although it is useful to be reminded, occasionally, of such natural disasters if only to have confirmed, yet again, the dynamic aspect of geomorphology, it is rather to the interrelation of the works of man with the form of the land that attention must be directed, and to the forces, both natural and mancontrolled, that have and still are shaping it. General statements or descriptions of large-scale examples will not be very profitable, and so this discussion will be confined to just two aspects of geomorphology and engineering—landslides and floodplains. Examination of a few cases will make clear that, for engineering purposes, geomorphological studies, while useful, by themselves are not enough. Correspondingly, if the significance of geomorphology to planning (in its broadest sense) is to be effective, its importance must be made clear far beyond closed geological-geotechnical circles, into the minds of public officials charged with the development and implementation of plans for urban and regional development and of the engineering works that serve the steadily growing cities of the world.

LANDSLIDES

The ancient and lovely city of Bath, in England, provides a good starting point. First developed by the Romans as a spa (Aquae Sulis) at the start of the Christian era (one can still see the Roman baths and the lead pipes they used), Bath was still a small walled city in the seventeenth century, although now a modern community of about 80,000. The splendid Georgian buildings of the crescents and terraces that grace its sloping building sites were erected, using the fine local Bath Stone, in the eighteenth century. As one walks the hills around the city, evidence of landsliding is to be seen in many places. It was in this area that William Smith, the father of British geology, and of engineering geology, did some of his notable work in controlling landslides with well-conceived drainage systems. It is not surprising, therefore, to find signs of slope instability within the city.

Camden Crescent, for example, can be seen to be unfinished; the Hedgemead slip made completion impossible. A wander round the pleasant streets of the main city area will show other evidences of earth movements. City authorities are well aware of the critical geological conditions with which they have to deal. The official City Development Plan allows for restriction of building in critical areas, being closely linked with the detailed geological studies that have been made of the city and its environs (Kellaway and Taylor, 1968). The city of Bath can well serve as an example in this to other cities, few of which have such difficult geological conditions with which to deal. It should also be a place of pilgrimage for all who are concerned with geomorphology and engineering and who have occasion to visit the United Kingdom.

In great contrast is the Cheakamus Dam of the British Columbia Power and Hydro Authority in western Canada. Set amid the mountains of the Coast Range, in magnificent scenery on a scale vastly different from that around Bath, the dam is about 1,500 ft long and 88 ft high, having been built between 1955 and 1957. Since it is an embankment dam with its exposed faces being of very coarse material, it blends well into the landscape. It has had the effect of raising the water level in the Cheakamus River, previously here enlarged into Stillwater Lake, by 65 ft. This enlarged reservoir-lake floods into Shallow Lake from which a 6.5-mi tunnel was driven to a steep slope on the northeastern side of the valley of the Squamish River. Pipelines convey the water diverted from the Cheakamus River to a power house on the Squamish River in which 190,000 hp are generated under a head of 1,125 ft, the power supplying the ever-growing metropolitan area of Vancouver.

Excellent though this example of modern water-power engineering is, why should it be mentioned here? Because the major part of the Cheakamus Dam is founded on landslide debris. In the planning of this necessary extension to the B.C. power system, this obviously desirable site was given special consideration. Through careful geological studies, notably by W. H. Mathews, it was known that, as a result of the failure of "The Barrier," a lava plug blocking the outlet from Lake Garibaldi, a massive landslide took place here about a century ago. About 20 million yd^3 of rock debris, including boulders up to 4 ft in diameter, flowed down the small creek that was blocked by "The Barrier" and filled the adjacent valley of the Cheakamus River for a distance of 2.5 mi up- and downstream to a depth of 150 ft, burying the forest that was previously there and forming Stillwater Lake when natural conditions had readjusted themselves.

Although topographically the damsite was ideal, geomorphologically the site was fraught with problems. No dam of this size had previously been built on top of a recent landslide. The B.C. power authorities therefore consulted Karl Terzaghi, internationally known geotechnical consultant, who thought that a dam could safely be built despite the difficulties. It was successfully completed under his expert guidance, Terzaghi making no less than 18 visits to the site while work was in progress, adjusting the design as excavation revealed new features of the heterogeneous landslide material. It was detailed study of the geomorphology of the damsite and of the surrounding area that permitted the dam to be so successfully designed and constructed, using some of the landslide debris as the main material for the dam. It provides a good example of "cooperating with the inevitable" when the nature of the "inevitable" is known (Terzaghi, 1960).

The same expression may well be used to describe, by way of contrast, another acceptance of an "inevitable" landslide, this being a slide that did not occur but which could have if measures had not been taken to prevent it. Coors beer is a familiar beverage to many, especially those who live in the West. This famous brand is brewed in a large plant on the outskirts of Golden, Colorado, located in a geologically sensitive area, of great interest to geomorphologists if only because of the famous twin mesas, North and South mountains. One of the roads leading from Golden to the east, toward Denver, is located at the foot of a steep slope near the top of which stands a large and valuable residence. When the plant had to be extended, a somewhat unusual solution was developed to maintain the stability of the slope and the integrity of the road. It can be best appreciated from a photograph, Fig. 2, showing what is possibly a unique association of beer with geomorphology.

Far removed from the delights of Colorado is another area featured by extensive sloping ground that I have studied. This is that part of the eastern coast of the south island of New Zealand in the Otago district between Palmerston and Dunedin. [If reference to New Zealand appears strange in this context, it should be explained that there are few, if any, countries of the world that exhibit more striking and widespread unusual geomorphological features than this distant British Dominion, as shown so well by the writings of Sir Charles Cotton (1926).] On a memorable visit I was guided along this fascinating coast by the venerable W. N. Benson, of the University of Otago, who has described some of its problems in some of his many geological papers (Benson, 1940).

The local bedrock is unusually complex, Pliocene basalts and phonolites overlying a Miocene peneplain, below which are Lower Cenozoic and Upper Cretaceous sandstones and mudstones and then a Cretaceous peneplain. The entire structure has been faulted and warped, with a general slope toward the sea. This is reflected in the relatively steep surface slopes along the coast, providing topographically desirable building sites, which are, unfortunately, geomorphologically dangerous. This can be easily seen by examining vegetation patterns, signs of earth movements downslope being well developed. Hedges planted when the region was first settled and the land cleared (1850–1860) serve as

FIGURE 2 *Main road near Golden, Colo., as it tunnels beneath an extension of the Coors Brewery, which serves as a support for the retaining wall on the right.*

FIGURE 3   *Abandoned Puketeraki Tunnel on New Zealand Government Railways, South Island, N.Z., showing the new location of the main line south in cutting.*

excellent indicators. Unfortunately, in earlier years the geomorphic significance of these slopes was not fully appreciated.

Pipelines to a small water power plant have had to be replaced, following movement of their pedestal supports, by new pipes in a tunnel, the excavation of which showed that the decomposed rock in the hillside was nearly 200 ft thick. At least one building was demolished by differential ground movements and others have been damaged. Roadways have been displaced, and the main railway line to the south moved outward toward the sea a distance of 18 ft in one location in the course of 3 yr. Even more serious was the enforced abandonment of the Puketeraki Tunnel (on this same rail line) due to serious damage to the tunnel lining as a result of successive ground movements. It had to be replaced by a curved rail alignment in a deep cut, as shown in Fig. 3.

In all these cases it was not merely the form of the land that caused the troubles, but the "lubricating effect" (to use an inaccurate but generally descriptive term) of water in association with the landforms, emphasing the dynamic aspect of geomorphology and its concern with the processes that have led to present-day land structure. Some of the water, as in New Zealand, was due to snowmelt, but more usually was the result of heavy rains. This is common experience with potentially unstable ground. It has probably never been more dramatically demonstrated than in the great landslide at Handlova in the center of Slovakia. Handlova is an important coal-mining town in this part of Czechoslovalkia. In December 1960 it was threatened with possible destruction as a direct consequence of the misbehavior of a local geomorphic feature.

To the south of the town, across the small Handlovka River, an extensive well-grassed slope is a dominant feature of the landscape. The slope consists of a mixture of sandy gravels, with clay and tuffites overlying a sloping surface of Paleogene marly shales, which are part of a flysch development. The existence of springs and of old drainage ditches (which had not been maintained) indicated possible instability of the slope, which had long been used for sheep grazing. A new power station in the town had emitted much fly ash; prevailing winds carried some of this on to the slope, with the result that grazing was interfered with to such an extent that part of the slope was taken "out of grass" and tilled for cultivation. Rainfall during the year 1960 totaled 1,045 mm (41 in.), the long-term annual average being only 689 mm (27 in.). This excessive precipitation, combined with the removal of some of the protective grass cover, disturbed the delicate groundwater equilibrium beneath the attractive sloping meadowland. Early in December of that year a slight earth movement at the foot of the big slope was noticed. Accurate measurements were started on December 22. It was found that at one point a movement of almost 150 m (492 ft) had taken place within 1 month; 20 million m^3 (700 million ft^3) of soil were on the move, threatening to engulf the town. The critical emergency can well be imagined. The remedial measures immediately brought into action were equal to the occasion. Local fire brigades brought batteries of their pumps into play and removed all standing and shallow water. Six drainage tunnels were excavated (at considerable risk) into the base of the slope, the tunnels and all other measures being based on geological advice from the group of engineering geologists assembled at the site. With the "capture" of the excess water, the movement was stopped, but only after 150 houses had been destroyed and much other damage done (Zaruba and Mencl, 1969). Today, the permanent drains installed in the inverts of the tunnels carry a very small flow but provide the drainage essential to maintain the integrity of the slope (Fig. 4).

There have been other major slides caused by the interaction of water and critical landforms, most of them naturally unrecorded. There are, however, brief records of a few such disasters in North America. One of the least known of these occurred on the Thompson River, British Columbia, near the small town of Ashcroft. The scars left by a

FIGURE 4   *Town of Handlova, Czechoslovakia, as seen from a point near the center of the sloping ground, which, when moving in December 1960, threatened the town.*

series of major slides in the latter part of the nineteenth century may still be seen by travellers to the west on the Canadian Pacific Railway or on the Trans Canada Highway.

In the early days of settlement of this relatively dry part of the interior of British Columbia, small irrigation projects were developed by individual settlers. Mountain streams were tapped for irrigating land on the first of the main terraces on the east bank of the river, here a very prominent geomorphic feature of the valley. The water thus spread on ground being used for agriculture naturally percolated into the soils beneath the surface. These are glacial deposits into which the river has downcut to its present position; they include much "white silt" (to use a local term), clean sand, stratified gravel, and some till. The percolating water found its way to the foot of the steep slopes below the irrigated terrace, disturbing their natural stability. When excavation for the Canadian Pacific Railway started along this section of the Thompson River, massive slides took place in consequence, seven occurring in a distance of 5 mi in the vicinity of Black Canyon. Further slides took place after the line was opened, causing major maintenance problems. It was found that even greater slides had taken place before railroad construction had started, the largest being known as the North Slide.

This took place in October 1881 after water in a small reservoir in the hills above the terrace was released, through failure of a primitive dam. The consequent excessive percolation from the terrace caused collapse of the riverside slopes to such an extent that a width of 2,000 ft was actually forced across the river on to the west bank. The total volume was later estimated to be about 60 million yd^3. It dammed the river with an "embankment dam" 160 ft high, which stayed in place until overtopped by the water being stored behind it. It then collapsed, resulting in a flood down the Thompson River that today can only be imagined. Figure 5 shows the location of these man-initiated geomorphic disasters, reprinted (by permission) from the only account of these slides so far located, a short paper in the *Minutes of Proceedings of the Institution of Civil Engineers* (already up to its 132nd volume in 1897), a source of information on early civil engineering works distinguished not only by its longevity, but also by the attention that the early engineer—authors paid to geology (Stanton, 1897).

The examples cited are typical of many. They demonstrate that an appreciation of the geomorphology of sites proposed for building, whether in city or in open country, *must* be the start of the planning process. Study of landforms, however, is but the start of this most important of all applications of geomorphology. The processes that have created present landforms must equally be appreciated. This is so easy to forget or to disregard, as some of the examples just cited clearly show. A good case can be made for the suggestion that all studies of sites proposed for building should include at least one visit by those responsible for planning in heavy rain, uncomfortable though it may be! Bright "sunny days in the country" can be so misleading. And it is rain—water—that can have so profound an effect once man attempts to use landforms that seem to be well suited topographically to his purposes.

FLOODPLAINS

Nowhere are the adverse effects of water more clearly demonstrated than in man's unthinking use of floodplain lands. Topographically, these most obvious of landforms appear to be ideal—generally level ground, formed usually of fertile soil, pleasantly contiguous to running water and often of easy access. And yet what dangers they present,

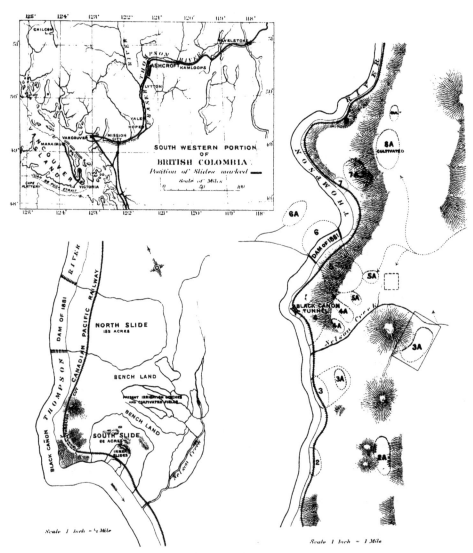

FIGURE 5 *Location of major slides on the Thompson River, B.C., Can. (Reproduced from Stanton, 1898, by permission of the Institution of Civil Engineers.)*

if any thought be given to their formation, to the fact that they are what they are because of the flooding to which the relevant river will occasionally be subject. There is probably no geomorphic feature that is more widely recognized by the "lay" public, that has in consequence been more widely used by man for his building purposes, and yet which can so often carry with it most serious potential danger to human lives because of its unthinking "development."

If these statements seem to be somewhat extreme, consider the results of a 1971 survey of floodplain land within cities and towns of the United States. No less than 5,200 were found to have some parts of their developed areas located on floodplains. Less than 1,000 have yet adopted regulations limiting the use of such floodplains (Goddard, 1971). A much smaller number had taken preventive measures for limiting the extent of damage when the "100-yr flood" does come. The situation in Canada is probably similar. As an extreme example, it has been officially stated that

> The Lower Mainland Regional Planning Board (of B.C.) has estimated that by 1968 the floodplain population (along the Fraser River, near Vancouver) will have increased to 210,000 and, by 2001, to approximately 290,000. With further expansion the potential danger from a major flood in the Lower Fraser Valley will become increasingly ominous (Canada Water Year Book, 1975).

There was flooding of some of these floodplain lands again the fall of 1975. Most fortunately, this was not nearly so extensive as the flood of the early summer of 1948, which I experienced and will never forget. It inundated 55,000 acres and destroyed 2,000 homes.

The experience of Rapid City, a well-settled community of about 50,000 in western South Dakota, has lessons of wide import, even though on a different scale in size and in tragedy. On the night of June 9, 1972, a phenomenal meeting over the adjacent Black Hills of a cold front and warm moist air resulted in extreme rainfall, as much as 15 in. in 6 h in some places. Rapid Creek, from which the city may be said to take its name, reached a flood flow of 50,000 ft^3/s. The Pactola Reservoir, 14 mi upstream of the city, gave a false sense of security; the worst of the rain fell between the reservoir and the city. Building had been permitted on the floodplain; 238 people lost their lives that night. Property damage was estimated to be $128 million. The area flooded was almost identical with that covered by a flood in 1907. The U.S. Geological Survey had published 1 : 24,000 maps of the Rapid City area before the flood; the coincidence of the flooded area with the alluvium plotted on the maps is significant. It is shown in Fig. 6, reproduced by permission from a recent paper (Rahn, 1975); this shows also the location of bodies recovered after the flood; five were never found.

This is a recent example; unfortunately, it is far from being alone. On October 15, 1954, Hurricane Hazel penetrated as far as the Toronto area in southern Ontario. Again through a phenomenal combination of meteorological features, extreme rainfall took place over a small area—as much as 1.06 in. in 1 h at one gage, a total of 7.02 in. in 24 h at Brampton. This was, however, only slightly more than a previous maximum of 6.90 in. in 1915. The small Humber River with adjacent streams, such as Etobicoke Creek, rose to catastrophic flood heights. Residential building had been permitted in floodplain lands in the township of Etobicoke. Eighty people lost their lives that night. Property damage was estimated at $25 million (Boughner, 1955).

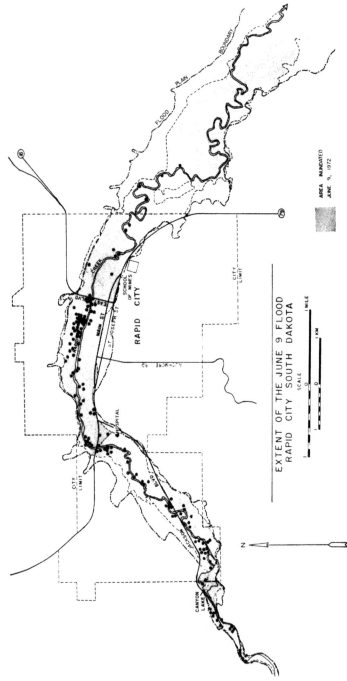

FIGURE 6 *Map of Rapid City, S.D., showing the area inundated in the 1972 flood, and the area mapped as alluvium by the U.S. Geological Survey: black dots show the location of bodies recovered after the flood. (Reproduced by permission of P. H. Rahn and the Association of Engineering Geologists.)*

All buildings were soon cleared from the floodplain lands in Etobicoke, now a part of Metropolitan Toronto. Regulations were promulgated prohibiting further building on this critical land. Despite all the pressures that have arisen, quite naturally, for the development of some of this "desirable land," the integrity of the regulations has been maintained; no more building has taken place.

Don Barnett, former mayor of Rapid City, reviewed the 1972 experience in a key address to a 1975 Engineering Foundation Conference on Flood Plain Management. He urged "community leaders to begin immediately developing an effective floodplain management program and to establish a long range program now rather than after a big flood." He coined a phrase that should become part of every planner's credo: "It is stupid to sleep in the flood plain" (Wall, 1975).

Reference to the Engineering Foundation may confirm an impression that all too many geologists and geomorphologists entertain—that floodplain problems are engineering matters only. On the contrary, it is difficult to imagine any activity in which geomorphology is, or should be, more closely allied with engineering than in the protection and proper use of floodplains in urban areas. Not only are there the obvious reasons for this, already indicated, but there is another neglected factor, geological in character, essential to sound floodplain planning. Study of catastrophies such as those just cited usually reveals either (1) that the catchment areas onto which phenomenal rains have fallen prior to extreme flooding were such that runoff took place immediately over solid rock at or very close to the surface, or (2) that the near-surface soil conditions were such that prolonged earlier rainfall had brought the water table up to or close to ground level, thus giving the same instant runoff as the impervious surface. Attention to groundwater equilibrium conditions in areas liable to flooding (and also in many other similar situations) has been neglected far too long, both by geomorphologists and engineers. It is essential and can only be studied against an understanding of the local geology. The pioneer and vitally useful work of Thornthwaite has yet to be given the recognition, and use, in geomorphic studies that it deserves. When one notes that the average annual flood losses in the greater Chicago area amount to $31 million, with damage to an average of 50,000 homes, the urgency of public attention to floodplain lands (as is being given in Chicago) will be apparent (Lanyon, 1974).

SUBSURFACE STUDIES

Studies of groundwater associated with geomorphic problems naturally necessitate detailed knowledge of the geology of the local subsurface. The same is true, however, whenever man changes or utilizes landforms for his own purposes. The New Zealand experience showed this well. Study of landforms is the starting point; determination of the natural processes allied with the present landforms is an essential complement. The engineer, however, must usually go beyond this and determine, without doubt, what subsurface conditions really are, despite the careful predictions from geomorphic studies.

An example from the Canadian Arctic may be cited as a final example. Admittedly, the locale involved is atypical, but the natural processes at work do not change. The government of Canada decided in 1953 that a new site for the settlement of Aklavik, in the Northwest Territories, must be found, local terrain and permafrost conditions making impracticable any improvement of the existing small town. It is located within the great delta of the Mackenzie River. A survey team was assembled. They studied, in Ottawa,

FIGURE 7   *Ground view of coalescing alluvial fans on the west side of the delta of the Mackenzie River, N.W. T., Can., looking west toward the Richardson Mountains.*

aerial photographs of the entire delta (4,700 mi^2), selecting 12 possible sites for a new town on the basis of the landforms as determined from the photographs (Merrill et al., 1960). Field studies of these sites in the summer of 1954 reduced the potential sites to four, all geomorphologically desirable. "Obviously," the best site was on sloping ground, formed by coalescing alluvial fans, between the Richardson Mountains and the Husky Channel on the west side of the delta. This site is seen in Fig. 7; sand and gravel were confidently anticipated beneath the surface since the disintegration of the mountain rocks, the products of which form the fans, could be seen in action in gullies less than 0.5 mi away.

As a check, in keeping with due engineering caution, test drilling was carried out. This revealed that at the "chosen" site there was no gravel at all but only sand and silt with high organic content to depths of almost 40 ft, exactly the type of subsurface condition that had been one of the factors discrediting the original town site. The reasons for this unusual condition have been discussed (Legget et al., 1966). Another site, on the east side of the delta, was therefore selected since excellent subsurface conditions were found beneath it. Today the new town of Inuvik stands on this site, with a population of over 3,000, all its buildings having performed excellently despite the underlying permafost condition.

RECOMMENDATIONS

The finding of the site for Inuvik and the development of the new town were entirely under the control of the government of Canada. The authorities responsible were fully appreciative of the necessary site studies, the minister directly in charge making the long journey from Ottawa in order to see the sites for himself before the final decision was made. More usually, site conditions are studied without any such governmental *imprimatur,* as in all private enterprise development, whereas strictly scientific field investigations may sometimes reveal conditions that would be dangerous should the area in question ever be developed. What can be done if dangerous geomorphic conditions are revealed?

In the case of private works, owners must be persuaded of possible dangers and guided, by their professional advisers, into an appreciation of possible ways of avoiding trouble while achieving a sound development. A residential housing development in San Clemente, California, was planned for a canyon site. The floodplain hazard was recognized, the floodplain being utilized for a most carefully designed nine-hole golf course; the course was planned to serve as an emergency floodway in the case of heavy floods. It has performed successfully in a recent 80-yr flood (Hood, 1972).

More difficult are the situations that arise when potentially dangerous ground conditions are encountered in the course of independent study, unrelated to any building proposals. The now famous case of the report by Miller and Dobrovolny (1959) on the surficial geology of Anchorage, Alaska, comes at once to mind (Dobrovolny and Schmoll, 1968). Publication of the Rapid City maps by the U.S. Geological Survey suggests a parallel situation. As better building sites are steadily used up, and less desirable sites have to be accepted, the problem will not disappear; it will probably become more widespread. I remember that when some of my (former) colleagues determined, in the course of research studies on some old landslides, that the slope of one area, known to be slated for development, was an old landslide scarp with a factor of safety against sliding of about 1, I wrote to the municipality in which the site was located advising of this finding and suggested that, before any development was authorized for this area, professional advice should be obtained to check the safety of the slope. The objections to any other course, such as approaching a developer, will be obvious.

Geomorphologists, no more than geologists, can no longer take refuge in their ivory towers. Their help is needed in ensuring that all urban and regional planning is carried out safely and efficiently. This means convincing all planning authorities that knowledge of local geology, and geomorphology, is an essential prerequisite for all town and country planning, just as it is for the planning of all engineering works. All university departments of geology face a special challenge in this regard in their own cities, but the need exists in all municipalities, large and small, whether or not they are the homes of universities or colleges. It is a need that must be met.

REFERENCES[1]

Benson, W. N. 1940. Landslides and allied features in the Dunedin District in relation to geological structure, topography, and engineering: *Trans. Roy. Soc. New Zealand,* v. 70, pt. 3, p. 249–263.

Boughner, C. C. 1955. Hurricane Hazel: *Weather,* v. 10, p. 200.

Canada Water Year Book. 1975. Ottawa, 232 p., p. 33.

Cotton, C. A. 1926. *Geomorphology of New Zealand; Part I, Systematic:* Wellington, N.Z. (and many later publications).

Dobrovolny, E., and Schmoll, H. R. 1968. Geology as applied to urban planning: an Example from the Greater Anchorage Area Borough, Alaska: *23rd International Geologic Congress, Proceedings,* Section 12, p. 39–56.

[1] Since *Civil Engineering* now paginates each issue separately, with no volume numbers given, it seemed best to list the papers from that journal as shown.

Goddard, J. E. 1971. Flood-plain management must be ecologically and economically sound: *Civil Eng.*, Sept. 1971, p. 81–85.

Hood, C. C. 1972. Flood channel doubles as golf course: *Civil Eng.*, May 1972, p. 49–50.

Kellaway, G. A., and Taylor, J. H. 1968. The influence of landslipping on the development of the city of Bath, England: *23rd International Geologic Congress, Proceedings,* Section 12, p. 65–76.

Lanyon, R. 1974. Flood plain management in metropolitan Chicago: *Civil Eng.*, May 1974, p. 79–81.

Legget, R. F., Brown, R. J. E., and Johnston, G. H. 1966. Alluvial fan formation near Aklavik, Northwest Territories, Canada: *Bull. Geol. Soc. America,* v. 77, p. 15–30.

Merrill, C. I., Pihlainen, J. A., and Legget, R. F. 1960. The new Aklavik: search for the site: *Eng. Jour.,* v. 43, p. 52–57.

Miller, R. D., and Dobrovolny, E. 1959. Surficial geology of Anchorage and vicinity, Alaska: *U.S. Geol. Survey Bull. 1093,* 128 p.

Rahn, P. H. 1975. Lessons learned from the June 9, 1972 flood in Rapid City, South Dakota: *Bull. Assoc. Eng. Geol.,* v. 12, p. 83–97.

Stanton, R. B. 1897–1898. The great landslides on the Canadian Pacific Railway in British Columbia: *Minutes Proc. Inst. Civil Engr.,* v. 132, p. 1–20.

Terzaghi, K. 1960. Storage dam founded on landslide debris: *Jour. Boston Soc. Civil Engr.,* v. 47, p. 64–94.

Wall, G. R. 1975. EF conference focuses on flood plain management: *Civil Eng.*, Oct. 1975, p. 20, 22.

Zaruba, Q., and Mencl, V. 1969. *Landslides and Their Control:* Prague, 205 pp.

16

RIVER MANAGEMENT CRITERIA
FOR OREGON AND WASHINGTON

Leonard Palmer

INTRODUCTION

In western Oregon and Washington, steep mountain streams descend into populated lowlands and empty into tidewaters or major rivers. Many of the rivers are, as yet, only moderately influenced by man's structures and landuse practices. Though landuse demands are expanding, strong pressures for recreational and scenic river use are also increasing and encourage preservation of natural conditions. The quality of a natural flowing stream is of growing value in public recognition.

The definition of natural stream values requires an understanding of the total stream system and the interaction of its parts. Geomorphic studies contribute significantly in these definitions. Single-purpose flood control or channelization design alone is often not compatible with the equilibrium system and often contributes to an escalating chain of river-control expenditures and resource losses.

In the studies reported here, rivers are classified into geohydraulic zones by combining evidence of the long-term (thousands of years) geomorphic processes with short-term channel hydraulic processes. The geohydraulic zones are used to define the river processes and the landuses that can be supported in each zone. Management guidelines oriented to sustaining the equilibrium of natural processes in the river as an entire system are proposed.

Review of the present river-control practices shows that flood protection works built to modify the river processes are not reducing flood damage (Leopold and Maddock, 1954). In Oregon, the annual cost of structures for flood protection over the last 30 yr is approximately equal to the present annual flood damage ($450 million/30 yr equals $15 million/yr, compared to $17 million/yr flood damage, not including other state, local, and federal expenditures or tributary stream and bank erosion losses). Estimates project a quadrupling of annual flood damage costs (1965 dollar values) in the next 50 yr (Oregon State Water Resources Board, 1972, p. 2).

To find the total costs, loss of natural resource values destroyed by the construction of flood protection works must be included. It appears that the past practice of landuse and river modification in Oregon has resulted in a loss of natural resources and a gain of bigger flood disasters for the future.

GEOHYDRAULIC ZONES

For a river-management classification system to work, it must be understandable and acceptable to a majority of those involved in the decision-making process. It must also be responsive to the wide range of mechanisms that occur in the river system.

The natural characteristics of most rivers in Oregon and Washington provide the basis for their subdivision into one or more of four geohydraulic zones. The zones are defined by the characteristic combinations of valley cross section, channel pattern, gradient, and bed load size. The Bauer geohydraulic river environment zones are summarized in Figs. 1 and 2. Bauer (who is a shore-resource engineering consultant in Seattle) considers these zones as energy-level zones dependent upon the height and velocity of the fluctuating river currents.

A brief list of the characteristics of the Bauer river zones shown in Fig. 1 are as follows:

The *boulder zone* is typified by nongraded mountain rivers with coarse bouldery sediment. The narrow channel runs in a bedrock gorge with valley walls rising to each side. Vertical erosion of the channel maintains a V-shaped valley in cross profile. Steep gradients exceed 5 m/km (25 ft/mi). Rapids are common.

The *floodway zone* is formed where coarse (sand to cobble sized) sediment load is equal to or greater than the transporting capacity of the river (balanced or aggrading). Braided and shifting multiple channels form an alluvial valley floor, with or without a floodplain. Moderate gradients range from about 1 to 5 m/km (5 to 25 ft/mi). Pools, ripples, eddies, and point bar sand and gravel beaches occur in the channel. Channel islands are common.

In the *pastoral zone*, fine bedload material (silt and sand) form cohesive bank material and deeper meandering single channels. Valley floors are typically wide with natural levees commonly bordering the channel. Floodplains may be submerged during flood stage once or twice per year. Sediment loads are usually balanced with transporting power of the river. Gradients range from 0 to 1 m/km (0 to 5 ft/mi).

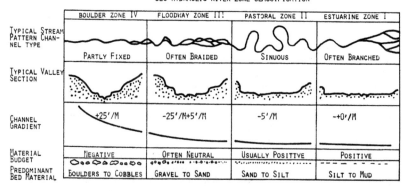

GEO-HYDRAULIC RIVER ZONE CLASSIFICATION

	BOULDER ZONE IV	FLOODWAY ZONE III	PASTORAL ZONE II	ESTUARINE ZONE I
Typical Stream Pattern Channel Type	Partly Fixed	Often Braided	Sinuous	Often Branched
Typical Valley Section				
Channel Gradient	+25'/M	-25'/M+5'/M	-5'/M	-+0'/M
Material Budget	Negative	Often Neutral	Usually Positive	Positive
Predominant Bed Material	Boulders to Cobbles	Gravel to Sand	Sand to Silt	Silt to Mud

FIGURE 1 *Bauer geohydraulic river zones. The action and natural function of rivers vary considerably with the valley-channel gradient and resulting current energy potential. These river reaches in different gradients display characteristic physical and ecologic streamway environments that will dictate constraints and opportunities in streamway planning and management, whether concerned with flood control and diking, recreation, fisheries, wildlife, pollution, or zoning.*

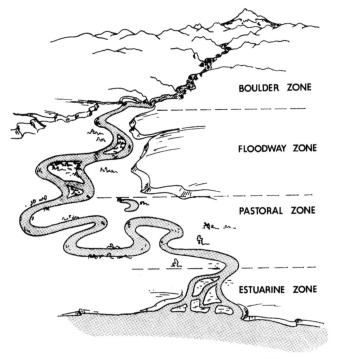

BOULDER ZONE

FLOODWAY ZONE

PASTORAL ZONE

ESTUARINE ZONE

FIGURE 2 *Graphic summary of the channel forms of the Bauer geo-hydraulic river zones.*

The *estuarine zone* is in the area of periodic river gradient reversal (plus or minus 0 m/km). Rivers discharging into tidal water produce branching channels in tidal marshes. Similar, but less-pronounced channel branching may occur at stream junctions, producing estuarine environment where the fluctuating base level seems to stimulate a measurable increase in upstream channel islands (Fig. 3). Sediment is typically fine (silt and mud) and is being deposited. The surge plain may be inundated by tides once or twice per day.

Although many additional varieties of river type could be differentiated, the four geohydraulic zones proposed by Bauer serve to segregate river forms with greatest utility in specifying the landuses that each environment can support. Landuse values of each zone are strongly influenced by the height, area, frequency, and velocity of flooding, plus the relative stability of the channel position. The environments for vegetation, fish, and wildlife are also significantly similar within each zone and different between zones.

Progression from coarse to fine sediment and steep to low gradient is not necessarily continuous and orderly. Many reversals in order of geohydraulic zone may occur along a river course. Changes in bedrock conditions and in sediment contribution from the river basin often cause changes in the zone. The progression of zones shown in Fig. 2 is only for convenience in demonstrating the four zone characteristics. Figure 4 shows the interchanging of boulder and floodway zones along the Snoqualmie River, Washington.

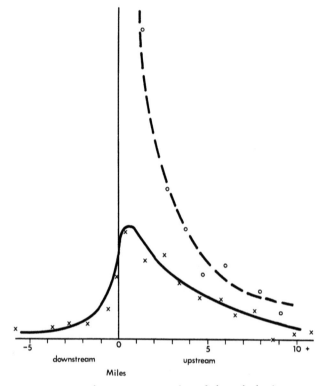

FIGURE 3 *Cumulative average number of channel islands per up-stream mile from Puget Sound tidewater is shown by dashed line. Islands per mile upstream and downstream from river junctions are shown by the solid line. (Data taken from U.S. Geological Survey topographic maps of eastern Puget Sound rivers.)*

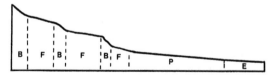

FIGURE 4 *Longitudinal profile of the Snoqualmie River showing the alternation of boulder and floodway zones: B, boulder zone; F, floodway zone; P, pastoral zone; E, estu-arine zone. (After Bauer, in Rivers Sub Committee, 1972.)*

Geomorphologists may recognize in the four zones characteristic equilibrium regimes, with independent variables of discharge and sediment load, supplied by the drainage basin, and resultant channel forms and patterns developed by the river. Among geomorphologists it is known that the long-term discharge–sediment supply controls the gradient. Engineers often assume that gradient conditions control the sediment by short-term channel processes. Integration of this viewpoint difference is important in river management where long-term equilibrium must often be maintained by the prudent application of short-term engineering works. Development of the technology of long- and short-term river processes is demonstrated by papers collected by Schumm (1972).

RIVER MANAGEMENT

Man's attraction to floodplain settlement is due to the availability of rich soils, low gradient transportation routes, level, well-drained construction sites, water supply, and the prestige of shoreline property. But settlement on the river corridor also causes exposure to losses by floods and channel shifting. Adjustment to these losses has been variously accommodated by (1) accepting the loss, (2) public relief, (3) emergency action, (4) structural changes, (5) floodproofing, (6) regulation of landuse, (7) flood insurance, and (8) flood control (Sewell, in Chorley, 1969, p. 432).

Integral to present local and national codes is the concept that a river can be viewed as a fixed-location ditch, in which the floodway is determined only by hydraulic measures, without consideration of sediment transport in meandering equilibrium channels (Fig. 5). The floodway is designated as the area that can carry the passage of water of the 100 yr flood (1% chance of recurrence) without increasing the water-surface elevation of that floodway more than 1 ft at any point (Fig. 6). The effect on flood height is all that many land managers are concerned about.

From the geomorphologist's viewpoint, however, the operation of a river is not so simple (Leopold et al., 1964; Morisawa, 1968; Cooke and Doornkamp, 1974). In addition to protecting property, the processes and environments of the river itself are of concern. The land managers, from a geomorphic viewpoint, should consider preservation of the balance of river processes while determining landuses compatible with various human needs. It should not be acceptable to permit construction of single-purpose projects that are destructive to the safety of adjacent properties and to the resources of the river.

At present, it is typical to find a mixed agglomeration of river controls: regulation of construction design in the river by building code; landuse type regulated by planning; land subdivision involving a third agency, railroad crossings a fourth, and bridges and pipelines still other agencies. Water quality may involve Public Health, Fish and Game, Department of Environmental Quality, and various water users such as cities, towns, industries, and agriculture. River-erosion activities may involve soil conservation agencies, state and local highways, the Corps of Engineers, and local governments. Flood control is another separate, special-interest, single-purpose, multiple-input problem. Fish and game is still another group, as is scenic and recreation interests. When so many different activities are imposed in uncoordinated special site actions, it is no surprise that so many rivers are being progressively turned into drainage ditches by one group or into primitive areas by other groups.

In rivers that have developed long-term equilibrium, the river itself displays the natural boundaries of the territory required for channel operation. Given reasonable

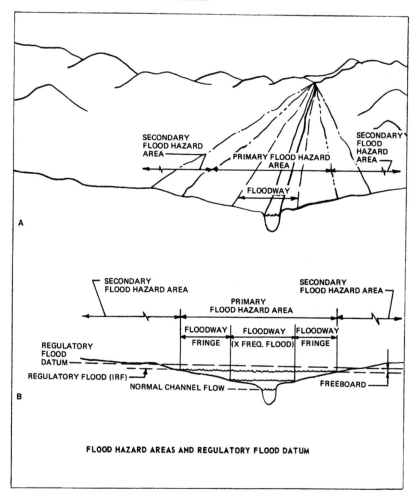

FIGURE 5 *Typical datum used for flood-hazard management. (From U.S. Army Corps of Engineers, 1972.)*

freedom to continue its normal functions, the river will provide for the discharge of water and sediment, and will sustain an environment of high quality for wildlife habitat and for man. The natural equilibrium in the river system operates by a complex interaction of variables. The more flexible systems adjust to the less flexible ones. The external (discharge and sediment) input to the river system control the channel form and are in long-term balance. The long-term balance is represented by a characteristic meander pattern, form and process of the channel which accommodate the conduct of water, the sediment transport, and the energy dissipation within the river reach. Therefore, if resources are to be preserved, management must provide a sufficient corridor to allow continuation of meander belt progression (Fig. 7).

Although the meander belt might shift over the entire floodplain in a period of several hundred years, the shifting of individual bends is usually accomplished in a few tens of years. Thus, the meander belt or streamway may be confined to a fixed position on the floodplain over the period of time considered for planning and management. In rivers with active channel shifting, attempts to confine the channel to a single position and to stop bank erosion or meander progression will often destroy the habitat and trigger erratic response by the river.

The amplitude and frequency of meanders in a natural stream will be related to the usual controls, such as discharge, sediment, and bank conditions. They in turn can be

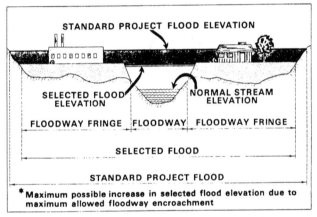

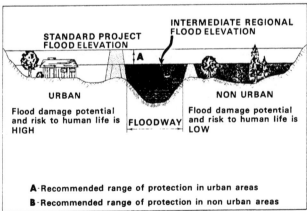

FIGURE 6 *Typical datum used for flood-hazard management. (From Oregon State Water Resources Board, 1972).*

FIGURE 7 *Digrammatic illustration of the natural features of a river channel (compare to Fig. 5): a, point bar; b, chute or lagoon; c, shallow still water; d, deep, fast current; e, eddy; f, cut bank; g, vegetation in floodway; h, streamway boundary.*

modified by natural changes, as in weather, watershed, bank vegetation, or driftwood. Interference with the channel form will also be caused artificially by man-imposed changes. Straightening and narrowing with dikes, levees, and revetments is particularly disruptive. Modifications instigate erratic erosion—accretion changes at the point of impact and cause up- and downstream upsets.

The job of geomorphic and engineering management is to avoid such modifications or minimize their disruptive effects. The geomorphologist can determine the streamway corridor (width) that can be maintained for the planning objectives (50 to 100 yr) and can provide interrelated data on equilibrium channel and meander geometry.

According to Bauer, the rewards to be gained by sustaining the river process include

> unique and important public resources . . . generated in the streamway corridor. These include accommodation of surface and storm runoff; recharge of adjacent aquifers; waste transport and partial waste recycling; gravel classification and stratification; and gravel habitat for fish spawning and larva. "The stream resources" also include processes of eddy cavitation—hydraulics that aerate the

water, increase water depths, and dig stilling and hold-up pools. "Hydraulic plowing" is also produced which, by scouring and accretion backfilling distributes washed-out vegetation, encourages organic and detritus processes, produces seed beds, and assures silt-mud environments for benthic, insect, and marshpond communities. In addition normal meander progression is produced which, by cyclic erosion—accretion, maintains a mineral, chemical, and organic input budget of water and soil in the various streamway reaches. Finally, the river resources include the dissipation of an immense quantity of hydraulic energy into the "useful" tasks of building and maintaining floodplains, wildlife and aquatic habitat, recreational beaches, and the aesthetic elements of our riverscapes. With each passing year, the preservation and enhancement of these many resource functions of the streamways become more appreciated and justifiable as economic assets to the community and the region (Bauer, in Palmer and Bauer, 1974, p. 32).

It is particularly important that these geomorphic processes be understood by the public. Near total ignorance of river process has generated a people who consider the river a hazard, to be controlled. Just as the private property owner may be required to remove snow from a walkway in front of his home, private property ownership adjoining a river imposes a responsibility upon the owner for maintenance of the floodway process. If the owner is unable to maintain the floodway, the corridor should be taken over by a government agency (as in street dedication) who can better manage the stream as a whole system. Thus, the four river zones provide a framework to segregate varying management needs in the river. Each zone has quite different properties, and provides quite different resource values; therefore, although the river must be considered as a total system, the type of river zone will guide the management needs in each segment.

A brief summary of the characteristics, values and management needs of each zone is presented next.

The *boulder zone* is a river environment that is adjusted to conduct water and sediment through steep, resistant terrain. The water is cold, pure, and has high potential energy. The river can transport all sediment discharged from the land and is down-eroding its channel. A bedrock channel in a V-shaped canyon conducts the water with little lateral spreading and without channel shifting during flood stage. The river is usually able to maintain its channel by clearing away landslides, rockfalls, and other less catastrophic sediment supply.

The value of the boulder zone river to man is in pure water supplies, fish and wildlife habitat, and scenic and recreational value. The dynamics of rapids and waterfalls is of special visual value and a favorite of rafters and kayakers.

Management needs of the boulder zone are few. The channel maintains itself and intrusions are soon flushed out, including inadequate bridges. Obstructions do not pond water far upstream owing to the steep gradient. Water-quality preservation requires consideration of the maintenance of near-channel tree cover, which affects the water temperature and the stability of valley walls. Clearcut logging can instigate serious mass wasting of sediment and warming of the water, which affect fish health.

The *floodway zone* develops a valley floor by lateral channel shifting, where the sand and gravel supply is balanced with the ability of the stream to transport sediment. A moderately steep gradient is developed to provide the energy required to transport the coarse sediment. A variety of channel features of high value for wildlife habitat and scenic and recreational use are formed. The alluvial channel has low bank resistance and forms wide-braided channels, which shift rapidly. The shifting channel provides accretional point bars and backwater pools with clean washed bottom gravel. The channel area includes a variety of vegetation and fauna. The unique channel area of the floodway zone is termed the "streamway" by Bauer. A diagram of some streamway features is shown in Fig. 7. The value of the floodway zone for scenic, recreational, and wildlife aspects is greater than that of any other river zone.

Management of the floodway zone is perhaps the most problematical, but the most rewarding if stream values are preserved. High energy and low bank resistance combine to make a dynamic channel. Efforts to straighten and narrow the channel will intensify the velocities and the aggressiveness of the river to shift its channel, and will upset the balance that the river has acquired over thousands of years. Such channelization will also destroy the habitat, recreational, and scenic values.

This river zone is best managed by protecting the natural streamway corridor. Careful definition of the streamway boundaries and strict landuse zoning to prevent development are very important. Landuse development that is incompatible with the streamway processes will not only disrupt the river balance, but will be subject to the highest risk of channel shifting and high-velocity floodwater damage. Cutting of downfall trees into short segments to prevent logjams is good housekeeping, since sudden shifting of flow by logjam obstructions at bridges or other locations can promote serious damage. Thus, it is important to provide bridge spans and approaches with sufficient clearance to allow debris passage. Dikes and revetments must be set back to provide streamway operation. Where necessary, owing to existing developments or natural obstructions in the channel, movable channel-training devices may help guide the current into narrows of the channel.

The *pastoral zone* channel is deeper, narrower, and more stable owing to the cohesive strength of finer sediment load. A single meandering channel flowing in a wide floodplain is typical. Banks are often steep and tree-lined on natural levees. Although not so desirable as the floodway zone, a limited amount of fishing, boating, and swimming occur in the pastoral zone. Commercial water transportation may be present, depending upon channel depth and water velocity. The cost of transport is attractive in energy consumption, being 5 times cheaper than railroad, 20 times cheaper than trucking, and 100 times cheaper than air (Cooke and Doornkamp, 1974, p. 99).

Management is concerned with isolating channel flooding and shifting from floodplain farming and building development. The rate and size of meander bend shifting will determine the area that should be reserved for the floodway corridor. Sustaining the uniform migration rate of the meanders may be assisted by preserving the bank vegetation. If dikes are required to protect adjacent landuses, they should be aligned tangent to the outer meander loops to provide shorter dike lengths and to preserve the meander belt. Cattle access and polluted effluent discharge is of greater concern because of the slower moving, less aerated water and the greater extent of floodplain area development. Preservation of any riffles in this zone is of considerable value to maintain dissolved oxygen in the water.

Estuarine zone channels are divided by low, marshy interfluves composed of weak-saturated fine-grained sediments. Channel overflow is much more frequent than in the upper segments of the river. Biotic activities include many interrelated flora and fauna. Food-chain and habitat value to fish and wildlife is high. Foundation conditions are a serious problem to development. Roadways, cattle grazing, and other development are inhibited by terrain and water conditions, and make the area of low value for man's use.

The estuarine zone is best left undisturbed to preserve the high habitat values. The resources of this type of environment in Oregon and Washington are uncommon and are needed for bird and fish migration.

RIVER-MANAGEMENT PROBLEMS

Almost all river problems could be avoided or minimized if the river corridor were considered as public domain and protected as a vital resource. Fragmented "ownership" of a flowing river is perhaps really only an enormous joke, or fraud, which could only be improved upon by selling private deeds to tomorrow's sunshine.

Examples are now presented to demonstrate some of the common types of river-management problems encountered in Oregon and Washington. Most are the result of single-purpose river-control attempts. The examples are taken from the Clackamas and Molalla rivers near Portland, Oregon, but are representative of problems encountered throughout the Northwest. They show the high cost incurred when the whole stream process is not preserved. The four examples demonstrate (1) the effect of landuse development into the streamway corridor, (2) artificial narrows created by bridge approaches, (3) channel diversion and obstruction for industrial land encroachment, and (4) a natural streamway.

Example 1: Landuse Development in the Streamway Corridor. The Molalla River flows from the Cascade range onto the Willamette lowland (Fig. 8). Near the upstream end of the floodway (4 river mi downstream from the boulder zone) the streamway occupies the entire floodplain (Fig. 9). Coarse cobble gravel is transported by discharges that range to 43,600 ft^3/s and average 1,165 ft^3/s.

The gradient of the river is steep (25 ft/mi) and adjusted to transport the coarse sediment with the moderate discharge available. Low cohesive strength of the alluvial channel banks and the high energy gradient combine to produce a braided, shifting streamway, which has been occupied by the channel on both sides within the last 30 yr.

A park without permanent facilities has been made to preserve an excellent stand of trees beside the river at Dickey Prairie bridge (river mile 20.8). Flood-stage overflow normally drained through the entire streamway, including the park, without obstruction. The Dickey Prairie bridge has been well placed with one abutment against the steep, east riverbank. The west bridge abutment has an earthfill approach obstructing flood-stage overflow; however, an adequate bypass channel exists through which floodwater could cut across the peninsula bend.

A real estate subdivision, called Shady Dell, has developed land parcels within the streamway adjacent to park property, and the bypass channel has been blocked, diverting all the flood runoff to the bridge passage. Obstruction of the flood relief channel has eliminated the countercurrent flow at the point of its rejoining the active channel (see Fig. 10A). A revetment has been built by a government agency to stop the resultant bank erosion at Shady Dell (Fig. 10B). The river channel is now crowded into a narrow corridor, and there is concern to build more river-control structures to stop bank erosion at the park. This would create an artificial barrier between the park and the river, destroying a large part of the natural value of the park.

Existing river and landuse management has converted a self-maintaining streamway corridor into a real estate commodity. Expensive river-control structures are being built

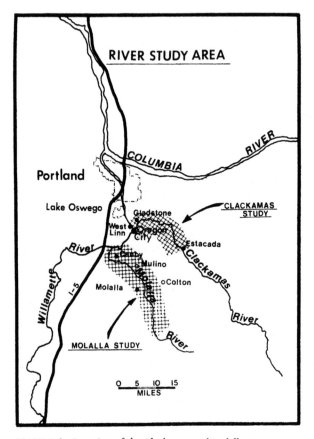

FIGURE 8   *Location of the Clackamas and Molalla rivers.*

with public funds to protect private property encroachment. Shady Dell is still subject to flood damage, which will probably be born by the general public through insurance. This exemplifies why flood protection and flood damage costs are both escalating.

Proper management would have excluded development within the streamway corridor. Adjacent land outside of the streamway could have been developed, preserving the scenic and recreational value of the river for more people at less cost (Fig. 10C).

Example 2: Artificial Narrows Created by Bridges.   At river mile 10 on the Molalla River, a railroad and a private logging highway have crossed the streamway corridor in an area where the floodplain is 1 mi wide (Fig. 11). Pasture and cropland farming without buildings utilize the floodplain. Gravel bedload is transported by moderate discharge, as given in Example 1. Low cohesive strength of the alluvial channel banks and the fairly steep gradient (15 ft/mi) combine to produce a semibraided, shifting channel.

Migration of the river has continued to a point so that the bridges are no longer in line with the channel. Abrupt changes in direction of channel flow now create large eddies at several points, which accelerate erosion and form a hazard to bathing and swimming. Steady channel shifting has been replaced by sudden erratic channel adjustments, which increase flood levels upstream. The natural meander progression has been disrupted.

Management has been directed to stopping bank erosion by use of dikes and revetments upstream from the bridges. Ponding above and high-velocity flow through the bridges are the result. The pinch-valve effect of the narrow bridge passage could be ameliorated by channel-training works designed to allow straight-through flow at the bridges. Also, where possible, bridge spans should have been designed with larger openings related to the natural channel needs. In general, management could work with the river, rather than react to stop the river.

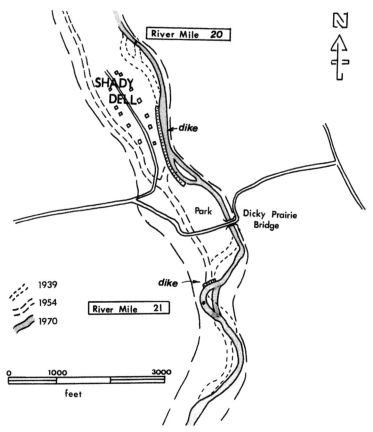

FIGURE 9 *Map of the Molalla River at river mile 20.*

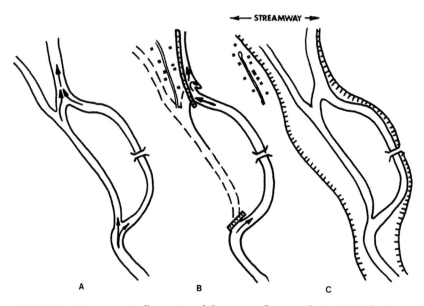

FIGURE 10 *Diagrammatic illustration of the current flow at mile 20.5, Molalla River: (A) original channel before development; (B) channel flow after development and overflow obstruction; (C) ideal location of development out of the streamway corridor.*

Example 3: Channel Diversion and Obstruction for Industrial Encroachment. The Clackamas River between river miles 13 and 16 is forming a new floodplain by entrenching into terraces (Fig. 12). Coarse cobble-gravel sediment from the Cascade Mountains 10 mi upstream is transported by a discharge of 336 to 120,000 ft³/s, averaging 3,785 ft³/s.

A moderately steep gradient of 10 to 18 ft/mi is generated to accommodate sediment transport with the range of discharges. (This is less steep than the Molalla River, as the same caliber of bedload is being moved with about three times the discharge.) The braided channel forms a streamway that occupies 50 to 100% of the floodplain. Channel shifting sweeps across the entire streamway in about each 50 yr.

Encroachment in the channel: at mile 13.3 a highway bridge has been located on the river at a natural narrow point where channel flow is relatively straight. The south bridge abutment is attached to a steep embankment and the other (northeast side) has been designed with a long, open trestle approach. This represents an excellent design and location to accommodate the natural river characteristics. The main channel is unobstructed, and the flood-bypass overflow still functions through the trestle approach ramp.

At river mile 13.7 a park with river beaches and permanent restroom facilities has been developed in the area of the 1914 channel. Park "protection" from river shifting has been attempted by use of bank revetments made of discarded concrete slabs. Channel-diversion structures have been attempted using channel gravel. A boat ramp has been constructed and destroyed repeatedly. The location of the park is well chosen on the inside bend of the channel. Accretion of point bar gravels sustains easy access to the river in a site with low water velocity.

However, permanent restroom facilities have been repeatedly damaged by floods, and should have been placed outside the active streamway. This popular park site was chosen to take advantage of the natural river values, but those same natural values are being destroyed to preserve toilets. A more flexible park management policy that includes natural river processes could enhance the site by calling attention to the working river functions through educational displays.

Private landowners on the west bank at river mile 15 are reported to have engaged in repeated river modifications to "protect" their property. Side channel diversions and bank revetments have been used to encroach into the streamway and to shift the river toward the opposite bank. A dike has been installed, repaired after flood damage, strengthened, and extended. A 1955 river channel has been blocked, concentrating the flow to the east bank channel. The intensified flow has accelerated and extended a large landslide at mile 14.3, increased bank erosion at Barton Park (mile 14.0), and instigated a series of current deflections

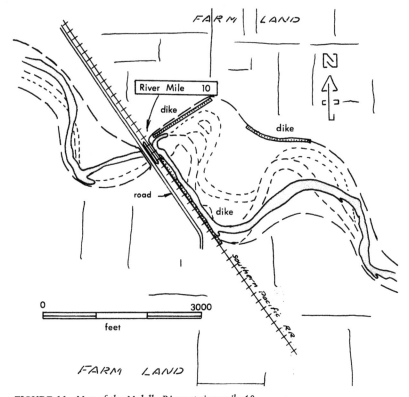

FIGURE 11 *Map of the Molalla River at river mile 10.*

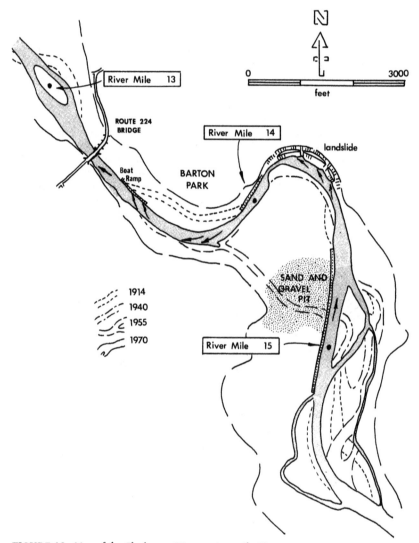

FIGURE 12 *Map of the Clackamas River at river mile 14.*

that bounce from bank to bank pass the park. The current deflections accelerate bank erosion, including removal of the series of boat ramps mentioned above, and also threaten the west bridge abutment. The intensified current deflections extend almost 2 mi downstream from the diversion dike at river mile 15. The closure of the 1955 channel has eliminated the countercurrent at its outlet (mile 13.8), which formerly interacted with the present channel flow to form relatively straight flow past the park and under the bridge. The mile 15 westbank landowner is now operating a sand and gravel pit in the former streamway. He volunteered to remove channel gravel from in front of Barton Park as a river correction action, and has asked for federal help to strengthen the dike protecting his property.

Proper management, to work with the natural river at this time, would require a restoration of flood overflow through the 1955 channel at river mile 15. The proposed excavation of the point bar channel accretion at Barton Park would probably not redirect the strong current that exits from the meander at mile 14, but would cause much water turbidity, and would form a dangerous hole where a safe beach now exists.

If the streamway encroachment at mile 14 to 15 were irreversible, as with a town development with high economic investment, one might consider the merit of river training to preserve a stable meander flow regime.

Example 4: Nonencroachment of a Natural Streamway.   At a site called Bonnie Lure at mile 17 on the Clackamas River, the braided channel streamway forms wooded islands with a rich variety of river forms and habitat. A request for a building permit to develop a mobile home park facility on a streamway island was made and refused. There has been no subsequent flood damage, no costly flood control works have been built, and the site is still a very attractive natural site enjoyed by many who raft, boat, and swim down the Clackamas River.

The residents of the Clackamas River have overwhelmingly voted for preservation of natural values in a large section of the Clackamas River. The county river legislation was later defeated in court, but replaced by state legislation.

Geomorphology of river processes can document many advantages of preserving natural conditions. Public opinion is favorable, but must have technical help to counter-act special interests and the riprap syndrome of most flood "protection" projects.

SUMMATION

This paper is based on the assumption that natural rivers attain conditions of equilibrium adjusted to many interdependent variables, including discharge, sediment, and channel characteristics. Disruption of river equilibrium will initiate complex counter-actions by the river. Therefore, river management that regulates landuse to sustain the minimum disruption of the river will preserve the maximum natural values and require the least maintenance cost.

The natural equilibrium of rivers should be a major concern in their management. The rewards of preserving the natural channel are many, including the natural accommodation of floods and saving of habitat and aesthetic values. Determination of the long-term equilibrium by geomorphic methods is needed to preserve those values.

The Bauer river classification system differentiates four geohydraulic zones within which similar management conditions exist. Use of this classification can help outline the landuse potential and management requirements of a river. The floodway zone is particularly valuable for scenic, recreational, and habitat uses. It is also the zone in which greatest losses are incurred from encroachments.

Strict landuse controls and preservation of the river process are believed to provide a solution to escalating flood damage and flood-control cost. The present practice of single-purpose landuse development and rigid channel controls is building bigger and more damaging flood losses.

Geomorphology can determine the long-term equilibirum conditions. Geomorphology with engineering can design the controls to maintain the most natural conditions and to accommodate unavoidable river modifications with least disruption to the equilibrium balance.

ACKNOWLEDGMENTS

The geohydraulic river classification system and management operations presented here are the result of combining geomorphology, engineering, and some intimacy with natural river hydraulics gained through wild-water kayaking. Wolf Bauer developed the concepts and classification in Washington State, where the method was first described. (Rivers Sub Committee, 1972). Bauer's witty and knowledgeable evangelism on behalf of rivers is appreciated by the public as well as by resource scientists and government officials. It is through working with Bauer on Oregon streams that I have gained much awareness of the practical application of geomorphology to rivers. Any credit is to him and to the geomorphology discipline; any errors and shortcomings are mine.

REFERENCES

Chorley, R. J. 1969. *Water, Earth, and Man:* Methuen, London, 588 p.

Cooke, R. U., and Doornkamp, J. C. 1974. *Geomorphology in Environmental Management:* Oxford University Press, New York, 413 p.

Leopold, L. B., and Maddock, T., Jr. 1954. *The Flood Control Controversey:* Ronald Press, New York, 278 p.

_____, Wolman, M. G., and Miller, J. P. 1964. *Fluvial Processes in Geomorphology:* W. H. Freeman, San Francisco, 522 p.

Morisawa, M. 1968. *Streams, Their Dynamics and Morphology:* McGraw-Hill, New York, 175 p.

Oregon State Water Resources Board. 1972. Oregon's flood plains; a status report and proposed flood plain management program: the Board, Salem, Ore., 40 p.

Palmer, L., and Bauer, W. 1974. Geo-hydraulic classification of the Clackamas and Molalla rivers: *Porland State University Publications in Earth Science No. 741*, 46 p.

Schumm, S. A., ed. 1972. *River Morphology:* Dowden, Hutchinson & Ross, Inc., Stroudsburg, Pa., 448 p.

U.S. Army Engineers. 1972. Flood proofing regulations: Document number EP 1165 2 314, U.S. Army Engineers, Washington, D.C., 79 p.

Wild, Scenic, and Recreational Rivers. Rivers Sub-Committee of the Interagency Committee for Outdoor Recreation, Olympia, Washington, 1972.

APPENDIX
TABLE OF CONVERSIONS

1 in = 2.54 cm
1 ft = 0.305 m
1 mi = 1.609 km
1 mi^2 = 2.589 km^2
1 acre = 0.405 hectare
1 gallon = 3.785 liters
1 ton = 1.016 tonnes (metric tons)
1 ft^3 = 0.0283 m^3
1 cm = 0.3937 in
1 m = 3.280 ft
1 km = 0.621 mi
1 hectare = 2.47 acres = 0.0039 mi^2
1 m^3 = 35.314 ft^3
1 tonne = 2,204.62 lb
1 kg = 1,000 g = 2.2046 lb

INDEX

Milton Keynes UK
Ingram Content Group UK Ltd.
UKHW021637071024
449327UK00020BA/1331